Applied Plant Virology

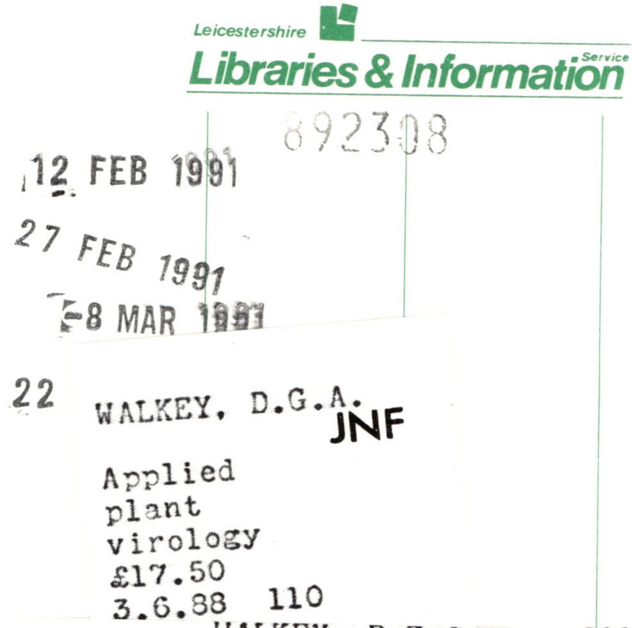

Applied Plant Virology

D. G. A. WALKEY

BSc, PhD, MI Biol

Principal Scientific Officer, National Vegetable Research Station, Wellesbourne
Honorary Lecturer, Plant Science Department, University College, Cardiff
Part-time Lecturer, Department of Biological Sciences, University of Warwick

HEINEMANN : LONDON

William Heinemann Ltd
10 Upper Grosvenor Street, London W1X 9PA

LONDON MELBOURNE TORONTO
JOHANNESBURG AUCKLAND

First published 1985

Walkey, D. G. A.
Applied plant virology.
1. Virus diseases of plants.
632'.8 SB736
ISBN 0–434–92225–0

Filmset by Deltatype, Ellesmere Port
Printed in England by
Redwood Burn Ltd, Trowbridge

Preface

For the past twenty years I have worked as an applied plant virologist, attempting to identify and control virus diseases in field crops. During the last ten years it has been my privilege to present short courses in plant virology to final-year students studying plant pathology, microbiology and general botany. Throughout the period I have been lecturing, it has been possible to recommend several excellent 'library' books for further reading in plant virology, but there has been no publication covering applied plant virology that a student might consider purchasing. With teaching requirements in mind, this book has been written to provide a concise introduction to applied plant virology based on the experiences I have gained working on virus diseases, both in an applied laboratory and in the field.

The text concentrates on introducing the reader to aspects of plant virology that would be encountered every day by an applied virologist, trying to identify viruses and develop control measures for virus diseases of crop plants. Although a brief introduction to virus structure and its terminology is given in the opening chapter of the book, no attempt is made to cover in detail the more fundamental aspects of virus structure, biochemistry and replication. Similarly, the symptoms caused by individual viruses are not described, although the various types of symptoms that plant viruses cause and, which might be encountered by a student or research worker, are described.

Each chapter contains key references that have been selected to illustrate the information cited in the text, and a number of selected references for further reading are given at the end of each chapter. These reviews and general articles or books, will allow the reader immediate access to more comprehensive treatments of specific subjects.

In the final chapter, detailed information is given of practical methods that are likely to be required by an applied virologist, together with a number of practical class exercises which could be undertaken by undergraduate students. Also included in this chapter, is an up-to-date list of plant viruses that have been described in the Commonwealth Mycological Institute/Association of Applied Biologists' publication *Descriptions of Plant Viruses*. This set of descriptions is essential to the work of any applied plant virus laboratory.

This book should be of value to the undergraduate in plant virology, plant pathology, microbiology and general botany, and to postgraduate students in applied plant virology or plant pathology, during the initial stages of their research experience.

DAVID WALKEY

Acknowledgements

I am indebted to my colleagues Dr R. T. Burchill, Dr I. R. Crute and Dr R. S. S. Fraser for reading the final manuscript of this book. Their critical and constructive comments were greatly appreciated.

I am grateful to Professor J. K. A. Bleasdale for encouraging me to write the book and for allowing me the use of the library facilities at the National Vegetable Research Station. I would also like to thank the following plant virologists and research organizations for kindly providing illustrations for the book: Dr A. A. Brunt, Glasshouse Crops Research Institute, Littlehampton; Dr M. F. Clark and Dr J. M. Thresh, East Malling Research Station; Dr M. Conti, Laboratorio di Fitovirologia, Turin; Professor D. J. Hagedorn, University of Wisconsin; Mr G. J. Hills, John Innes Institute, Norwich; Professor G. P. Martelli and Dr M. Russo, Istituto di Pathologia Vegetale, Bari; Dr J. A. Tomlinson and Mr M. J. W. Webb, National Vegetable Research Station, Wellesbourne; Dr W. M. Robertson, Mr I. M. Roberts and Dr C. E. Taylor, Scottish Crops Research Institute, Dundee; Agriculture Canada, Research Station, Vancouver and the National Vegetable Research Station, Wellesbourne.

Finally, but not least, I wish to thank my wife Heather, for her help in editing and checking the draft manuscripts of the book.

Contents

CHAPTER 4 MECHANICAL TRANSMISSION AND VIRUS ISOLATION

CHAPTER 5 VIRUS PURIFICATION

CHAPTER 6 VIRUS IDENTIFICATION

CHAPTER 7 VIRUS TRANSMISSION BY BIOLOGICAL MEANS

Plates

Plant Virology: An Introduction

Although this book is primarily concerned with applied aspects of plant virology, an understanding of basic plant virus structure and its terminology is essential for any newcomer to the subject. In this chapter the development of plant virology as a science is outlined, the worldwide economic importance of plant viruses is illustrated and the basic structure and composition of plant viruses are described.

1.1 A Definition

The meaning of the word virus has changed considerably during the last century. In Roman times the word meant poison and even during the eighteenth century a dictionary referred to a virus as *a poison, venum, also a rammish smell as of the armpits, also a kind of watery matter, whitish, yellowish, and greenish at the same time, which issues out of ulcers and stinks very much; being induced with eating and malignant qualities* (Phillips, 1720). During the nineteenth century it came to denote *the poisonous element by which infection is communicated* or simply *a micro-pathogen* (Gibbs and Harrison, 1976).

Since the beginning of this century, the modern concept of the word *virus* and its study *virology*, has taken on a more specific meaning to denote a group of extremely small (not usually visible in the light microscope, *see* Figure 1.1), obligately parasitic, pathogenic agents. In 1950, a virus was described by Bawden as an *obligately parasitic pathogen with dimensions of less than 200 nm*, but this and other early definitions (Lwoff, 1957; Pirie, 1962) were based on the small size of the particle, pathogenicity, possession of nucleic acid and an inability to multiply outside a living cell. As knowledge of viruses and associated disease agents increased, it became clear that these definitions were not entirely satisfactory. They failed to distinguish between viruses and other disease agents, such as mycoplasma and rickettsia, and excluded large animal viruses such as the pox viruses.

These anomalies were covered by Gibbs and Harrison (1976), when they defined a virus as a *transmissible parasite whose nucleic acid*

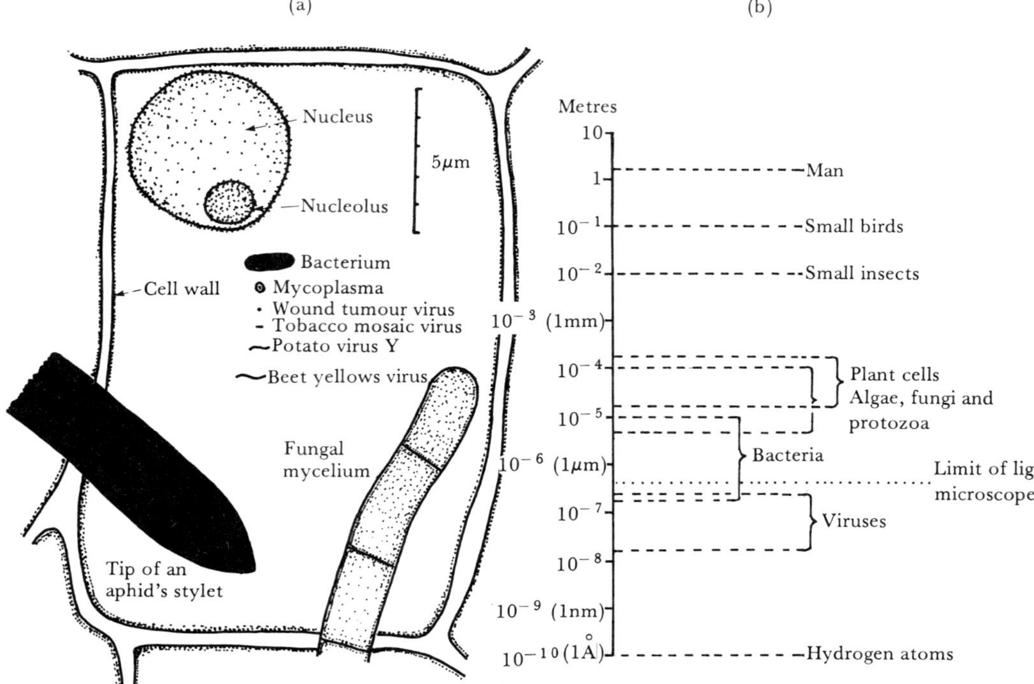

Fig. 1.1 The scale of 'things'. (*a*) The size of plant pathogenic agents in relation to a plant cell; (*b*) the size range of living organizms (logarithmic scale) (based on Marshall, 1976).

genome is less than 3×10^8 daltons in weight and that need ribosomes and other components of their host cells for multiplication. The value of 3×10^8 for the nucleic acid molecular weight was large enough to include bacteriophages (viruses infecting bacteria), and the pox and iridoviruses infecting animals (Matthews, 1979). This definition was not entirely satisfactory, however, for it included the disease agents known as plant viroids (*see* Section 2.5.1), such as potato spindle tuber viroid, whose nucleic acid has a molecular weight of 10^5 daltons or less, but which unlike a typical virus, is not contained within a protein shell (*see* Section 1.4.1).

Recently, Matthews (1981) has more specifically defined a virus as *a set of one or more nucleic acid template molecules, normally encased in a protective coat, or coats of protein or lipoprotein, which is able to organise its own replication only within suitable host cells. Within such cells virus production is (a) dependent on the host's protein synthesising machinery, (b) organised from pools of the required materials rather than by binary fission, and (c), located at sites which are not separated from the host cell contents by a lipoprotein, bilayer membrane.* Such a definition clearly distinguishes a virus from other plant disease agents such as viroids, mycoplasmas and the rickettsia group of bacteria, which cause virus-like symptoms in diseased plants (*see* Section 2.5).

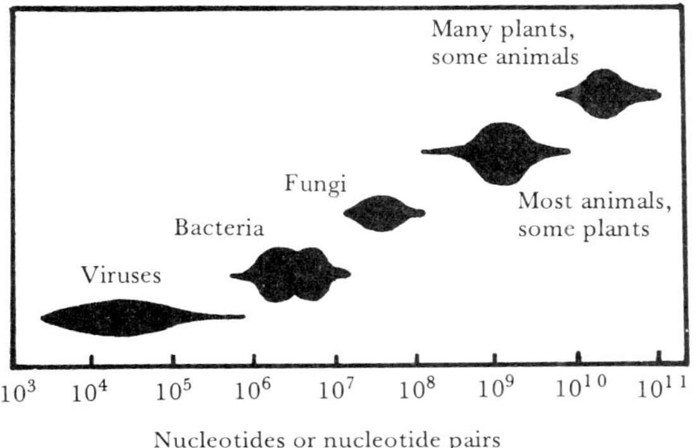

Fig. 1.2 A diagram of organisms classified according to their genome size. The vertical axis indicates the approximate number of species (or viruses) within the size range of each group (based on Hinegardner, 1976).

Figure 1.2 illustrates the relative size of the virus nucleic acid genome compared with those of other major groups of living organisms.

1.2 The Development of Plant Virology as a Science

The first reference to a symptom in plants, that we now know to be caused by a virus, occurred in a poem composed by the Empress Koken in the year 752 (Inouye and Osaki, 1981). In this anthology the symptoms of the yellow leaf disease of *Eupatorium* were described. Later, during the seventeenth century in Holland, colour variegation or striping of tulip petals (*see* Plate 1.1) in plants infected with tulip mosaic virus, was much prized by Dutch tulip growers. Pictures painted by Dutch masters during the seventeenth century, often illustrated the distinct symptoms of the petal *break* disease, and the demand for infected bulbs was so great during this period, that the craze became known as *tulipomania*. One report tells how a single bulb was bartered for oxon, pigs, sheep, tons of grain and 1,000 lb of cheese (Dubos, 1958). Although the cause of the tulip petal symptoms was unknown at the time, some growers knew that the condition could be grafted to a normal flowered bulb. It was not until 1926 that tulip 'breaking' was associated with a virus and shown to be transmitted by infected sap or aphids (McKay and Warner, 1933).

Another early reference to the grafting of a condition that was later

Plate 1.1 A tulip flower showing petal break symptoms caused by tulip breaking virus (*right*) and a healthy flower (*left*).

shown to be caused by a virus disease, was the transmission in 1692 of yellow-striped *Jasminium* (caused by jasminium mottle virus) to a normal flowered plant (Cane, 1720).

In 1886 Mayer, an agricultural chemist working at Wageningen in Holland, was investigating a mosaic disease of tobacco whose cause was unknown. He found that the disease could be transmitted to healthy tobacco plants in juice extracts taken from infected plants. A few years later, Ivanowski (1892) worked on two diseases of tobacco in the Crimea. He recognized these as two distinct diseases in the same plant and described one as a pox disease, and the other as a mosaic disease similar to the one reported by Mayer. He confirmed Mayer's report that the mosaic disease could be sap transmitted and showed that the sap was still infectious after it has been passed through a Chamberland filter, which was known to retain bacteria. Mayer suggested that the mosaic disease might be caused by a toxin produced by a bacterium, or to a new small bacterium that could pass through a filter. Further agar diffusion experiments carried out with this mosaic disease, led Beijerinck (1898) to conclude that the disease was not caused by a microbe, but by a *contagium vivum fluidum*. He thought that the contagium could reproduce itself in the living plant and used the word virus to describe it.

Between 1894 and 1895, Hashimoto showed that the dwarf disease of rice was transmitted by a leafhopper, although this was not reported until 1911 (Fukushi, 1934). In 1900 the leafhopper was identified as *Nephotettix apicalis*, and between 1906 and 1908 it was shown that only leafhoppers from certain areas of Japan could transmit the disease and that *N. apicalis* was not the causal agent of the disease, but only the vector.

Although numerous studies were reported between 1900 and 1935 on the symptoms of virus diseases (Corbett, 1964), little progress was made on the nature of the virus agent itself. Perhaps one of the most significant discoveries during this period, was that the discreet lesions (referred to as *local* lesions) that developed on tobacco leaves when they were inoculated with tobacco mosaic virus (TMV) sap (*see* Section 3.2.1), could be used as a quantitative assay for determining the amount of virus present in the infected sap. The number of local lesions produced could be correlated with the concentration of virus in the inoculum (Holmes, 1929). This method still forms the basis for the quantitative assay of many plant viruses today (*see* Section 4.4). Also of importance during this period was the work of Purdy (1929), who showed that plants infected with virus contained antigenic materials that were capable of inducing antibody formation in mammals. The use of virus antibodies in various serological reactions, has undoubtedly played a major role in the development of plant virology, both for diagnostic purposes and quantitative assay (*see* Section 6.6).

It was not until 1935, however, that the most important discovery concerning the nature of the plant virus itself was made. Working with TMV, Stanley reported the isolation and characterization of the virus as a crystallizable protein. Although this study did not recognize the distinct nucleoprotein nature of the virus, it was, nevertheless, the beginning of modern plant virology. Two years later, in 1937, Bawden and Pirie reported that TMV consisted of 95% protein and 5% nucleic acid, and in 1938 they purified tomato bushy stunt virus and showed that it contained 18% ribonucleic acid. Another major step forward, in respect of virus function, was made when the viral nucleic acid was shown to be involved in virus infectivity (Markham, 1953; Gierer and Schramm, 1956), and during the next twenty years numerous other discoveries were made concerning virus chemistry and structure (Markham, 1977).

During the early years of plant virology it was impossible to visualize the virus agent itself. Although it is possible to see plant virus inclusion bodies (*see* Section 3.3.2) and some of the larger animal viruses, such as vaccinia, in the light microscope, it was not until the development of the electron microscope, that the smaller plant viruses could be seen and photographed. The first electron microscope picture of a plant virus was taken of TMV by Kausche, Pfankuch and Ruska (1939), but it was not until the technique of metal shadowing was developed by Williams and Wycoff (1945) that details of the particle could be seen. Even the shadowing method had many limitations, however, and the basis of modern-day electron microscopy techniques developed as a result of the negative staining procedures introduced by Brenner and Horne (1959).

Another major technical advance in plant virology, was the development of density gradient centrifugation (Brakke, 1951). The idea of using solutions of different density to separate viruses was not new, but Brakke's method allowed the liquid column to be stabilized in a centrifuge tube and the virus specimen placed on its surface. The individual components in the virus solution could then be sedimented into bands in the tube during centrifugation, according to the order of their sedimentation coefficients (*see* Section 5.4.1). From these bands, the infectious portions could be removed, and correlations made between the physical and biological properties of the virus concerned.

In the field of plant virus control, a major step forward was made in 1952 by Morel and Martin. Using meristem-tip culture they showed that virus-free plants could be obtained from totally infected parents. Later, in 1954, Kassanis demonstrated that viruses could be eradicated from infected plants by high temperature treatments. These two techniques, on their own, or combined, have played an important role in producing virus-free clones of numerous vegetatively propagated crop plants (*see* Chapter 11).

Finally, other major advances have undoubtedly been the discovery of viruses with multi-component genomes (Lister, 1968; Sanger, 1968) which are described in detail in Section 1.4.2; the development of an internationally accepted system for plant virus classification (*see* Chapter 2); and the recognition of diseases with virus-like symptoms that are caused by distinct, separate agents such as viroids (Diener, 1971), mycoplasma (Doi *et al.*, 1967) and rickettsia (Nienhaus and Sikora, 1979) (*see* Section 2.5).

1.3 The Economic Importance of Plant Viruses

Most, if not all economically important crop plants may become infected with viruses. In most cases the virus (or viruses) will cause a reduction in yield or quality of the infected crop, but the extent of the economic loss can vary greatly. In highly developed countries which have relatively uncontrolled markets, it is difficult to assess such losses, for frequently shortages of a particular crop may result in higher prices and no financial loss to horticulture as a whole. Individual growers may suffer, but others with healthy crops will benefit considerably because of the shortage, and higher prices. Such factors do not apply of course, if complete crop losses occur throughout a wide geographical area, when all growers are likely to suffer. Nor do such factors apply in 'undeveloped countries' in which maximum crop yields are often required to maintain even basic food supplies. In such countries, even a relatively small yield loss will be important and a complete crop loss disastrous.

Crops can be divided into three classes, annuals, perennials and vegetatively propagated plants. Annual crops, such as vegetables and cereals, are usually grown from seed, and virus infection in such crops may be serious and result in complete crop loss within a particular season. Provided that the factors causing the disease can be controlled however (*see* Chapters 9 and 10), the grower may be able to raise a healthy crop on the same site the following season. The epidemiology of the virus concerned in such outbreaks is critical, and factors, such as the population size of an insect vector during the early part of the crop's growth, may determine whether a crop will be diseased or not (*see* Chapters 8 and 9). Consequently, serious outbreaks of virus disease in annual crops, such as those caused by beet yellows virus in sugar beet, tend to be spasmodic and differ in their intensity from season to season (*see* Table 1.1).

Table 1.1 Seasonal variations in losses caused by beet yellows virus in British sugar beet crops

Year	*% Infection*				*%*	*Estimated cash loss*
	June	*July*	*August*	*Sept.*	*Yield loss*	*(£)**
1970	0	0	2	4	<1	447,000
1971	0	0	1	3	<1	331,000
1972	0	1	2	5	1	520,000
1973	0	6	11	14	3	3,111,000
1974	2	42	66	76	18	14,859,000
1975	0	6	37	59	9	6,128,000

* Based on 1974 prices of £14/ton, Heathcote (1978).

Outbreaks of virus disease in perennial crops, such as trees, are often more serious, for once a tree is infected, it will remain infected for life. The symptoms and crop losses caused by the virus in a particular tree may vary from season to season, but in the case of severe infections, or in order to prevent the virus from spreading to adjacent healthy trees, the grower may have to remove the tree. In such crops, not only must the immediate loss be considered, but also the loss of income that occurs before the healthy replacement tree becomes productive. Examples of such losses are seen in virus infected citrus trees in North and South America (Wutscher, 1977) and in the swollen shoot disease of cocoa trees in West Africa (Brunt, 1975).

Virus infection of vegetatively propagated plants can also be serious (Walkey, 1980) and unless special measures are taken (*see* Chapter 11), all the propagules removed from an infected plant will be infected (*see* Section 7.8). Infection of vegetatively propagated crops is widespread and often every plant of a particular cultivar may be infected with one or more viruses. This was the case with the rhubarb crop in Britain (Tomlinson and Walkey, 1967). Sometimes symptoms are relatively mild or the viruses may be latent (e.g. infecting without symptoms) in the infected cultivars. Also plants may have been inadvertently selected for virus resistance by growers over a long period of time (Walkey *et al.*, 1982). In other instances, infected clones have been lost to horticulture because of a serious decline in yield and quality.

Various workers have tried to assess crop losses caused by viruses in monetary terms (Bos, 1982), and others have devised techniques for measuring crop losses (James, 1974), but because of the difficulties in obtaining accurate information there are few precise estimates. Some examples are given below.

In cereal crops, serious losses were caused by viruses in wheat in Kansas, U.S.A., in 1953 and 1954. Based on 1955 values it was estimated that losses of $3,000,000 and $14,000,000 were caused by soil-borne wheat mosaic and wheat streak mosaic viruses, respectively (Sill *et al.*, 1955). Between 1972 and 1974 soil-borne wheat mosaic virus was also reported to cause severe losses in Nebraska (Palmer and Brakke, 1975) and wheat streak mosaic virus caused considerable losses in southern Alberta in 1963 and 1964 (Atkinson and Grant, 1967). Serious losses in cereals caused by barley yellow dwarf (BYDV), oat mosaic and wheat spindle streak mosaic viruses have also been reported in North America (Slykhius, 1976) and in recent years in Britain, BYDV has resulted in yield losses in barley, wheat and oat crops (Plumb, 1981). Heavy losses also occurred in rice crops in the Philippines in 1964, where rice mosaic virus reduced grain yield by 47% in the worst affected area of Okayama (Ling, 1972).

Sugar beet yellows has caused serious losses in sugar beet crops in both Europe and America for many years. In 1925, 75% of the beet crop was lost in the Yakima valley in the state of Washington, and serious losses have been reported in England (Watson *et al.*, 1946; Hull, 1958), West Germany (Heiling, 1953), and Sweden (Bjorling, 1949). Other heavy losses have been reported in France, The Netherlands, Hungary, Rumania, Turkey, China, Japan and other areas of the United States (Duffus, 1973). During the period 1970–5, it was estimated that sugar beet growers in England lost about 5% of their crop per year as a result of yellows infection (*see* Table 1.1). This was equivalent to £4.2 million per annum based on the price of beet in 1974 (Heathcote, 1978).

Major losses due to viruses have also occurred in potato crops. Potato virus X was estimated to have caused losses in Australia amounting to $1,750,000 per annum at 1941 values (Bald and Norris, 1941). More recently, field trials in the United States have shown that potato leaf roll virus can reduce yields of the cultivar Netted Gem by between 65% and 92% (Harper *et al.*, 1975) and in other studies at three sites in the United States and one in Canada, potato viruses Y and S reduced yields in the cultivars Netted Gem and White Rose by 14% to 37% (Wright, 1974).

Viruses of other solanaceous crops are also of economic importance. Tomato vein mottling virus for instance, was estimated to have reduced yields in the North Carolina tobacco crop by over 400,000 lb in 1978, with a loss of about $5.2 million (Gooding and Main, 1981) and tomato mosaic virus has been estimated to cause between 15 and 25% loss of yield in infected tomato crops (Broadbent, 1976).

There are also many reports of field studies designed to determine yield losses caused by viruses. Some recent examples of such studies are shown in Table 1.2. These illustrate the importance of virus infection, both in terms of the range of crops infected and the geographic occurrence of disease.

Another consideration in respect of the economic importance of plant viruses, is seen not only in the value of the crop losses they cause, but in the high cost of preventative or control measures required to avoid infection. Such measures include chemical sprays to control insect vectors, the provision of virus-free seed to prevent seed-borne virus diseases, breeding for disease resistance, and certification schemes to provide healthy planting stock for vegetatively propagated crops.

Finally, not all virus infections cause economic losses, and in a few instances virus infection can actually be beneficial to the grower. Latent infection, in which the virus causes no symptoms and little or no loss of yield in the infected plant (*see* Section 3.4), has already been mentioned. The symptoms of flower-break in tulips caused by tulip mosaic virus were greatly valued by the Dutch in the seventeenth century (*see* Section 1.2), and mild strains of a virus may be used to inoculate a host plant to protect it against later infection by a more severe strain (*see* Section 9.7). This technique has been used successfully to control tomato mosaic virus in commercial tomato crops (Rast, 1972.) Other examples of virus protection using mild strains have been reviewed recently by Cohen (1981).

Table 1.2 Examples of yield reduction caused by various plant virus diseases

Crop	*Virus*	*Yield reduction (%)*	*Country*	*Reference*
Beans (*Phaseolus vulgaris*)	Bean yellow mosaic	33	United States	Hampton (1975)
	Bean common mosaic	64	United States	Hampton (1975)
Cabbage	Turnip mosaic	36	England	Walkey & Webb (1978)
Cassava	Cassava mosaic	24–75	Kenya	Seif (1982)
Lettuce	Lettuce mosaic	56	United States	Zink & Kimble (1960)
	Cucumber mosaic	8–50	England	Walkey & Ward (1983)
	Beet western yellows	7–58	England	Walkey & Ward (1983)
Pepper	Various	9–67	United States	Villalon (1981)
Potato	Potato leaf roll	65–92	United States	Harper *et al.* (1975)
Maize	Maize streak	25–60	Kenya	Guthrie (1978)
Wheat	Barley yellow dwarf	9–29	Australia	Smith & Sward (1982)
Apple	Various	39–51	Germany	Schmidt (1972)
Pear	Ring pattern	35	United States	Waterworth (1976)
Raspberry	Raspberry mosaic	50	United States	Converse (1963)
		11–14	Canada	Freeman & Stace-Smith (1970)
Strawberry	Various	14–38	Belgium	Aerts (1977)
Sweet cherry	Prunus ringspot and prune dwarf	70	England	Cameron (1977)
Tobacco	Tobacco etch	3–18	United States	Gooding (1970)
	Tobacco mosaic	5–16	United States	Gooding (1970)

1.4 The Composition of Plant Viruses

1.4.1 Morphology and structure

Basically the plant virus particle consists of infectious nucleic acid referred to as the *genome*, which is *encapsidated* (enclosed) within a protective protein coat or shell (Crick and Watson, 1956). The genome, which carries the genetical information necessary for the virus's replication, is composed of *ribonucleic acid* (RNA) in most groups of plant viruses, but consists of *deoxyribonucleic acid* (DNA) in members of the caulimovirus and geminivirus groups (*see* Section 2.2). The RNA and DNA may be single (*ss*) or double (*ds*) stranded.

The nucleic acid genome of the caulimovirus group and a number of the RNA plant virus groups, is composed of a single molecular species or molecule of nucleic acid, and is often referred to as a monopartite genome (*see* Tables 2.2 and 2.4). Other groups with RNA genomes and some members of the DNA geminivirus group have two or more molecular species of nucleic acid, usually, though not always, encapsidated within separate protein shells. Such genomes are referred to as bi-, tri- or multi-partite. Genomes with more than one molecular species of RNA, usually require the presence of all their major genomic components for complete infectivity. Viruses of the tomato spotted wilt and reovirus groups also have multi-partite, segmented genomes, but these are encapsidated within the same protein shell.

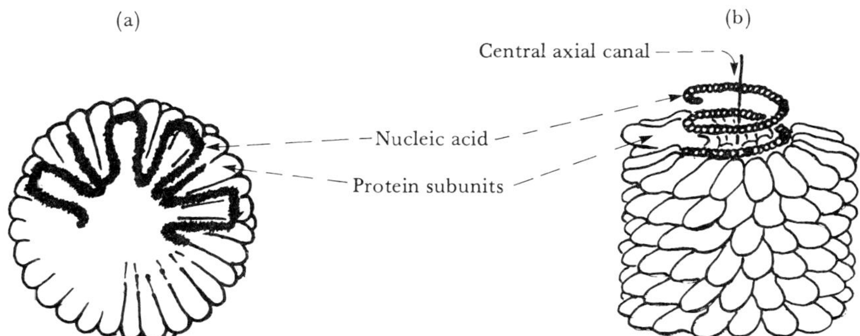

Fig. 1.3 Diagrams of the arrangements of the protein sub-units and ribonucleic acid in an isometric (1.3a) and rod-shaped virus (1.3b). (*a*) Shows the RNA closely associated with the inner surfaces of the protein sub-units of the isometric virus turnip yellow mosaic virus (based on Klug *et al.*, 1966); (*b*) shows the structure of tobacco mosaic virus in which the helical chain of RNA is arranged on the inner surface of the helically arranged protein sub-units, around a central axial canal (based on Caspar, 1964).

The complete, mature virus particle is sometimes referred to as a *virion* and the protective protein shell the *capsid*. Each capsid is made up of a number of individual protein sub-units (polypeptide chains) (*see* Figures 1.3*a* and 1.3*b*). The arrangement of the sub-units varies in different viruses and these morphological units may be seen in the electron microscope on the surface of the virion, and are referred to as *capsomeres*. In membrane-bound viruses, such as those of the rhabdovirus and tomato spotted wilt virus groups (*see* Figure 2.1), the inner nucleoprotein core is often called the *nucleocapsid*.

Plant viruses may be isometric (e.g. spherical), bacilliform (e.g. bullet-shaped) or rod-shaped. Isometric particles vary from 17 nm (the satellite virus of tobacco necrosis virus) up to 70 nm (the reoviruses) in diameter. Particles of bacilliform viruses, such as those of the rhabdovirus group, measure up to 300 nm in length × 95 nm in width, and the elongated, rod-shaped plant viruses range from short rigid rods measuring 114–215 nm in length × 23 nm in width (the tobraviruses), to long flexuous particles up to 2,000 nm in length × 10 nm in width (the closteroviruses).

Crick and Watson (1956) were the first to suggest that the protective protein coat of a virus, was built up from a number of identical protein molecules or sub-units, packed together in a regular pattern. It might be thought, that in view of the wide range of virus particle shapes and sizes, there must be many ways in which these protein sub-units could be arranged. In fact, this is not the case, and it has been shown that there are only a few possible efficient designs by which the numerous protein sub-units can be arranged to form the capsid (Caspar and Klug, 1962; Caspar, 1964). The principles of construction of isometric particles proposed by these workers, and confirmed by such methods as X-ray diffraction and electron microscopy (Matthews, 1981), dictate that the protein sub-units must be arranged in shells having a basic icosahedral symmetry (i.e. having 20 plane faces). Variations occurring only in the way the individual sub-units are packed in lattices to make up the shell, and these are based on sub-divisions of the icosahedral structure, as shown in Figure 1.4 (Gibbs and Harrison, 1976; Matthews, 1981). Studies on the structure of various viruses with icosahedral symmetry, indicate that the nucleic acid genome is usually closely associated with the internal surface of the protein sub-units (*see* Figure 1.3*a*).

Intensive studies have been made of the structure of the rod-shaped particles of tobacco mosaic virus (TMV). X-ray diffraction pictures have shown that the TMV rod consists of protein sub-units built up in a regular, helical array, with the RNA chain compactly coiled in a corresponding helix on the inside of the protein sub-units (*see* Figure 1.3*b*). The protein coat and RNA genome, surround an axial hole or

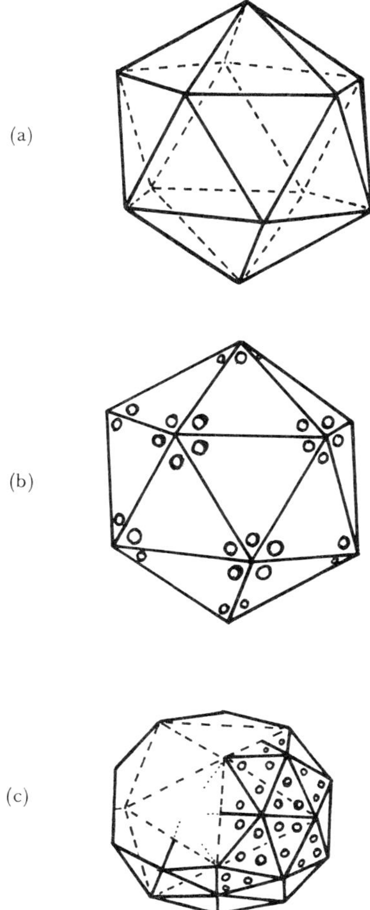

Fig. 1.4 Diagram of a regular icosahedron (*a*), one possible arrangement of its protein subunits (*b*), and a sub-divided icosadeltahedron (*c*). (*a*) The structure of many of the smaller isometric viruses is based on an icosahedron arrangement. A regular icosahedron has 20 identical triangular faces and each face may be composed of 3 structural protein sub-units of any shape in a regular position (*see* Caspar and Klug, 1962), making a total of 60 sub-units; (*b*) one possible arrangement is for the sub-units to be clustered in groups of five at the vertices (pentamer clusters), as is the case in particles of satellite virus; (*c*) many isometric viruses are composed of more than 60 sub-units and their arrangement is based on sub-division of the basic icosahedron. A common arrangement for many isometric viruses is for the faces of the basic icosahedron to be divided into 9 to form an icosadeltahedron consisting of 180 sub-units. Different viruses vary in the way these sub-units are clustered to form each face of the sub-division.

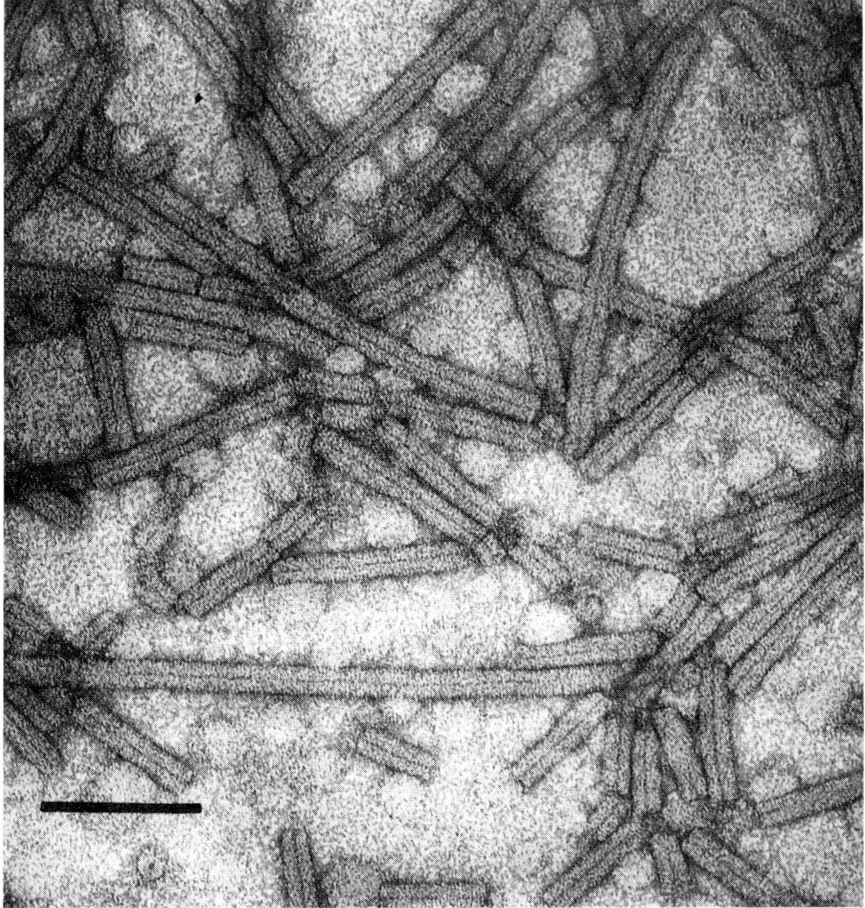

Plate 1.2 Electron micrograph of particles of tobacco mosaic virus showing the central axial canal, magnification bar = 100 nm (courtesy of M. J. W. Webb).

canal that is clearly visible in particles observed in the electron microscope (*see* Plate 1.2). The rod-shaped particles of tobacco rattle virus have been shown to have a basic structure, similar to those of TMV, but the structure of other rod-shaped viruses has not been so extensively studied. The structure of rod-shaped viruses has been reviewed by Hull (1976).

Plant viruses vary considerably in the relative amounts of nucleic acid and protein they contain (*see* Table 1.3). Viruses with isometric particles may contain between 15% and 45% nucleic acid, whereas viruses with rod-shaped particles have only about 5% nucleic acid.

Although nucleic acid and protein are the main chemical constituents of most viruses, they have also been shown to contain cations, polyamines and enzymes (Matthews, 1981). Some viruses, such as tomato spotted wilt, also contain significant quantities of lipids (19%, Matthews, 1975). All virus particles also contain various amounts of water, which may represent up to 50% of the total weight of a virus in suspension, but the water content is normally disregarded when the relative amounts of protein and nucleic acid are described.

Table 1.3 Examples of the relative amounts of nucleic acid and protein in particles of different plant viruses

Virus	% *Nucleic acid**	% *Protein*	*Particle shape and size* (nm)
Tobacco mosaic	5	95	Rigid rod 300 × 18
Potato virus X	5–7	93–95	Flexuous rod 515 × 13
Carnation latent	6	94	Flexuous rod 650 × 12
Potato virus Y	5	95	Flexuous rod 730 × 11
Beet yellows	5	95	Flexuous rod 1250 × 10
Tomato bushy stunt	17	83	Isometric 30
Cauliflower mosaic	17†	83	Isometric 50
Wound tumour	22	78	Isometric 70
Cowpea mosaic	23 & 34	66–77	Isometric 24
Turnip yellow	35	65	Isometric 28–30

* Information based on Matthews (1979).
† Deoxyribonucleic acid (DNA), all other examples have ribonucleic acid (RNA) genomes.

1.4.2 Multicomponent plant viruses

The division of the genetic information into two or more parts, is a feature found exclusively in certain groups of RNA viruses (*see* Section 2.2). In such *multi-component* viruses, the individual components are not infectious alone and when a host plant is inoculated with virus the two or more genomic elements must be present in the inoculum, if virus replication is to occur.

Typical multi-component viruses may have two (e.g. tobacco ringspot virus – nepovirus group), or three (e.g. brome mosaic virus – bromovirus group) major molecular species of RNA encapsidated in separate protein shells. The situation is even more complex in tobacco streak virus (ilarvirus group), where in addition to the three major separately encapsidated RNA molecules, a fourth, smaller RNA molecule is also required for infection. This fourth RNA is encapsidated together with one of the three major RNA molecules. Individual component particles, even if they are the same size, may differ in their sedimentation rates in a sucrose density gradient (*see* Section 6.8) because of the different quantities of nucleic acid they contain. This allows them to be physically separated from one another, and enables virologists to demonstrate that more than one type of particle is required for infection, and even to construct hybrid viruses by mixing the necessary genomes in a test-tube (Jaspars, 1974). Laboratory mixing also makes it possible to discover which part of the genome carries the genetic determinants for such characters as vector transmission, protein coat serology type and symptom pattern (Fulton, 1980).

The RNA genomic components of some multi-component viruses, such as those of cucumber mosaic virus (cucumovirus group), are so similar in their molecular weight that they are difficult to separate by sedimentation (Fulton, 1980), but the components of others, such as tobacco streak virus (ilarvirus group), are readily separated, because the genomic elements are contained in isometric particles of different diameter.

In tobacco rattle and pea early browning viruses (tobravirus group), the two genomic RNA molecular species are contained in rod-shaped particles of two different lengths (Lister, 1968), and both sized rods are required for complete infection. On their own the shorter rods are not infectious, and the longer rods alone can produce RNA, but not a protein coat. The reason for this is that the long rod alone, lacks the genetic coding for the capsid (coat) protein, whereas the shorter rods contain the coat protein gene, but lack other genes essential for replication.

Fulton (1980) suggested that one advantage of the divided genome in multi-component viruses, is that it allows the virus particle to remain small, and to be easily protected by its protein capsid. For if the elements of the genome were combined the amount of protein required to protect it might be inadequate. A more likely explanation for the origin of the divided genome, however, has been provided by Reanney (1982). He suggests that if the genome is divided into a number of short RNA strands rather than one long one, lethal errors (referred to as *noisiness*) that may occur during replication will be less important. In the replication of a single long-stranded genome, for instance, a single error would be lethal, but if the genome is multipartite, the same error would only eliminate one strand of the genome, leaving the other parts intact and capable of participating in the replication process.

The ability of virologists to mix genomic material from different virus strains in the laboratory to create new hybrid strains of a virus, also indicates that the divided genome may be of evolutionary advantage to the virus. This fact demonstrates the possibility, if not probability, that such hybrids may occur naturally under suitable conditions.

The nature and importance of multi-component plant viruses has been reviewed by Jaspars (1974), Reijnders (1978) and Fulton (1980).

1.4.3 Satellites of plant viruses

In addition to the multi-component genome that has been described in Section 1.4.2, other complex genomic systems occur which control or modify the infection process of some plant viruses. The most important of these are plant virus '*satellites*', the subject of a recent review by Murant and Mayo (1982).

The term virus satellite was first used to describe the small spherical virus associated with tobacco necrosis virus (TNV) (Kassanis, 1962). In this relationship the larger TNV particles (30 nm in diameter) often occur quite independently, and are able to replicate normally from their own nucleic acid genome. Sometimes, however, the small satellite virus (STNV) particles (17 nm in diameter) occur in joint infection with TNV in the host plant, and are dependent upon the TNV genome for their own replication. TNV and STNV are not serologically related (*see* Section 6.6).

In addition to describing complete virus-like particles that are dependent on a '*helper*' virus for their replication, the term satellite is also used to describe certain nucleic acid molecules that are unable to multiply in the host cell without the aid of a specific helper virus. In all these cases the helper virus itself can exist quite independently.

The best documented occurrence of a satellite RNA is that associated with cucumber mosaic virus, which is referred to as CMV-RNA 5 or CARNA-5 RNA (Kaper and Waterworth, 1977). The basic CMV genome is multi-component, consisting of three separately encapsidated RNA molecules (RNA 1, RNA 2 and RNA 3). All three are necessary for virus replication (N.B. RNA 4 is encapsidated with the three major RNA molecules and does not carry basic genetic information necessary for the replication of the CMV genome). The additional CMV-RNA 5 is not necessary for the replication of CMV, but is a true satellite in that it is dependent upon the remainder of the CMV genome for its own replication. The satellite CMV-RNA 5 is encapsidated together with the other CMV genomic elements, and is produced in varying amounts in different host species (Kaper and Tousignant, 1977).

A satellite RNA-5 nucleic acid is also associated with tomato aspermy and peanut stunt viruses, two other members of the cucumovirus group. Other viruses reported to produce satellite RNAs include the nepoviruses tobacco ringspot (Schneider, 1969; 1977), tomato black ring (Murant *et al.*, 1973), arabis mosaic (Clark and Davies, 1980), strawberry latent ringspot (Mayo *et al.*, 1974), myrobalan latent ringspot (Gallitelli *et al.*, 1981) and chicory yellow (Quacquarelli *et al.*, 1973); turnip crinkle virus a possible member of the tombusvirus group (Altenbach and Howell, 1981) and panicum mosaic virus (Buzen *et al.*, 1977).

Satellite RNAs should not be confused with genomic fragments of RNA which may occur during normal virus replication (Murant and Mayo, 1982), or with nucleic acids that multiply independently in plants with the production of infective particles, but which rely on other viruses for some biological functions such as transmission by aphids (Falk and Duffus, 1981, and Section 7.4.2).

1.5 References

Aerts, J. (1977). De betekenis van virusvrij plantmateriaal voor de glasaardbeienteelt. *Meded Fac. Landbouwwet Gent*, **42**, 1135–9.

Altenbach, S. B. and Howell, S. H. (1981). Identification of a satellite RNA associated with turnip crinkle virus. *Virol*, **112**, 25–33.

Atkinson, T. G and Grant, M. N. (1967). An evaluation of streak Mosaic losses in winter wheat. *Phyt*, **57**, 188–92.

Bald, J. G. and Norris, D. O. (1941). Obtaining virus-free potatoes. *J Council Sci Ind Res Aust* **14**, 187–90.

Bawden, F. C. (1950). *Plant viruses and virus diseases*. Chronica Botanica: Waltham, Massachusetts

Bawden, F. C. and Pirie, N. W. (1937). The isolation and some properties of liquid crystalline substances from solanaceous plants infected with three strains of tobacco mosaic virus. *Proc R Soc Lond Ser B Biol Sci* **123**, 274–320.

Bawden, F. C. and Pirie, N. W. (1938). Crystalline preparations of tomato bushy stunt virus. *Br J Ex Pathol* **19**, 251–63.

Beijerinck, M. W. (1898). Over een contagium vivum fluidum als oorzaak van de vlekziekte der tabaksbladen. *Versl Gewone Vergad Wis Natuurk. Afd. K.Akad. Wet. Amsterdam* **7**, 229–35.

Bjorling, K. (1949). Virus yellows of beet. Symptoms and influence on the Swedish sugar beet yield. *Socker Handl* **5**, 119–40.

Bos, L. (1982). Crop losses caused by viruses. *Crop Protect* **1**, 263–82.

Brakke, M. K. (1951). Density gradient centrifugation, a new separation technique. *J Am Chem S* **73**, 1847–48.

Brenner, S. and Horne, R. W. (1959). A negative straining method for high resolution electron microscopy of viruses. *Bio Biop Acta* **34**, 103–10.

Broadbent, L. (1976). Epidemiology and control of tomato mosaic virus. *Ann Rev Phyto* **14**, 75–96.

Brunt, A. A. (1975). The effects of cocoa swollen-shoot virus on the growth and yield of Amelonado and Amazon cocoa (*Theobroma cacao*) in Ghana. *Ann Appl Biol* **80**, 169–80.

Buzen, F. G., Niblett, C. L. and Hooper, G. R. (1977). A possible satellite virus of panicum mosaic virus. *Proc Am Phytopathol Soc* **4**, 132.

Cameron, H. R. (1977). Effects of viruses on deciduous fruit trees. *Hortic Sci* **12**, 484–87.

Cane, H. (1720). *Roy Soc Lond Philos Trans B* **31**, 102.

Caspar, D. L. D. (1964). Structure and function of regular virus particles. In *Plant Virology* (ed. Corbett, M. K. and Sisler, H. D.). University of Florida Press: Florida.

Caspar, D. L. D. and Klug, A. (1962). Physical principles in the construction of regular viruses. *Cold Spr Harb Symp Quant Biol* **27**, 1–24.

Clark, M. F. and Davies, D. L. (1980). Arabis mosaic virus and hop nettlehead disease. *Ann Rep East Malling Res Stn* 1979, pp. 101–2.

Cohen, M. (1981). Beneficial effects of viruses for horticultural plants. *Hortic Rev* **3**, 394–410.

Converse, R. H. (1963). Influence of heat-labile components of the raspberrry mosaic virus complex on growth and yield of red raspberrries. *Phyt* **53**, 1251–4.

Corbett, M. K. (1964). Introduction. In *Plant Virology* (ed. Corbett, M. K. and Sisler, H. D.). University of Florida Press: Florida.

Crick, F. H. C. and Watson, J. D. (1956). Structure of small viruses. *Nature* **177**, 473–5.

Diener, T. O. (1971). Potato spindle tuber 'virus'. IV. A replicating, low molecular weight RNA. *Virol* **45**, 411–28.

Doi, Y., Teranaka, M., Yora, K. and Asuyama, H. (1967). Mycoplasma or PLT group-like micro-organisms found in the phloem elements of plants infected with mulberry dwarf, potato witches broom, aster yellows, or Paulownia witches broom. *Ann Phytopathol Soc Jpn* **33**, 259–66.

Dubos, R. J. (1958). In Pollard, p. 291.

Duffus, J. E. (1973). The yellowing virus disease of beet. *Adv Virus Res* **18**, 374–86.

Falk, B. W. and Duffus, J. E. (1981). Epidemiology of helper-dependent persistent aphid transmitted virus complexes. In *Plant diseases and vectors, ecology and epidemiology* (ed. Maramorosch, K. and Harris, K. F.). Academic Press: London, pp. 161–79.

Freeman, J. A. and Stace-Smith, R. (1970). Effects of raspberry mosaic viruses on yield and growth of red raspberry. *Can J Plant Sci* **50**, 521–7.

Fukushi, T. (1934). Studies on the dwarf disease of the rice plant. *J Fac Agric Hokkaido Univ* **78**, 41–164.

Fulton, R. W. (1980). Biological significance of multicomponent viruses. *Ann Rev Phyto* **18**, 131–46.

Gallitelli, D., Piazzolla, P., Savino, V., Quacquarelli, A. and Martelli, G. P. (1981). A comparison of myrobalan latent ringspot virus with other nepoviruses. *J Gen Virol* **53**, 57–65.

Gibbs, A. J. and Harrison, B. D. (1976). *Plant virology, the principles*, Edward Arnold: London.

Gierer, A. and Schramm, G. (1956). Infectivity of ribonucleic acid from tobacco mosaic virus. *Nature* **177**, 702.

Gooding, G. V. (1970). Tobacco etch virus and tobacco mosaic virus in *Nicotiana tabacum*. In *Crop loss assessment methods* (ed. Chiarappa, L.). Commonwealth Agricultural Bureaux: Slough.

Gooding, G. V. and Main, C. E. (1981). Estimating losses caused by tomato vein mottling virus in Burley tobacco. *Plant Dis R* **65**, 889–91.

Guthrie, E. J. (1978). Measurement of yield losses caused by maize streak disease. *Plant Dis R* **62**, 839–41.

Hampton, R. O. (1975). The nature of bean yield reduction by bean yellow and bean common mosaic viruses. *Phyt* **65**, 1342–6.

Harper, F. R., Nelson, G. A., and Pittman, U. J. (1975). Relationship between leafroll symptoms and yield in Netted Gem potato. *Phyt* **65**, 1242–4.

Heathcote, G. D. (1978). Review of losses caused by virus yellows in English sugar beet crops and the cost of partial control with insecticide. *Plant Path* **27**, 12–17.

Heiling, A. (1953). *Zucker* **6**, 27.

Hinegardner, R. (1976). Evolution of genome size. In *Molecular evolution* (ed. F. J. Ayaya), Sinauer: Sunderland, Massachusetts. pp. 179–99.

Holmes, F. O. (1929). Local lesions in tobacco mosaic. *Bot Gaz* **87**, 39–55.

Hull, R. (1958). Sugar beet yellows. The search for control. *Agr (Gt Brit)* **65**, 62–5.

Hull, R. (1976). The structure of tubular viruses. *Adv Virus Res* **20**, 1–32.

Inouye, T. and Osaki, T. (1981). *The earliest record of a possible plant virus disease.* Abst. p. 22/05, 5th. Int. Cong. of Virology, Strasbourg, p. 238.

Ivanowski, D. (1892). Ueber die mosaikkrankheit der tabakspflanze. *Bull Acad Imp Sc St. Petersb (New Ser.)* **3**, 65–70.

James, W. C. (1974). Assessment of plant diseases and losses. *Ann Rev Phyto* **12**, 27–48.

Jaspars, E. M. J. (1974). Plant viruses with a multipartite genome *Adv Virus Res* **19**, 37–149.

Kaper, J. M. and Tousignant, M. E. (1977). Cucumber mosaic virus – associated RNA 5. I. Role of host plant and helper strain in determining amount of associated RNA 5 with virions. *Virol* **80**, 186–95.

Kaper, J. M. and Waterworth, H. E. (1977). Cucumber mosaic virus – associated RNA 5: Causal agent for tomato necrosis. *Sci* **196**, 429–31.

Kassanis, B. (1954). Heat therapy of virus-infected plants. *Ann Appl Biol* **41**, 470–4.

Kassanis, B. (1962). Properties and behaviour of a virus depending for its multiplication on another. *J Gen Micro* **27**, 477–88.

Kausche, G. A. Pfankuch, E. and Ruska, H. (1939). Die sichtbormachung von pflanzlichem virus im ubermikroskop. *Naturwissen* **27**, 292–99.

Klug, A., Longley, W. and Leberman, R. (1966). Arrangement of protein subunits and the distribution of nucleic acid in turnip yellow mosaic virus. 1. X-ray diffraction series. *J Mol Biol* **15**, 315–43.

Ling, K. C. (1972). Rice virus diseases. *Int Rice Res Inst* (*Los Banos*) *Tech Bull*, pp. 1–142.

Lister, R. M. (1968). Functional relationships between virus-specific products of infection by viruses of the tobacco rattle type. *J Gen Virol* **2**, 43–58.

Lwoff, A. (1957). The concept of virus. *J Gen Micro* **17**, 239–53.

Markham, R. (1953). Nucleic acids in virus multiplication. *Soc Gen Micro Symp.* **2**, 85–98.

Markham, R. (1977). Landmarks in plant virology. *Ann Rev Phyto* **15**, 17–39.

Marshall, K. C. (1976). *Interfaces in microbial ecology.* Harvard University Press: Cambridge.

Matthews, R. E. F. (1975). A classification of virus groups based on the size of the particle in relation to genome size. *J Gen Virol* **27**, 135–49.

Matthews, R. E. F. (1979). Classification and nomenclature of viruses. 3rd report of the International Committee on Taxonomy of Viruses. *Intervirol* **12**, 131–296.

Matthews, R. E. F. (1981). *Plant Virology*. Academic Press, pp. 1–897.

Mayer, A. (1886). Ueber die mosaikkrankheit des tabaks. *Landwirtsch Vers Stn* **32**, 451–67.

Mayo, M. A., Murant, A. F., Harrison, B. D. and Goold, R. A. (1974). Two protein and two RNA species in particles of strawberry latent ringspot virus. *J Gen Virol* **24**, 29–37.

McKay, M. B. and Warner, M. F. (1933). Historical sketch of tulip mosaic or breaking. The oldest known plant virus disease. *Natl Hortic Mag* **3**, 179–216.

Morel, G. M. and Martin, C. (1952). Guérison de dahlias atteints d'une maladie a virus. *C R Hebd Séances Acad Sci* **235**, 1324–5.

Murant, A. F. and Mayo, M. A. (1982). Satellites of plant viruses. *Ann Rev Phyto* **20**, 49–70.

Murant, A. F., Mayo, M. A., Harrison, B. D. and Goold, R. A. (1973). Evidence for two functional RNA species and a 'satellite' RNA in tomato blackring virus. *J Gen Virol* **19**, 275–8.

Nienhaus, F. and Sikora, R. A. (1979). Mycoplasmas, spiroplasmas, and rickettsia-like organisms as plant pathogens. *Ann Rev Phyto* **17**, 37–58.

Palmer, L. T. and Brakke, M. K. (1975). Yield reduction in winter wheat infected with soilborne wheat mosaic virus. *Plant Dis R* **59**, 469–71.

Phillips, E. (1720). *New world of words or universal English dictionary*. 7th edn, Kings Arm in St. Paul's Church Yard: London.

Pirie, N. W. (1962). *Perspectives in Biology and Medicine* **5**, 446.

Plumb, R. T. (1981). Chemicals in the control of cereal virus diseases. In *Strategies for the control of cereal disease* (ed. Jenkyn, J. K. and Plumb, R. T.). pp. 135–45, Blackwell Scientific Publications.

Purdy, H. A. (1929). Immunologic reactions with tobacco mosaic virus. *J Exp Med* **49**, 919–35.

Quacquarelli, A., Piazzolla, P., Vovlas, C. and Martelli, G. P. (1973). Chicory yellow mottle; a new multicomponent plant virus. *Acta Biol Yugosl Ser B Mikrobiol* **10**, 15–25.

Rast, A. T. B. (1972). MI.1–16, an artificial symptomless mutant of tobacco mosaic virus for seedling inoculation of tomato crops. *Neth J Plant Pathol* **78**, 110–12.

Reanney, D. C. (1982). The evolution of RNA viruses. *Ann Rev Microbiol* **36**, 47–74.

Reijnders, L. (1978). The origin of multicomponent small ribonucleoprotein viruses. *Adv Vir Res* **23**, 79–102.

Sanger, H. L. (1968). Characteristics of tobacco rattle virus 1. Evidence that its two particles are functionally defective and mutually complementing. *Mol Gen Genet* **101**, 346–67.

Schmidt, H. (1972). The effect of latent virus infections on the yield of maiden trees on 20 apomictic apple seedling rootstocks. *J Hort Sci* **47**, 159–63.

Schneider, I. R. (1969). Satellite-like particle of tobacco ringspot virus that resembles tobacco ringspot virus. *Sci* **166**, 1627–9.

Schneider, I. R. (1977). Defective plant viruses. In *Beltsville Symposia in Agr.Res* (ed. Romberger, J. A., Anderson, J. D. and Powell, R. L.) Allanheld, Osmun & Co:

Seif, A. A. (1982). Effect of cassava mosaic virus on yield of cassava. *Plant Dis R* **66**, 661–2.

Sill, W. H., Fellows, H. and King, C. L. (1955). Kansas wheat mosaic situation (1953–4). *Plant Dis R* **39**, 29–30.

Slykhius, J. T. (1976). Virus and virus-like diseases of cereal crops. *Ann Rev Phyto* **14**, 189–210.

Smith, P. R. and Sward, R. J. (1982). Crop loss assessment studies on the effect of barley yellow dwarf virus in wheat in Victoria. *Aust J Agric Res* **33**, 179–85.

Stanley, W. M. (1935). Isolation of a crystalline protein possessing the properties of tobacco mosaic virus. *Sci* **81**, 644–5.

Tomlinson, J. A. and Walkey, D. G. A. (1967). The isolation and identification of rhubarb viruses occurring in Britain. *Ann Appl Biol* **59**, 415–27.

Villalon, B. (1981). Breeding peppers to resist virus diseases. *Plant Dis R* **65**, 557–62.

Walkey, D. G. A. (1980). Production of virus-free plants. *Acta Hort* **88**, 23–31.

Walkey, D. G. A., Creed, C., Delaney, H. and Whitwell, J. D. (1982). Studies on the reinfection and yield of virus-tested and commercial stocks of rhubarb cv. Timperley Early. *Plant Path* **31**, 253–61.

Walkey, D. G. A. and Ward, C. M. (1983). Unpublished results.

Walkey, D. G. A. and Webb, M. J. W. (1978). Internal necrosis in stored white cabbage caused by turnip mosaic virus. *Ann Appl Biol* **89**, 435–41.

Waterworth, H. E. (1976). Effect of some pome fruits viruses, mycoplasma-like, and other pests on Bradford ornamental pear trees. *Plant Dis R* **60**, 104–5.

Watson, M. A., Watson, D. J. and Hull, R. (1946). Factors affecting the loss of yield of sugar beet caused by beet yellows virus. *J Agric Sci* **36**, 151–66.

Williams, R. C. and Wycoff, R. W. G. (1945). Electron shadow-micrography of virus particles. *Proc Soc Exp Biol Med* **58**, 265–70.

Wright, N. S. (1974). Combined effects of potato viruses X and S on yield of Netted Gem and White Rose potatoes. *Am Potato J* **47**, 475–8.

Wutscher, H. K. (1977). Citrus tree virus and virus-like diseases. *Hort. Sci.* **12**, 478–84.

Zink, F. W. and Kimble, K. A. (1960). Effect of time of infection by lettuce mosaic virus on rate of growth and yield of Great Lakes Lettuce. *Proc Am Soc Hort Sci* **76**, 448–54.

1.6 Further Selected Reading

Ansa, A. O., Bowyer, J. W. and Shepherd, R. J. (1982). Evidence for replication of cauliflower mosaic virus DNA in plant nuclei. *Virol* **121**, 147–56.

Bos, L. (1982). Crop losses caused by viruses. *Crop Prot* **1**, 263–82.

Gordon, K. H., Gill, D. S. and Symons, R. H. (1982). Highly purified cucumber mosaic virus induced RNA-dependent RNA polymerase does not contain any of the full length translation products of the genomic RNAs. *Virol* **123**, 284–95.

Hall, T. C. and Davies, J. W. (1979). *Nucleic acids in plants.* C.R.C. Press: Florida.

Matthews, R. E. F. (1981). *Plant Virology.* Academic Press: London.

Murant, A. F. and Mayo, M. A. (1982). Satellites of plant viruses *Ann Rev Phyto* **20**, 49–70.

Reddy, D. V. R., Rhodes, D. P., Lesnaw, J. A., McLeod, R., Banerjee, A. K. and Black, L. M. (1977). In vitro transcription of wound tumor virus RNA by virion-associated RNA transcriptase. *Virol* **80**, 356–61.

Toriyama, S. and Peters, D. (1980). In vitro synthesis of RNA by dissociated lettuce necrotic yellows virus particles. *J Gen Virol* **50**, 125–34.

2 *Plant Virus Classification*

2.1 Introduction

For all types of organism some system of naming and grouping is required, if order is to be created out of chaos. In this respect the viruses which infect the higher plants (Angiospermae) are no exception. In classification schemes, however, if a system is going to stand the test of time, it is essential for individuals to be grouped according to consistent and accurately determined characteristics and relationships.

In considering plant virus classification, it must be remembered that the first plant virus was only purified and partially characterized as recently as the mid 1930s (Stanley, 1935; Bawden and Pirie, 1936). Until then most plant virologists usually gave a virus a name based on the host plant in which it was found, and the disease symptoms it caused. For example, the virus inducing mosaic symptoms in tobacco was called *tobacco mosaic virus*.

In 1927, Johnson proposed a system for naming and grouping plant viruses, on the basis that a virus should take the common (vernacular) name of the host, with an appropriate number added. Thus, tobacco mosaic virus would be called *tobacco virus 1.* Later in 1937, Smith proposed that the name should be Latinized and the generic name of the host used, so tobacco virus 1 would become *Nicotiana virus 1.* This was followed by attempts to use a Latin binomial system, under which tobacco mosaic virus would be called *Marmor tabaci* (Holmes, 1939; 1948). These complex schemes gained little support amongst plant virologists and for the next two decades most workers preferred to use a virus's vernacular name, with the name of the host in which the virus was first described taking precedence over hosts discovered later. This is irrespective of the host's economic importance. Thus, arabis mosaic virus which causes important diseases in strawberry, raspberry and other commercial crops (Murant, 1970) retains the name of the economically unimportant *Arabis* species from which it was first isolated.

The vernacular names of plant viruses are frequently long and virologists soon started to abbreviate them to initial letters, for example *TMV* for tobacco mosaic virus. These abbreviations have not been officially standardized, however, and confusing situations can arise with viruses such as cucumber mosaic and cauliflower mosaic which have the same initial letters. Consequently, authors always write the full vernacular name of the virus followed by its abbreviation, on the first occasion the name is used in a publication, e.g. cucumber mosaic virus (CMV). Elsewhere in the publication only the abbreviation is used. Most virologists now try to standardize abbreviated names based on previous, well-used abbreviations, and so cucumber mosaic virus is usually referred to as CMV and cauliflower mosaic virus as CaMV.

In addition to the vernacular name, a system of cryptograms was introduced to give concise information on the properties of individual viruses (Gibbs *et al.*, 1966; Gibbs and Harrison, 1968). The cryptogram besides giving the reader an immediate summary of a virus's specific characteristics, also helped to relate the particular virus to others with similar properties. An example of a cryptogram and an explanation of its use is given in Table 2.1, and the cryptogram for the type member of each plant virus group is given in Section 2.2. Cryptograms are still frequently encountered in plant virus literature, but in general their use is decreasing (Matthews, 1981).

Table 2.1 Key to plant virus cryptograms

Example of a cryptogram

Tobacco mosaic virus (TMV) =	R/1 :	2/5 :	E/E :	S/O.
	(1st)	(2nd)	(3rd)	(4th)

Key to terms

1st term:	type of nucleic acid/number of nucleic acid strands	
	R = RNA	1 = single
	D = DNA	2 = double
2nd term:	Molecular weight of nucleic acid in millions/% of nucleic acid in infective particle	
3rd term:	outline of particle shape/outline of nuclear capsid	
	S = spherical (isometric)	
	E = elongate (rod shaped)	
	B = bacilliform.	
4th term:	type of host infected/type of vector	
	B = bacterium	Ap = aphid
	F = fungus	Au = leafhopper
	I = invertebrate	Cl = beetle
	S = seed plant (Angiosperm)	Fu = fungus
		Ne = nematode
		Th = thrips
		W = whitefly
		O = spreads without vector
		Se = seed transmitted

*Information unknown or unconfirmed.

The requirement for a sound classification system for plant viruses, based on well defined characteristics, became imperative during the 1950s and 60s as more and more information on individual plant viruses was accumulated. This requirement was met by the appointment, at the International Congress of Microbiology held in Moscow in 1966, of a committee to investigate virus taxonomy. The committee later became known as the International Committee on Taxonomy of Viruses (ICTV) and their objective was to develop an internationally agreed classification and nomenclature for all viruses, including those infecting plants. As a result of the work of this committee and its Plant Virus Sub-Committee, a system for plant virus classification has been introduced, based on such characteristics as virus particle morphology, type and quantity of nucleic acid, genome structure and type of vector. This classification system has now become widely accepted by most plant virologists.

The characteristics of the plant virus groups which have received international approval by the ICTV are summarized in Section 2.2

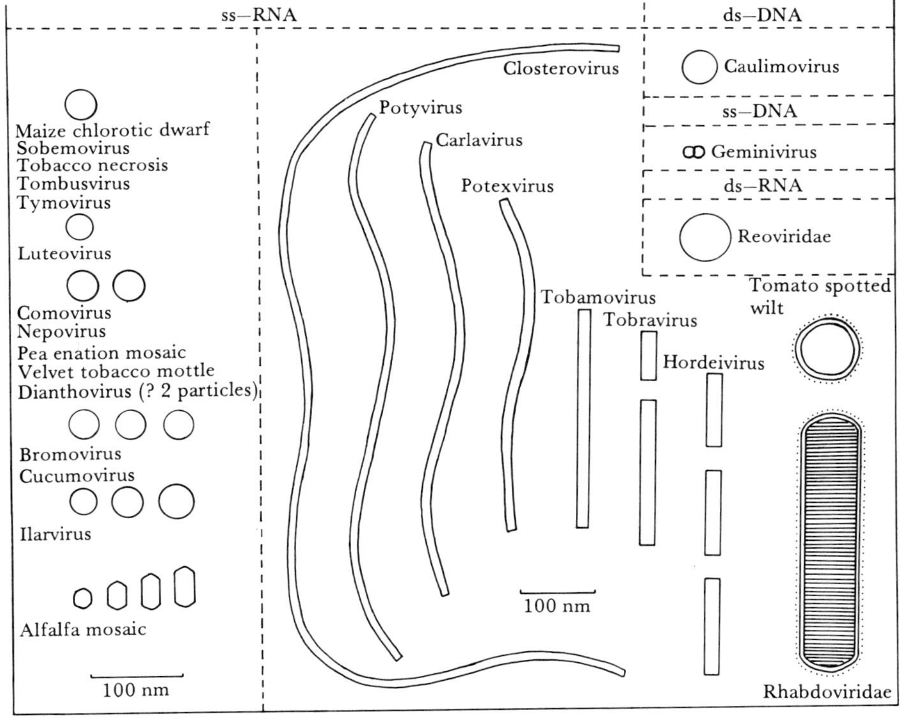

Fig. 2.1 Plant virus groups and families.

Table 2.2 Plant virus groups with ss-RNA genomes and isometric particles without envelopes

Virus group	*Type member*	*Genome structure*	*Particle size*	*Reference*	*CMI/AAB No.*
Luteovirus	Barley yellow dwarf (BYDV)	Monopartite	25	Rochow & Duffus (1981)	–‡
Maize chlorotic dwarf*	MCDV	Monopartite	30	Gingery *et al.* (1981)	–
Sobemovirus	Southern bean mosaic (SBMV)	Monopartite	28	Sehgal (1981)	–
Tobacco necrosis*	TNV	Monopartite	26–28	Uyemoto (1981)	–
Tombusvirus	Tomato bushy stunt (TBSV)	Monopartite	30	Martelli (1981)	–
Tymovirus	Turnip yellow mosaic (TYMV)	Monopartite	28–30	Koenig & Lesemann (1981)	214
Comovirus	Cowpea mosaic (CPMV)	Bipartite	24	Stace-Smith (1981)	199
Dianthovirus	Carnation ringspot (CRSV)	Bipartite	31–34	Matthews (1982)	21
Nepovirus	Tobacco ringspot (TRSV)	Bipartite	28	Murant (1981)	185
Pea enation mosaic*	PEMV	Bipartite	30	Hull (1981)	–
Alfalfa mosaic*†	AlfMV	Tripartite	18–58 × 18	Van Regenmortel & Pinck (1981)	–
Bromovirus	Brome mosaic (BMV)	Tripartite	26	Lane (1981)	215
Cucumovirus	Cucumber mosaic (CMV)	Tripartite	28	Kaper & Waterworth (1981)	–
Ilarvirus	Tobacco streak (TSV)	Tripartite	26–35	Fulton (1981)	275
Velvet tobacco mottle	VtMoV	–§	30	Matthews (1982)	–

* Not yet given an approved ICTV group name

† Particles are bacilliform but thought to be isometric in origin.

‡ Group not yet described in CMI/AAB descriptions of plant viruses, *see* Section 12.8 for description No.'s of type members.

§ Genome differs from those of other virus groups in this Table in that the nucleic acid consists of linear and circular ss-RNA (*see* Matthews, 1982).

Table 2.3 Plant virus groups with ss-RNA genomes and rod-shaped particles

		Particle			
Virus group	*Type member*	*Shape*	*Size (nm)*	*Reference*	*CMI/AAB No.*
Tobravirus	Tobacco rattle (TRV)	rigid rod†	180–215 / 46–114 × 22	Harrison & Robinson (1981)	–*
Tobamovirus	Tobacco mosaic (TMV)	rigid rod	300 × 18	Van Regenmortel (1981)	184
Hordeivirus	Barley stripe mosaic (BSMV)	rigid rod†	100–150 × 20	Jackson & Lane (1981)	–
Potexvirus	Potato virus X (PVX)	flexuous rod	470–580 × 13	Purcifull & Edwardson (1981)	200
Carlavirus	Carnation latent (CLV)	flexuous rod	620–700 × 13	Wetter & Milne (1981)	259
Potyvirus	Potato virus Y (PVY)	flexuous rod	680–900 × 11	Hollings & Brunt (1981)	245
Closterovirus	Beet yellows (BYV)	flexuous rod	600–2000 × 10	Lister & Bar-Joseph (1981)	260

* Group not yet described in CMI/AAB descriptions of plant viruses, *see* Section 12.8 for description number of TRV.
†Virus has a multi-component genome.

Table 2.4 Other plant virus groups

Virus group	Type member	Genome type	Particle: Shape	Particle: Size (nm)	Reference	CMI/AAB. No.
Rhabdoviridae	Lettuce necrotic yellows (LNYV)	ss-RNA monopartite	Bacilliform (within envelope)	160–380 × 50–90	Francki *et al.* (1981)	244
Tomato spotted wilt	TSWV	ss-RNA multipartite	Isometric (within envelope)	85	Francki & Hatta (1981)	–
Reoviridae						
i. Phytoreovirus	Wound tumour (WTV)	ds-RNA multipartite	Isometric	70	Shikata (1981)	–*
ii. Fijivirus	Fiji disease (FDV)					
Geminivirus	Maize streak (MSV)	ss-DNA monopartite†	Isometric (in pairs)	18–20	Goodman (1981)	–
Caulimovirus	Cauliflower mosaic (CaMV)	ds-DNA monopartite	Isometric	50	Shepherd & Lawson (1981)	–

* Group not yet described in CMI/AAB descriptions of plant viruses, *see* Section 12.8 for description numbers of type members.
† The structure of this genome has still to be confirmed. Some group members (e.g. cassava latent virus) are reported to be bipartite.

and have been described in more detail by Matthews (1979 and 1982) and in the '*Handbook of Plant Virus Infection*' (Kurstak, 1981). At the present time plant viruses have been classified into twenty-seven groups of which two are referred to as families (*see* Figure 2.1 and Tables 2.2–2.4). Unlike the animal virus groups, the plant virus groups are not normally referred to as families, but the two families, the *rhabdoviridae* and the *reoviridae*, include viruses that can infect plants, arthropods and vertebrates. Of the remaining twenty-five groups, nineteen have been given ICTV approved names and the other six groups are at present known by the name of the type member of the group. In this classification scheme the viruses have been grouped according to their known physical, chemical and biological characters. Frequently only a few essential characters need to be known in order to identify a virus as belonging to a particular group. In many cases the most characteristic feature of a virus is the nature of its genome. For example, only two plant virus groups (*see* Table 2.4) have a genome consisting of deoxyribonucleic acid (DNA), one of these, the caulimovirus group, has double stranded (ds) DNA and the other, the geminivirus group, has single stranded (ss) DNA. These characters, together with the distinct morphologies of the virus particles (*see* Figure 2.1 and Plate 2.1), immediately identifies members of these groups.

The majority of plant virus groups have a genome consisting of ribonucleic acid (RNA) and additional properties of the genome or particle morphology must usually be known, before the group can be identified. Viruses of one group, the family *reoviridae*, have a genome consisting of ds-RNA, but the remaining twenty-four groups have ss-RNA genomes. Two of these groups, the family *rhabdoviridae* and the tomato spotted wilt virus group, have particles contained within an envelope, but the particle morphology of the two groups is quite distinct (*see* Figure 2.1).

The particles of the remaining ss-RNA groups are not surrounded by envelopes. Seven groups have elongated, rod-shaped particles (*see* Figure 2.1 and Table 2.3), and are classified by their particle length and other characters (*see* Section 2.2). The remaining fifteen have isometric particles (N.B. the alfalfa mosaic virus group has some particles which are bacilliform in shape, but these are thought to have evolved from an isometric form (Hull, 1969)). Some of these fifteen groups have a genome consisting of a single molecular species of ss-RNA and are said to be *monopartite*, others have *bipartite* genomes consisting of two molecular species of RNA, and others a *tripartite* genome of three major RNA species (*see* Table 2.2). The characteristics of some viruses with a multicomponent genome are further complicated by the presence of additional minor RNA molecules (Matthews, 1981; Kurstack, 1981).

There are still many plant viruses that have not yet been classified as

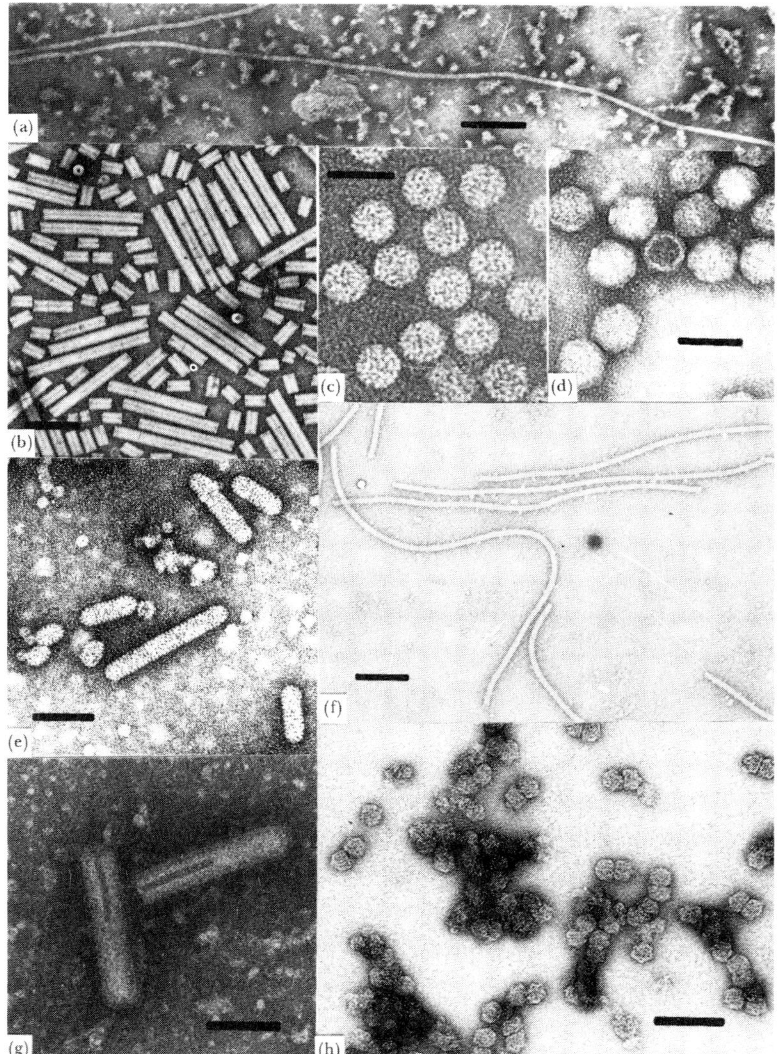

Plate 2.1 Electron micrographs of various types of virus particles.
(*a*) Beet stunt virus (closterovirus group), magnification bar = 100 nm; (*b*) tobacco rattle virus (tobravirus group), magnification bar = 100 nm; (*c*) cucumber mosaic virus (cucumovirus group), magnification bar = 40 nm; (*d*) tobacco ringspot virus (nepovirus group), magnification bar = 40 nm; (*e*) alfalfa mosaic virus, showing particles of different lengths, magnification bar = 35 nm; (*f*) turnip mosaic virus (potyvirus group), magnification bar = 100 nm; (*g*) lettuce necrotic yellows virus (rhabdovirus group, magnification bar = 100 nm; (*h*) cassava latent virus (geminivirus group), magnification bar = 50 nm.
(*a*, *c*, *d*, *f* and *g* courtesy of M. J. W. Webb, *b* and *h* courtesy of I. M. Roberts and *e* courtesy of G. J. Hills).

belonging to any particular group (*see* Commonwealth Mycological Institute/Association of Applied Biologists (CMI/AAB) descriptions of plant viruses in Chapter 12). This may either be because they do not fit any of the twenty-seven approved groups, or because too little is known of their properties to allow them to be classified. In the future, the present classification scheme will be enlarged to accommodate these distinct viruses, and others that will undoubtedly be discovered.

2.2 The Plant Virus Groups Infecting Angiosperms

A summary of the characteristics of each of the twenty-seven ICTV approved groups is given in this section, together with some probable and possible members of each group. Where a CMI/AAB description exists for a group member or the group itself, the description number is given following the virus's name (e.g. maize chlorotic dwarf virus (MCDV, 194)). The full list of viruses described in the CMI/AAB descriptions of plant viruses is given in Chapter 12.

2.2.1 Viruses with a ss-RNA genome, non-enveloped

(a) With isometric particles

Luteovirus Group
Type member: *barley yellow dwarf virus* (BYDV, 32).
R/1 : 2/28 : S/S : S/Ap.

Name derived from the Latin 'luteus' meaning yellow, because of the yellowing symptoms produced in infected hosts.

The genome consists of monopartite, ss-RNA within isometric particles of about 25 nm diameter. The host range of individual members of the group is restricted in some cases, but extensive in others. The viruses are not transmitted by sap inoculation, but are aphid transmitted in a persistent manner (*see* Section 7.4.2). Various strains of BYDV show marked vector specificity (Rochow and Duffus, 1981).

Other members of the group include *beet mild yellowing*, *beet western yellows* (89), *carrot red leaf* (249), *legume yellows*, *malva yellows*, *pea leaf roll* (syn; *bean leaf roll*), *potato leaf roll* (36), *rice giallume*, *soybean dwarf* (179), *tobacco necrotic dwarf* (234) and *turnip yellows virus*.

Possible group members include *banana bunchy top*, *beet yellow net*, *celery yellow spot*, *cotton anthocyanosis*, *filaree red leaf*, *groundnut rosette assistor*, *milk-vetch dwarf*, *physalis mild chlorosis*, *physalis vein blotch*, *raspberry leaf curl*, *solanum yellows*, *strawberry mild yellow edge*, *subterranean clover red leaf*, *tobacco vein distorting, tomato yellow net and tomato yellow top virus.*

Maize chlorotic dwarf virus Group
Type member: *maize chlorotic* dwarf virus (MCDV, 194).
R/1 : 3.2/36 : S/S : S.I /Au.
Not yet given an ICTV name.

The monopartite ss-RNA genome is contained in isometric particles approximately 30 nm in diameter. The virus has a narrow host range within the *Graminae* and is transmitted by leafhoppers in a semi-persistent manner (*see* Section 7.4.2). It is not transmitted by sap inoculation. MCDV occurs in the southern U.S.A. (Gingery *et al.*, 1981).

Rice tungro virus (67) is another possible member of this group.

Sobemovirus Group
Type member: *southern bean mosaic virus* (SBMV,57; 274).
R/1 : 1.4/21 : S/S : S/Cl, Se.
Name derived from *so*uthern *be*an *mo*saic.

Group members have monopartite, ss-RNA genomes with isometric particles approximately 28 nm in diameter. They have a restricted host range and may be seed transmitted. They are also transmitted by sap inoculation and beetles (Sehgal, 1981).

Turnip rosette virus (125) is also a probable member of this group and other possible members include *cocksfoot mottle* (23), *lucerne transient streak* (224), *rice yellow mottle* (149) and *sowbane mosaic* (64)

Tobacco necrosis virus Group
Type member: *tobacco necrosis virus* (TNV,14)
R/1 : 1.5/19 : S/S : S/Fu.
Not yet given an ICTV group name.

The isometric particles contain monopartite ss-RNA and are about 26–28 nm in diameter. The virus has a wide host range and its distribution is world-wide (Uyemoto, 1981). It is readily transmitted by sap inoculation and is also transmitted in the soil by zoospores of the chytrid fungus *Olpidium brassicae* (*see* Section 7.6).

Cucumber necrosis virus (82) is a possible member of this group.

Tombusvirus Group
Type member: *tomato bushy stunt* virus (TBSV,69).
R/1 : 1/4 : S/S : S/O.
Name derived from *tom*ato *bu*shy *s*tunt.

Group members have a monopartite, ss-RNA genome with isometric particles approximately 30 nm in diameter. The host range is wide and they can be transmitted by sap inoculation. Infection may also occur through virus particles in the soil (Martelli, 1981).

Artichoke mottle crinkle, *carnation Italian ringspot*, *pelargonium leaf curl* and *petunia asteroid mosaic* are other members of the group; possible members include *cymbidium* ringspot (178), *galinsoga mosaic* (252), *narcissus tip necrosis* (166), *pelargonium flower-break* (130), *saguaro cactus* (148), and *tephrosia symptomless virus* (256). *Turnip crinkle virus* (109) is a possible member of this group.

Tymovirus Group

Type member: *turnip yellow mosaic virus* (TYMV,2; 230).
R/1 : 2/35 : S/S : S, I/Cl.
Name derived from *t*urnip *y*ellow *mo*saic.

The monopartite ss-RNA genome is contained in isometric particles 28–30 nm in diameter. The host range of TYMV is mainly restricted to the *Cruciferae*, and other group members have narrow host ranges. They are readily transmitted by sap inoculation and field transmitted by flea-beetles (Koenig and Lesemann, 1981). The tymovirus group is described in CMI/AAB description 214.

Other members of the group include *Andean potato latent belladonna mottle* (52), *cacao yellow mosaic* (11), *clitoria yellow vein* (171), *desmonium yellow mottle* (168), *dulcamara mottle*, *egg-plant mosaic* (124), *erysimum latent* (222), *kennedya yellow mosaic* (193), *okra mosaic* (128), *ononis yellow mosaic*, *physalis mosaic*, *plantago mottle*, *scrophularia mottle* (113), *and wild cucumber mosaic virus* (105).

Comovirus Group

Type member: *cowpea mosaic virus* (CPMV,47; 197).
R/1 : 2/36 + 1.4/25 : S/S : S/Cl, Se.
Name derived from *co*wpea *mo*saic

The genome consists of bipartite ss-RNA. The two types of RNA are contained in separate isometric particles, each measuring about 24 nm in diameter. Both types of particle are required for infection. Other particles containing no RNA may also be present in the infected plant. The host range of each group member is relatively restricted and they induce mainly mottling and stunting symptoms in their hosts. All group members are readily transmitted by sap inoculation and by beetles in a persistent way. Some members are seed transmitted (Stace-Smith, 1981).

Group members include *Andean potato mottle* (203), *bean pod mottle* (108), *bean rugose mosaic* (246), *broad bean stain* (29), *broad bean true mosaic* (20), *cowpea severe mosaic* (209), *quail pea mosaic* (238), *radish mosaic* (121), *red clover mottle* (74) and *squash mosaic virus* (43).

Broad bean wilt virus is another possible member of the group.

Dianthovirus Group

Type member: *carnation ringspot virus* (CRSV,21).
R/1 : 1.5/20.5 (+0.5/20.5) : S/S : S/*.
Name derived from *Dianthus* the generic name of carnation.

Group members have a bipartite ss-RNA genome but it is not yet known if the two species are encapsidated separately or not. The isometric particles are between 31 and 34 nm in diameter. They are readily transmitted by sap inoculation and are reported to be transmitted in the soil (Matthews, 1982).

Other members of the group include red clover necrotic mosaic (181) and sweet clover necrotic mosaic.

Nepovirus Group

Type member: *tobacco ringspot virus* (TRSV,17).
R/1 : 2.8/46 + 1.3–2.4/27–40 : S/S : S/Ne, Se.
Name derived from the *ne*matode transmitted *po*lyhedral shaped virus particles which are characteristic of the group.

The angular, isometric particles are approximately 28 nm in diameter and the genome is bipartite, consisting of two ss-RNA molecular species in distinct particles. Both types of particle are required for infection. A third particle containing no RNA is frequently detected in infected plants. The viruses of the group have a wide host range causing ringspot and mottle symptoms. Latent infection is common in some hosts, especially in seedlings grown from infected seed. These viruses are transmitted by sap inoculation and by soil-inhabiting nematodes. Seed transmission is frequent in many hosts and often a very high percentage of the seed is infected. Pollen transmission has been shown for many members of the group (Murant, 1981). The group has been described in CMI/AAB description 185.

Other group members are *arabis mosaic* (16), *arracacha A* (216), *artichoke Italian latent* (176), *artichoke yellow ringspot* (271), *blueberry leaf mottle* (267), *cacao necrosis* (173), *cherry leaf roll*† (80), *cherry rasp leaf* (159), *chicory yellow mottle* (132), *grapevine Bulgarian latent* (186), *grapevine chrome mosaic* (103), *grapevine fan-leaf* (28), *hibiscus latent ringspot* (233), *lucerne Australian latent* (225), *mulberry ringspot* (142), *myrobalan latent ringspot* (160), *peach rosette mosaic* (150), *potato black ringspot* (206), *raspberry ringspot* (6; 198), *strawberry latent ringspot* (126), *tomato blackring* (38) and *tomato ringspot* (18).

†Nematode transmission has not yet been conclusively demonstrated for this virus.

Pea enation mosaic virus Group
Type member: *pea enation mosaic virus* (PEMV,25;257).
R/1 : 1.7/28 + 1.3/28 : S/S S/Ap.
Not yet given an ICTV group name.

PEMV has a bipartite ss-RNA genome encapsidated separately in isometric particles about 30 nm in diameter. Both types of particle are required for infection. PEMV is the only known member of the group and its host range is narrow. The virus is easily transmitted by sap inoculation and by aphids in a persistent manner (Hull, 1981). Infected pea plants develop mosaic and enation symptoms (*see* Section 3.2.2).

Alfalfa mosaic virus Group
Type member: *alfalfa mosaic virus* (AlfMV,46; 229)
R/1 : 1.1/16 + 0.8/16 + 0.7/16 : B/B : S/Ap, Se.
Not yet given an ICTV group name.

The tripartite genome is composed of the virus's three larger molecular species of ss-RNA. A fourth smaller species of RNA (the coat protein messenger RNA) is also present and each is separately encapsidated. The three larger RNA species are contained in the larger particles and these three, together with the fourth RNA or the coat protein are required for infectivity. Three of the particles are bacilliform in shape measuring 58 × 18, 48 × 18 and 36 × 18 nm in length and the fourth is isometric measuring 18 nm in diameter (Van Regenmortel and Pinck 1981) (*see* Section 2.1).

The virus is readily transmitted by sap inoculation and by aphids in a non-persistent manner. It is also seed transmitted in some hosts. The host range is wide (Hull, 1969), and it has a world-wide distribution causing economically important diseases in legumes and other crops.

There is only one member of the group.

Bromovirus Group
Type member: *brome mosaic virus* (BMV,3; 180).
R/1 : 1.1/23 + 1/22 + 0.7 + 0.3/22 : S/S : S/Cl, Ne.
Name derived from *bro*me *mo*saic.

The group contains viruses with a tripartite genome of ss-RNA, separately encapsidated in isometric particles about 26 nm in diameter. All three molecular species of RNA are required for infection and although the group members are of considerable biophysical and biochemical interest, they have not yet been reported to cause diseases of economic significance. Group members have a narrow host range and are readily transmitted by sap inoculation and sometimes naturally transmitted by beetles. In the laboratory they have also been transmitted by nematodes (Lane, 1981).

Other members of the group include *broad bean mottle* (101), *cowpea chlorotic mottle* (49) and *melandrium yellow fleck* (236).

Cucumovirus Group
Type member: *cucumber mosaic virus* (CMV,1; 213).
R/1 : 1.3/18 + 1.1/18 + 0.3/18 : S/S : S/Ap, Se.
Name derived from *cucum*ber *mo*saic.

The viruses of this group have tripartite genomes, encapsidated in three types of isometric particles of about 28 nm in diameter. All three components are necessary for infection. Numerous strains of CMV are known causing variable reactions in many hosts. Host reactions may be further complicated by the production of a virus-dependent satellite-like RNA (often referred to as RNA-5 or CARNA 5), that interacts with the normal genome RNA and the host (Kaper and Waterworth, 1977). Group members are easily transmitted by sap inoculation and by aphids in a non-persistent manner (*see* Section 7.4.2). Seed transmission also occurs in some hosts. CMV has a particularly wide host range and is of considerable economic importance in crops throughout the world, particularly in temperate regions (Kaper and Waterworth, 1981).

Other members of the group include *peanut stunt* (92), *robinia mosaic*, (65) and *tomato aspermy virus* (79).

Ilarvirus Group
Type member: *tobacco streak virus* (TSV,44).
R/1 : 1.1/14 + 0.9/14 + 0.7/14 : S/S : S/Th, Se.
Name derived from the *i*sometric *la*bile *r*ingspot characteristics of the viruses in this group

All members of the group have a tripartite ss-RNA genome each contained in separate isometric particles which vary from 26 to 35 nm in diameter. All particles have the same density, but their sedimentation rates (*see* Section 6.8) differ because of their different sizes. A mixture of the three largest RNA molecules is not infectious on its own, and infection only occurs when a fourth and smaller RNA molecule (mol. wt. 0.3), or the coat protein is present (Fulton, 1981). The group members have a wide host range and are easily sap transmitted. One member has been transmitted by a thrips and others are seed and pollen transmitted. The group is described in CMI/AAB description 275.

Group members include *apple mosaic* (83), *black raspberry latent* (106), *citrus leaf rugose* (164), *cherry rugose mosaic*, *citrus variegation*, *elm mottle* (139), *Danish plum line pattern*, *hop A*, *hop C*, *lilac ring mottle* (201), *prune dwarf* (19), *prunus necrotic ringspot* (5), *rose mosaic and tulare apple mosaic* (42).

Possible members include *American plum line pattern* and *spinach latent virus*.

Velvet tobacco mottle virus Group
Type member: *velvet tobacco mottle virus* (VtMoV).
R/1 : 1.5 + 1.2/* : S/S : S/Cl.
Not yet given an ICTV name or approval as a group.

The genome consists of one molecule of linear ss-RNA and a second molecule of circular ss-RNA with a viroid-like structure. Both are required for infectivity and in addition, there are smaller molecules of circular-like RNA which are also encapsidated. The particles are isometric and about 30 nm in diameter. Group members have a very narrow host range and are readily transmitted by sap inoculation and by beetles.

Other group members are *lucerne transient streak*, *solanum nodiflorum mottle* and *subterranean clover mottle viruses* (Matthews, 1982).

(b) With rod shaped particles

Tobravirus Group
Type member: *tobacco rattle virus* (TRV,12).
R/1 : 2.4/5 + 0.6 − 1.4/5 : E/E : S/Ne, Se.
Name derived from *to*bacco *ra*ttle.

The viruses of this group have rigid, rod-shaped particles of two lengths measuring 180–215 nm and 46–114 nm × 22 nm in width. The long and short particles each contain a separate ss-RNA molecular species. Both particles are necessary for infection. The host range of members of the group is wide and they are transmitted by nematodes. Seed transmission occurs in some hosts, but transmission by sap inoculation is difficult with some virus isolates (Harrison and Robinson, 1981).

Pea early browning virus (120) is another member of the group.

Tobamovirus Group
Type member: *tobacco mosaic virus* (TMV, 151).
R/1 : 2/5 : E/E : S/O, Se.
Name derived from *to*bacco *mo*saic.

The rigid, rod-shaped particles measure 300 nm × 18 nm and contain a single molecular species of ss-RNA. The protein sub-units of the virus are arranged on a helical RNA molecule and a distinct central hole is visible in electron micrographs (*see* Plate 1.2). Group members have a relatively wide host range, are transmitted by sap inoculation and are sometimes seed transmitted (*see* Section 7.7). The group has been described in CMI/AAB description 184 and by Van Regenmortel (1981).

Other members of the group are *cucumber green mottle mosaic* (154), *frangipani mosaic* (196), *odontoglossum ringspot* (155), *ribgrass mosaic* (152), *Sammon's opuntia*, *sunn-hemp mosaic* (153) and *tomato mosaic virus* (156).

Possible members include *beet necrotic yellow vein* (144), *broad bean necrosis* (223), *chara australis*, *hypochoeris mosaic* (273), *nicotiana velutina mosaic* (189), *peanut clump* (235), *potato mop top* (138) and *soil-borne wheat mosaic* (77).

Hordeivirus Group
Type member: *barley stripe mosaic virus* (BSMV,68).
R/1 : 1/4 : E/E : S/Se.
Name derived from the Latin *Hordeum* meaning barley.

Group members have rigid, rod-shaped particles measuring 100 to 150 nm in length x 20 nm in width. They have a ss-RNA genome within particles of two to four different lengths, depending on the virus strain. Particles of different lengths contain different RNA molecular species and two or three particle components are required for infectivity, indicating a bi- or tripartite genome (Jackson and Lane, 1981). The host range is narrow, but BSMV causes serious diseases throughout the world in barley crops. Group members are transmitted by sap inoculation and in seeds.

Other members of the group include *lychnis ringspot*, and *poa semi-latent virus*.

Potexvirus Group
Type member: *potato virus X* (PVX,4).
R/1 : 2.1/6 : E/E : S/O (Fu).
Name derived from *pot*ato *X*.

The particles are flexuous rods about 470 to 580 nm in length × 13 nm in width. The genome is a single species of ss-RNA. Individual group members frequently have a narrow host range although the host range of PVX is wide. All group members are readily transmitted by sap inoculation and there are no confirmed vectors (Purcifull and Edwardson, 1981). The group is described in CMI/AAB description 200.

Other members of the group are *cactus X* (58), *cassava common mosaic* (90), *clover yellow mosaic* (111), *cymbidium mosaic* (27), *daphne X* (195), *foxtail mosaic* (264), *hydrangea ringspot* (114), *narcissus mosaic* (45), *papaya mosaic* (56), *pepino mosaic*, *viola mottle* (247) and *white clover mosaic* (41).

Possible members include *artichoke curly dwarf*, *bamboo mosaic*, *dioscorea latent, hippeastrum latent, malva veinal necrosis, potato aucuba mosaic* (98) and *rhododendron necrotic ringspot virus.*

Carlavirus Group
Type member: *carnation latent virus* (CLV,61).
R/1 : 2.3/6 : E/E : S/Ap.
Name derived from *car*nation *la*tent.

The group has flexuous, rod-shaped particles measuring 620 to 700 nm in length × 13 nm in width, containing a single species of ss-RNA. Members of the group have a relatively narrow host range and with the exception of some of the legume virus members, do not cause economically important diseases. They are transmitted by sap inoculation and by aphids in a non-persistent manner (Wetter and Milne, 1981). The group is described in CMI/AAB description 259.

Group members include *alfalfa latent* (211), *American hop latent* (262), *cactus 2, chrysanthemum B* (110), *cowpea mild mottle* (140), *eggplant mild mottle, elderberry carlavirus* (263), *helenium S* (265), *hippeastrum latent, honeysuckle latent, hop latent* (261), *lilac mottle, lily symptomless* (96), *mulberry latent, muskmelon vein necrosis, narcissus latent* (170), *passiflora latent, pea streak* (112), *poplar mosaic* (75), *potato M* (87), *potato S* (60), *red clover vein mosaic* (22), and *shallot latent virus* (250).

Possible members include *cassia mosaic, chicory blotch, cole latent, cynodon mosaic, elderberry A, garlic mosaic, nasturtium mosaic* and *white bryony mosaic virus.*

Potyvirus Group
Type member: *potato virus Y* (PVY,37; 242).
R/1 : 3.1/6 : E/E : S/Ap.
Name derived from *pot*ato *Y.*

The flexuous, rod-shaped particles measure from 680 to 900 nm in length × 11 nm in width, and contain a single ss-RNA. Members of the group are of considerable economic importance. They infect a wide range of plants, but most individual members have a relatively narrow host range. They are transmitted by aphids in a non-persistent manner and by sap inoculation. Some group members are seed transmitted (Hollings and Brunt, 1981). The group is described in CMI/AAB description 245.

Group members include *bean common mosaic* (73), *bean yellow mosaic* (40), *beet mosaic* (53), *bidens mosaic* (161), *carnation vein mottle* (78), *carrot thin leaf* (218), *celery mosaic* (50), *clover yellow vein* (131), *cocksfoot streak* (59), *cowpea aphid-borne mosaic* (134), *dasheen mosaic* (191), *henbane mosaic* (95), *hippeastrum mosaic* (117), *iris mild mosaic* (116), *leek yellow stripe* (240), *lettuce mosaic* (9), *nothoscordum mosaic, onion yellow dwarf* (158), *papaya ringspot* (84), *parsnip mosaic* (91), *passionfruit woodiness* (122), *pea necrosis*, *pea seed-borne mosaic* (146), *peanut mottle* (141), *pepper mottle* (253), *pepper veinal mottle* (104), *Peru tomato* (255), *plum pox* (70), *pokeweed mosaic* (97), *potato* A (54), *soybean mosaic* (93), *sugar cane mosaic* (88), *tobacco etch* (55; 258), *tulip breaking* (71), *turnip mosaic* (8), *and watermelon mosaic virus* (63).

Possible members include *bearded iris mosaic* (147), *blackeye cowpea mosaic*, *cowpea mosaic*, *carrot mosaic*, *daphne Y*, *datura shoestring*, *dioscorea greenbanding*, *dock mottling mosaic*, *euphorbia ringspot*, *freesia mosaic*, *grapevine leafroll*, *Guinea grass mosaic* (190), *maclura mosaic* (239), *malva vein clearing*, *mungbean mosaic*, *mungbean mottle*, *narcissus yellow stripe* (76), *passion fruit ringspot*, *pepper severe mosaic*, *tamarillo mosaic*, *tobacco vein mottle*, *wild potato mottle and wisteria vein mosaic virus*.

Closterovirus Group

Type member: *beet yellows virus* (BYV,13).
R/1 : 2.3–4.3/5 : E/E : S/Ap.
Name derived from the Greek kloster meaning spindle, because of the group's long thread-like particles.

Group members have very long flexuous rod-shaped particles, measuring 600 to 2000 nm in length × 10 nm in width, with a genome consisting of a single molecular species of ss-RNA. Individual group members have a moderately wide host range and some cause important economic diseases. Transmission by sap inoculation is often difficult, but some members are transmitted by aphids in a semi-persistent manner (Lister and Bar-Joseph, 1981). The group is described in CMI/AAB description 260.

Other members of the group include *beet yellow stunt* (207), *burdock yellows*, *carnation necrotic fleck* (136), *carrot yellow leaf*, *clover yellows*, *citrus tristeza* (33), *festuca necrosis*, *lilac chlorotic leafspot* (202) and *wheat yellow leaf virus* (157).

Possible members include *apple chlorotic leaf spot* (30), *apple stem grooving* (31), *heracleum latent* (228) and *potato virus T*.

2.2.2 Viruses with a ss-RNA genome and particles enclosed in envelopes

Rhabdovirus Group

Probable type member: *lettuce necrotic yellows virus* (LNYV, 26)
R/1 : 4/2 : B/E : S, I/Ap, Au.
Name derived from the Greek *rhabdos* meaning rod.

This is one of the two plant virus groups that is classified as a family, the *rhaboviridae*; other members of the same family infect invertebrates and vertebrates (for example, *rabies* virus). The plant virus members are not grouped into genera. They have bacilliform or bullet shaped particles measuring 160 to 380 nm in length × 50 to 95 nm in width (*see* Plate 2.1g). The bullet-shaped particles are thought to arise through breakage of a bacilliform particle during preparation for electron microscopy. Characteristically, the particles consist of an outer membrane containing lipid and protein, and an inner helically constructed core containing protein and a single molecular species of ss-RNA.

Some plant rhabdoviruses, such as LNYV, are of economic importance. Generally they cannot be transmitted by sap inoculation, but are transmitted by aphids and leafhoppers. Transmission is persistent and the virus has been shown to multiply within the vector. Recently, other members of the group have been shown to be transmitted by a lacebug and a mite (Francki *et al.*, 1981). The group is described in CMI/AAB description 244.

Other members of the group include *barley yellow striate mosaic*, *beet leafcurl* (268), *bobone disease*, *broccoli necrotic yellows* (85), *carrot latent*, *cereal chlorotic mottle* (251), *coffee ringspot*, *cow-parsnip mosaic*, *digitaria striate*, *eggplant mottled dwarf* (115), *lucerne enation*, *maize mosaic* (94), *northern cereal mosaic*, *potato yellow dwarf* (35), *raspberry vein chlorosis* (174), *rice transitory yellowing* (100), *sonchus yellow net* (205), *sowthistle yellow vein* (62), *strawberry crinkle* (163), *wheat (American) striate mosaic* (99) and *wheat chlorotic streak virus*.

Various other rhabdoviruses have been reported (Francki *et al.*, 1981), and the list continues to grow, but as so few characteristics of many of these viruses have been studied, it is not known if many of these reports are of the same virus infecting different host species.

Tomato spotted wilt virus Group

Type member: *tomato spotted wilt virus* (TSWV,39).
R/1 : 2.6/* + 1.7/* + 1.3/* + 1.9/* : S/S : S/Th.
Not yet given an ICTV group name.

The isometric particles are 85 nm in diameter and are enclosed within a lipoprotein envelope. The ss-RNA genome is composed of at least four molecular species of RNA, but the RNA alone is not infectious and the nature of its relationship with the viral proteins has not yet been determined. TSWV is transmitted readily by sap inoculation and by thrips in a persistent manner. It has a very wide host range and is common in temperate and sub-tropical regions causing diseases of economic importance in tomato, tobacco, potato and various other crops (Francki and Hatta, 1981).

TSWV is the only known member of the group.

2.2.3 Viruses with a ds-RNA genome

Reovirus Group
The reoviruses comprise the second plant virus group classified as a family, the *reoviridae*. Besides viruses that infect plants, other members infect invertebrates and vertebrates. The plant virus members are divided into two sub-groups, the genus *phytoreovirus* and the genus *fijivirus*.

Phytoreovirus Genus
Type member: wound tumour virus (WTV,34).
R/2 : 16/22 : S/S : S, I/Au.
Name derived from *r*espiratory *e*nteric *o*rphan.

The double-stranded (ds) RNA genome has twelve segments encapsidated within an isometric particle measuring approximately 70 nm in diameter. WTV has no known natural plant host and was originally isolated from its leafhopper vector. Group members are not transmitted by sap inoculation, but are transmitted by leafhoppers in which they are propagative (*see* Section 7.4).

The other phytoreovirus is *rice dwarf virus* (102).

Fijivirus Genus
Type member: Fiji disease virus (FDV,119).
R/2 : 1.1–2.9/45 : S/S: S, I/Au.

The isometric particles are similar to those of the phytoreovirus genus. They are approximately 70 nm in diameter with a ds-RNA genome divided into ten segments. Group members infect only plants of the *graminae* family and the insect group known as the plant hoppers (*Flugoroidea*). Group members are transmitted only by plant hoppers in which they are propagative (Shikata, 1981).

Other group members include *arrhenatherum blue dwarf*, *cereal tillering disease*, *lolium enation*, *maize rough dwarf* (72), *oat sterile dwarf* (217), *pangola stunt* (175) and *rice black-streaked dwarf virus* (135).

Rice ragged stunt virus (248) is a possible member of the group.

2.2.4 Viruses with a single-stranded DNA genome

Geminivirus Group
Type member: *maize streak virus* (MSV,133).
D/1 : 0.7/* : S/S : S/Au, W.
Name derived from the latin gemini meaning 'twins'.

Group members have an unusual particle morphology in that their

18–20 nm isometric, particles occur mainly in pairs (*see* Plate 2.1*h*). The ss-DNA genome is so far unique among plant viruses. Members of the group infect a wide range of plant species, but individual members have a narrow host range. Some members of the group are transmitted by leafhoppers and others by whitefly (Goodman, 1981). They are found predominantly in tropical areas, where they are of considerable economic importance in food and fibre crops.

Other group members include *bean golden mosaic* (192), *beet curly top* (210), *cassava latent, chloris striate mosaic* (221) *euphorbia mosaic, tomato golden mosaic, tomato yellow dwarf, tobacco leaf curl* (232), *tobacco yellow dwarf, bean summer death* and *mungbean yellow mosaic virus*.

2.2.5 *Viruses with a double-stranded DNA genome*

Caulimovirus Group
Type Member: *cauliflower mosaic virus* (CaMV,24; 243).
D/2 : 4.8–5/17 : S/S : S/Ap.
Name derived from *cauli*flower *mo*saic.

Members of the group have a genome consisting of one molecule of ds-DNA which has three discontinuities, two in one strand and one in the other. The genome is encapsidated in isometric particles measuring about 50 nm in diameter. They have a restricted host range and are aphid transmitted in a non-persistent manner.

Some members, such as CaMV, are readily transmitted by sap inoculation, but others, such as strawberry vein banding virus, are not (Shepherd and Lawson, 1981).

Other group members include *carnation etched ring* (182), *dahlia mosaic* (51), *figwort mosaic, mirabilis mosaic* and *strawberry vein banding virus* (219).

Possible members include *cassava vein mosaic, blueberry red ringspot* and *petunia vein-clearing virus*.

2.2.6 *Other angiosperm viruses*

In addition to the plant virus groups listed, a number of additional viruses have been described in the CMI/AAB descriptions of plant viruses (*see* Chapter 12). Their characteristics do not conform with those of groups already named by the ICTV. Many of these viruses do have natural affinities with each other, and have been separated into groups in the CMI/AAB descriptions. It should be emphasized, however, that at the time of writing these additional groups are provisional and have not yet been considered by the ICTV committee. They include the following.

Viruses with rod-shaped particles

(*a*) ***Apple Stem Grooving Virus Group***: including *apple stem grooving* (31) and *potato virus T* (187).

(*b*) ***Barley Yellow Mosaic Virus Group***: including *barley yellow mosaic* (143), *oat mosaic* (145), *rice necrosis mosaic* (172) and *wheat spindle streak mosaic virus* (167).

(*c*) ***Wheat Streak Mosaic Virus Group***: including *agropyron mosaic* (118), *oat necrotic mottle* (169), *ryegrass mosaic* (86) and *wheat streak mosaic virus* (48).

Viruses with isometric particles

(*a*) ***Cocksfoot Mild Mosaic Virus Group***: including *cock's foot mild mosaic* (107) and *panicum mosaic virus* (177).

(*b*) ***Oat Blue Dwarf Virus Group***: including *maize rayado fino* (220) and *oat blue dwarf virus* (123).

In addition to these groups, a number of viruses which so far have not been classified in any group, have been described in the CMI/AAB descriptions.

2.3 Virus Strains

In common with all organisms, viruses are subject to processes which generate variation between individuals. Variants, which can be recognized by some characteristic of the phenotype (such as changes in the symptoms they induce in diseased hosts), may be classed as distinct strains, and it is probable that most virus populations actually consist of mixtures of genetically different individuals (Matthews, 1981).

Care should be taken when referring to individual virus populations to distinguish between the terms *strain* and *isolate*. When a virus is initially isolated from diseased plants and its characteristics are unknown, it should be referred to as an isolate of that particular virus. The term strain should only be used when sufficient characteristics of the isolate have been determined, to know if it can be classified as a distinct or existing strain of the virus concerned.

The presence of variants within a virus culture can be readily demonstrated with tobacco mosaic virus (TMV) (Kunkel, 1940). If TMV is passaged through tobacco (*Nicotiana tobacum*) for several generations by sap inoculation and then inoculated to *N. glutinosa*, it produces discrete local lesions on the inoculated leaves. Each individual lesion is composed of the progeny from one inoculated virus particle and if single lesions are sub-cultured; the population arising from the original individual particle can be cultured. Such cultures may be distinct strains, which can be shown to have differences in their biological, physical and chemical properties. Kunkel has shown that with TMV, between 0.5 and 2% of its lesions contain variants from the parent virus type. In normal cultures, however, these variants are hidden by the natural, parent type.

Sometimes the symptoms produced by a virus culture following many passages through a particular host, become milder, or differ from those observed when the virus was first isolated from the diseased field plant. In such circumstances the virus is said to have *attenuated*. One explanation for this change in symptoms, is that milder strains have been selected during the successive passages.

2.3.1 The importance of virus strains

The existence of virus strains creates not only a problem for classification and identification, but is also of considerable practical importance. Different strains of a virus can cause different kinds of symptoms, and therefore, different diseases in a crop plant. Strains may also differ in their host range and vector transmissibility, which in turn affects their epidemiology and eventual control. In addition, strains often vary in their serological affinities (*see* Section 6.6), which can complicate and confuse identification.

The occurrence and characteristics of virus strains are particularly important in breeding for virus resistance (*see* Chapter 10). Some strains may cause severe symptoms in a particular host plant, while others may induce only mild symptoms or even fail to infect. Therefore, when screening plants for resistance to a particular virus, great care must be taken to select suitable strains (Walkey and Innes, 1979; Walkey *et al.*, 1983).

In some circumstances, the existence of different strains of a particular virus may be useful for controlling the disease. For example, mild strains of TMV may be used to inoculate tomato plants, which will protect them from later infection by more severe strains of that virus (Rast, 1972, *see* Section 9.7).

2.3.2 Strain identification

There is often considerable debate over the criteria for deciding whether a previously unrecognized virus, is a new strain of an existing virus, or a new virus. Matthews (1981) has suggested that the following criteria be adopted for distinguishing between virus strains and a new virus.

1 The size and shape of the virus particle and the characteristics of any of its sub-structure that can be seen in the electron microscope.
2 Serological identity (*see* Section 6.6).

3 The disease symptoms and host range of the isolate when inoculated to a range of indicator plants (*see* Chapter 3).
4 The type of transmisison, especially in respect of insect, nematode or fungal vectors (*see* Chapter 7).

2.3.3 Mechanisms generating variability in viruses

The possible origins of virus strains have been described by Gibbs and Harrison (1976) and Matthews (1981).

In the case of a virus which has a single nucleic acid genome (*see* Section 1.4), strains may arise by chemical or physical action on the genome. An alteration in one or more of the bases in the nucleotide sequence of the genome results in a *mutant*. In the laboratory mutant strains may be produced artificially by treating virus with nitrous acid (Gierer and Mundry, 1958; Siegel, 1960).

New virus strains may arise in several different ways. First, they may be formed by the breakage and renewal of covalent links in the nucleic acid chain. This occurs readily between the DNA genomes of different strains of cauliflower mosaic virus. The process is called *recombination* and results in the creation of *recombinants* (Fenner, 1970; Cooper, 1969).

Secondly, and more commonly, dual or multiple strain infections may occur by viruses with divided genomes, which may produce new strains by reassortment of their nucleic acids during replication. The new virus forms that result from this procedure are sometimes referred to as *pseudo-recombinants*, to distinguish them from the true recombinants mentioned above (Gibbs and Harrison, 1976).

Finally, new strains may result from *complementation*. This occurs when a virus, which is unable to produce one or more necessary proteins, is assisted in this by another virus (or another strain of the same virus) during a mixed infection (Rochow, 1970). This association may then be perpetuated.

All the above procedures may be used by virologists to produce new strains in the laboratory. It is impossible to know, however, which of these processes are most frequently responsible for the natural variability that occurs in virus populations.

2.4 Viruses Infecting Other Plant Classes

In addition to the viruses that infect higher plants, there are reports of the presence of viruses or virus-like particles from most other major plant classes except the *Bryophyta* (Matthews, 1981). The majority of the non-angiosperm studies, however, have concentrated on viruses infecting fungi.

Table 2.5 Examples of viruses infecting fungi

Host	*Particle characteristics*		*Reference*
BASIDIOMYCETES			
Agaricus bisporus	Isometric	25 nm	Hollings (1962)
	Isometric	29 nm	
	Bacilliform	19 × 50 nm	
	Isometric	34 nm	Dieleman-van Zaayen & Temmink (1968)
Laccaria laccata	Isometric	28 nm	Blattńy & Králík (1968)
ASCOMYCETES			
Gaeumannonyces graminis	Isometric	29 nm	Lapierre *et al.* (1970)
	Isometric	27 and 35 nm	Rawlinson *et al.* (1973)
Peziza ostracoderma	Isometric	25 nm	Dieleman-van Zaayen *et al.* (1970)
	Tubular	350 × 12 nm	
Saccharomyces cerevisiae (yeast)	Isometric	30 nm	Herring & Bevan (1974)
PHYCOMYCETES			
Thraustochytrium sp.	Herpes-like		Kazama & Schornstein (1973)
FUNGI IMPERFECTI			
Penicillium stoloniferum	Isometric	34 nm	Banks *et al.* (1968)
	Isometric	34 nm	Buck & Kempson-Jones (1970)
	Isometric	40–45 nm	
Helminthosporium victoriae	Isometric	35–40	Sanderlin & Ghabrial (1978)
	Isometric	34–40 nm	

2.4.1 Fungal viruses (mycoviruses)

Since virus particles were first seen in 1962 in tissues of the cultivated mushroom, *Agaricus bisporus* (Hollings, 1962), viruses or virus-like particles (VLPs) have been reported from over one hundred different species, belonging to all the major fungal groups (Ghabrial, 1980). In fact Hollings (1978) has estimated that mycoviruses may be expected to occur in at least 5,000 species of fungi. The term virus-like particle rather than virus, is used by many workers in instances in which a mycovirus has not been isolated and characterised, and where evidence for its existence is based only on electron microscopy. Examples of viruses that infect various fungi are given in Table 2.5.

Mycoviruses frequently have isometric particles varying from 25 to 48 nm in diameter (Bozarth, 1979), but bacilliform (Hollings, 1962), rod-shaped (Dieleman-van Zaayen *et al.*, 1970) and herpes-like (Kazama and Schornstein, 1973) particles have been reported (*see* Table 2.5). Most have segmented ds-RNA genomes, with the different RNA segments usually encapsidated in separate, but otherwise identical particles (Bozarth, 1979).

Transmission of mycoviruses occurs most frequently during cell division, cell fusion and spore production in the fungal host. No efficient natural vector of mycoviruses is known and one of the major problems is the lack of routine infectivity tests, such as transmission by sap inoculation to a healthy host. Most mycoviruses are not serologically related and have a narrow host range among fungal species and genera. The reason for this apparent specificity may be the absence of known vectors or laboratory transmission methods, rather than their inability to infect different fungal species.

Although most of the mycoviruses reported appear to cause symptomless or latent infection in their fungal host (Hollings, 1978; Ghabrial, 1980), the pathogenic effects of virus infection in the cultivated mushroom are well documented. Various names have been given to diseases of cultivated mushrooms, including watery stipe, La France, brown, mummy and X-disease (Gibbs and Harrison, 1976). The sporophores of diseased mushrooms are small, distorted and may be shrivelled or watery (*see* Plate 2.2). The disease may cause a complete crop loss and may be spread on commercial farms by diseased mycelial fragments or in spores. Commercially, the disease may be controlled by the use of healthy mycelium (spawn), sterilized compost, sterilized growing trays and houses, and by using air filters to exclude virus-infected spores.

Virus infection of yeast, associated with 'killer' proteins, has been well documented (*see* review by Bruenn, 1980) and the economically important disease Chesnut Blight, caused by the fungus *Endothia*

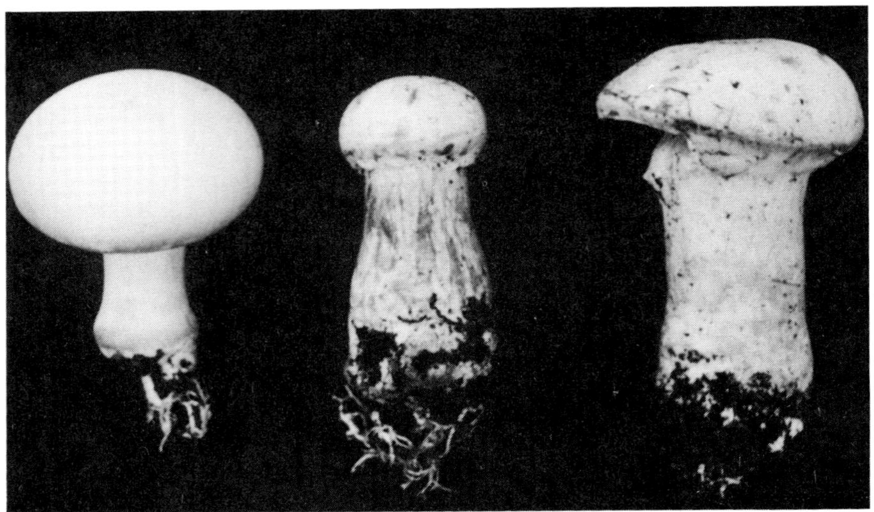

Plate 2.2 Cultivated mushrooms infected with watery-stipe disease. Diseased sporophores *right*, healthy *left* (courtesy of A. A. Brunt).

parasitica, may be controlled by hypovirulent strains of the fungus. These strains contain genetic determinants which are associated with ds-RNA (Anagnostakis, 1982).

Mycoviruses have been the subject of recent reviews by Lemke, 1976 and 1979; Hollings, 1978; Molitoris *et al.*, 1979; Ghabrial, 1980 and Hollings, 1982.

2.4.2 Viruses of algae, ferns and gymnosperms

Viruses have been shown to infect *Eukaryotic* algae, including *Chara corrallina* in which rod-shaped particles measuring about 530 nm in length were observed (Gibbs *et al.*, 1975). More recently, virus-like particles have been observed in sections of algal cells (Dodds, 1979), but like fungal viruses, algal viruses are difficult to study because suitable assay methods are not available.

In 1963 a virus was isolated from a fern (*Pteridophyta*) for the first time. Canova and Casalicchio reported virus-like symptoms in *Polypodium vulgare* and *Scolopendrium vulgare*. They succeeded in transmitting a virus from infected to healthy ferns by sap inoculation. Later Hull (1968) isolated viruses from the hart's tongue fern (*Phyllitis scolopendrium*), which had rod-shaped particles measuring 135 and 320 nm in length × 22 nm in width.

There have been several reports of virus-like diseases occurring naturally in spruce and pine trees (Class *Gymnospermae*), but the viral nature of these diseases has not been confirmed (Matthews, 1981). Viruses which commonly infect angiosperms have, however, been experimentally transmitted to *Pinus sylvestris* (Yarwood, 1959) and *Chamaecyparis lawsoniana* (Harrison, 1964).

2.5 Virus-like Diseases

A number of plant diseases which were once thought to be caused by viruses, because of the virus-like symptoms they induced, have in recent years been shown to be caused by other disease agents. These include diseases caused by viroids, mycoplasma and rickettsia-like organisms.

2.5.1 Viroids

Although some of the diseases now known to be caused by viroids have been known for many years, the true nature of the causal agent was not described until 1971. Then Diener, working with potato spindle tuber disease, showed that the causal agent was non-encapsidated RNA of low molecular weight, which he termed a viroid. Viroids are the smallest and simplest known agents of infectious disease, and despite their small size ($1.1–1.3 \times 10^5$ mol. wt.), they are able to replicate in susceptible host plants. The RNA of all viroids so far studied, has been shown to consist of a single molecular species which may occur in a circular or linear form.

Table 2.6 Examples of viroid diseases

Viroid	*Reference*
Chrysanthemum stunt	Diener & Lawson (1973)
Chrysanthemum chlorotic mottle	Romaine & Horst (1975)
Citrus exocortis	Sanger (1972)
Cucumber pale fruit	Van Dorst & Peters (1974)
Hop stunt	Sasaki & Shikata (1977)
Potato spindle tuber	Diener (1971)
Tomato apical stunt*	Walter (1982)
Tomato bunchy-top	Benson *et al.* (1965)
Tomato planta-macho	Galindo *et al.* (1982)
Possible viroid diseases	
Avocado sunblotch	Palukaitis *et al.* (1979)
Coconut cadang-cadang	Randles (1975)

* It is possible that more than one of these tomato diseases are caused by the same viroid (Diener, 1979).

All the viroids so far discovered have been isolated from higher plants (angiosperms) and include a number of economically important

Table 2.7 Examples of mycoplasma and rickettsia-like diseases

Disease	*Distribution*	*Vector*	*Reference*
MLOs			
Aster yellows	E., N.Am., Far East*	*Macrosteles fascifrons* (L)†	Kunkel (1926)
Clover phyllody	E., N.Am.	*Euscelis plebejus* (L)	Sinha & Paliwal (1970)
Corn stunt	N.Am., S.Am.	*Dalbulus maidis* (L)	Kunkel (1948)
Mulberry dwarf	Far East	*Hishimonus sellatus* (L)	Doi *et al.* (1967)
Papaya bunchy top	Caribbean	*Empoasca papayae* (L)	Story & Halliwell (1969)
Pear decline	N.Am.	*Psylla pyricola* (Ps)	Jensen *et al.* (1964)
Phormium yellow leaf	New Zealand	*Oliarus atkinsoni* (Pl)	Ushiyama *et al.* (1969)
Sugarcane white leaf	Taiwan	*Epitettix hiroglyphicus* (L)	Matsumoto *et al.* (1969)
Rice yellow dwarf	Far East	*Nephotettix apicalis* (L)	Nasu *et al.* (1967)
Rubus stunt	E.	*Macropsis fuscula* (L)	Murant & Roberts (1971)
Stolbur (tomato)	E., N.&.S.Am.	*Hyalestes obsoletus* (L)	Maramorosch *et al.* (1970)
RLOs			
Clover club leaf	E., N.Am.	*Agalliopsis novella* (L)	Black (1944)
Citrus greening	Africa, India, Far East	*Trioza erytreae* (Ps)	Moll & Martin (1973)
Peach phony	N.Am.	*Homalodisca coagulata* (L)	Turner & Pollard (1959)
Pierce's grapevine	N.Am.	*Draeculacephala minerva* (L)	Hewitt *et al.* (1942)

* E = Europe N.&.S.Am. = North and South America. † L = leafhopper. Pl = plant hopper, Ps = psyllid.

diseases (*see* Table 2.6). Potato spindle tuber viroid is common in potato growing areas of northern U.S.A. and Canada (Diener and Raymer, 1971) and can considerably reduce yields (Singh *et al.*, 1971). Citrus exocortis disease is common in all major citrus growing areas (Diener, 1978), chrysanthemum stunt disease occurs widely in the U.S.A., Canada, the Netherlands and England and coconut cadang-cadang disease causes serious losses in the Philippines (Randles, 1975). Some viroids are readily transmitted by sap inoculation, and they may also be transmitted by plant contact, or on contaminated budding knives.

It was suggested at first that viroids might be primitive or degenerate forms of conventional plant viruses, but increased knowledge of their biochemical nature and structure, suggests that they are quite distinct (Diener, 1980). The nature of viroids and viroid diseases has recently been described by Diener (1979, 1980, 1981 and 1982).

2.5.2 Mycoplasma and rickettsia-like plant diseases

A number of other plant diseases, with virus-like symptoms were also thought for many years to be viral induced, but no virus particles were ever observed by electron microscopy in tissues from these plants. Then in 1967, one of these diseases, mulberry dwarf disease, was shown by Japanese workers to be caused by a mycoplasma-like organism (Doi *et al.*; Ishiie *et al.*). During the next decade, a number of other similar diseases were also shown to be caused by mycoplasma-like organisms (MLOs) (*see* Table 2–7).

Mycoplasma belong to the group *Mycoplasmatales* and are characterized by having no cell wall, but they do have a confining unit membrane. They are pleomorphic in shape and are usually confined in the diseased plants to phloem or xylem cells. Although they are too small to be seen in the light microscope (0.1 to 1.0 μm in diameter), MLOs are readily seen in the electron microscope (*see* Plate 2.3). They reproduce by binary fission and are resistant to penicillin, but susceptible to tetracycline.

Plant MLOs contain ribosomal RNA and DNA in the form of a coil, in their nuclear region. The presence of both RNA and DNA clearly distinguishes them from plant viruses. Many cause diseases of considerable economic importance (Maramorosch *et al.*, 1970), inducing symptoms of yellowing, stunting and proliferations of the witches-broom type. They may also cause flowers to change to leaf-like structures, a condition referred to as phyllody. Transmission is not normally possible by sap-inoculation, but they are often transmitted by leafhoppers and occasionally by plant hoppers or psyllids (*see* Table 2.7). Diseases caused by MLOs may be controlled by antimycoplasma

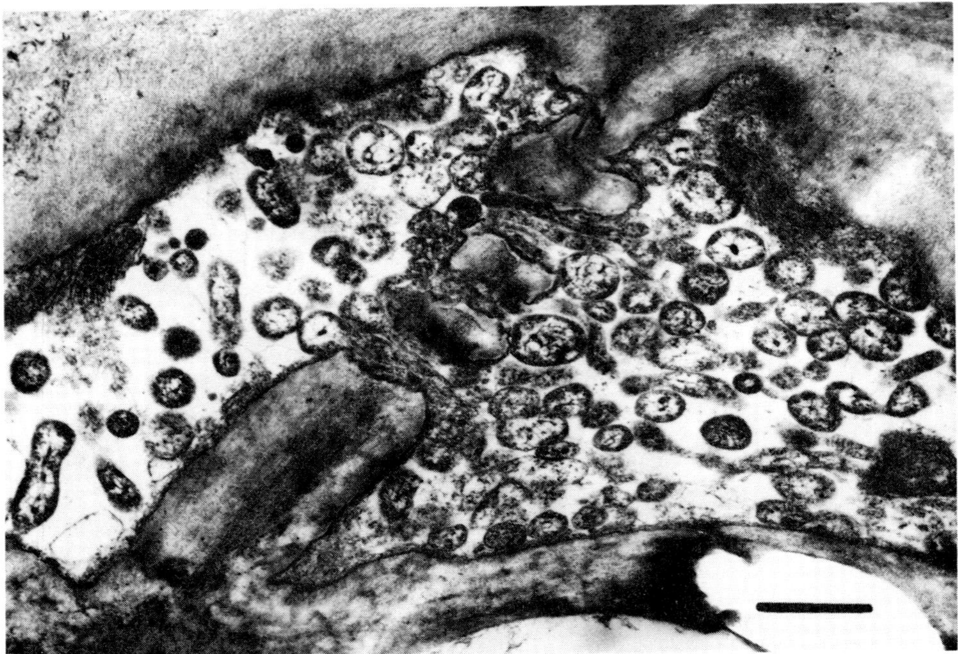

Plate 2.3 Electron micrograph of a section through the phloem elements of a clover leaf infected with the mycoplasma disease, clover phyllody. The rounded bodies of the disease agent are visible in the phloem vessel near a sieve plate, magnification bar = 1 μm (courtesy of M. J. W. Webb).

drugs, breeding for resistance, heat therapy or by control of their vectors. Details of these control measures have been described by Maramorosch (1982).

In recent years various MLOs have been cultured on artificial media and all those cultured have been placed in a new family called the *Spiroplasmataceae* and are commonly referred to as *spiroplasma* (Davis and Worley, 1973). MLOs such as *Spiroplasma citri*, the agent of citrus stubborn disease (Maramorosch, 1974), frequently produce a helical, filamentous, mobile form during a rapid growth phase in culture. It is from these spiral structures that the name spiroplasma was derived.

Up to the present time many authorities have considered plant-infecting mycoplasma and spiroplasma to be distinct, but since all MLOs appear to produce the distinct spiral bodies when they are finally cultured, Maramorosch (1982) considers that they all belong to the same group. He considers that they differ only in whether or not they can be cultured. The controversy concerning their classification has been covered in recent reviews (Maramorosch, 1981; Davis *et al.*, 1981).

In addition to MLOs, a further group of plant disease agents, the rickettsia-like organisms (RLOs) have been recognized in recent years. These organisms have distinct cell walls and belong to the group of bacteria called *Schizomycetes*. RLOs such as Pierce's disease of grapevine, occur in xylem and phloem vessels and measure about 0.4 μm in diameter and up to 3 μm in length (Goheen *et al.*, 1973). They are readily seen in the electron microscope, are leafhopper transmitted and unlike MLOs, are susceptible to penicillin. Other examples of RLOs are given in Table 2.7 and the group together with MLOs has been the subject of a review by Nienhaus and Sikora (1979).

2.6 References

Anagnostakis, S. L. (1982). Biological control of Chesnut Blight. *Sci* **215**, 466–71.

Banks, G. T., Buck, K. W., Chain, E. B., Himmelweit, F., Marks, J. E., Tyler, J. M., Hollings, M., Last, F. T. and Stone, O.M. (1968). *Nature* **218**, 542.

Bawden, F. C. and Pirie, N. W. (1936). Experiments on the chemical behaviour of potato virus X. *Br J Ex Pathol* **17**, 64–74.

Benson, A. P., Raymer, W. B., Smith, W., Jones E. and Monro, J. (1965). Potato diseases and their control. *Potato Handb* **10**, 32–8.

Black, L. M. (1944). Some viruses transmitted by agallian leafhoppers. *Proc Amer Phil Soc* **88**, 132–44.

Blattńy, C. and Králík, O. (1968). A virus disease of *Laccaria laccata* (Scop. ex. Fr.) Cooke and some other fungi. *Ceskà Mykol* **22**, 161–6.

Bozarth, R. F. (1979). Physicochemical properties of mycoviruses. An overview. In *Fungal Viruses* (ed. Molitoris, H. P., Hollings, M. and Wood, H. A.), Springer-Verlag, pp. 48–61.

Bruenn, J. A. (1980). Virus-like particles of yeast. *Ann Rev Microbiol* **34**, 49–68.

Buck, K. W. and Kempson-Jones, G. F. (1970). Three types of virus particles in *Penicillium stoloniferum*. *Nature* **225**, 945–6.

Cooper, P. D. (1969). In *The biochemistry of viruses* (ed. Levy, H. B.), Marcel Dekker: New York.

Canova, A. and Casalicchio, G. (1963). Infezioni da virus su *Polypodium vulgare* L., e *Scolopendrium vulgare* Sm. *Phytopathol Mediterr* **2**, 88–90.

Davis, R. E. and Worley, J. F. (1973). Spiroplasma; motile, helical micro-organism associated with corn stunt disease. *Phyt* **63**, 403–8.

Davis, R. E., Chen. T. A. and Worley, J. F. (1981). Corn stunt spiroplasma. In *Virus and virus-like diseases of maize in the United States* (ed. Gordon, D. T., Knoke, J. K. and Scott, G. E.) Southern Co-op. Ser. Bull. **247**, 43–53.

Dieleman-van Zaayen, A., Igesz, O. and Finch, J. T. (1970). Intracellular appearance and some morphological features of virus-like particles in an ascomycete fungus. *Virol* **42**, 534–7.

Dieleman-van Zaayen, A. and Temmink, J. H. M. (1968). A virus disease of cultivated mushrooms in the Netherlands. *Neth J Plant Pathol* **74**, 48–51.

Diener, T. O. (1971). Potato spindle tuber 'virus'. IV. A replicating, low molecular weight RNA. *Virol* **45**, 411–28.

Diener, T. O. (1978). In *Viruses and environment* (ed. Kurstak, E. and Maramorosch, K.), Academic Press: New York. pp. 113–15.

Diener, T. O. (1979). *Viroids and viroid diseases*, John Wiley and Sons: Chichester.

Diener, T. O. (1980). In *Plant disease etiology*, Abstr. Meet. Fed. Br. Plant Pathol/Soc. Gen. Microbiol., London, 1980.

Diener, T. O. (1981). Viroids. In *Handbook of plant virus infections* (ed. Kurstak, E.). Elsevier/North Holland: London. pp. 913–34.

Diener, T. O. (1982). Viroids and their interactions with host cells. *Ann Rev Microbiol* **36**, 239–58.

Diener, T. O. and Lawson, R. H. (1973). Chrysanthemum stunt; a viroid disease. *Virol* **51**, 94–101.

Diener, T. O. and Raymer, W. B. (1971). Potato spindle tuber 'virus'. CMI/AAB. Descriptions of plant viruses No. 66.

Dodds, J. A. (1979). Viruses of marine algae. *Experimentia*, **35**, 440–2.

Doi, Y., Teranaka, M., Yora, K. and Asuyama, H. (1967). Mycoplasma or PLT group-like micro-organisms found in the phloem elements of plants infected with mulberry dwarf, potato witches broom, aster yellows, or Paulownia witches broom. *Ann Phytopathol Soc Jpn* **33**, 259–66.

Fenner, F. (1970). The genetics of animal viruses. *Ann Rev Microbiol* **24**, 297–334.

Francki, R. I. B. and Hatta, H. (1981). Tomato spotted wilt virus. In *Handbook of plant virus infections* (ed. Kurstak, E.), Elsevier/North-Holland: London. pp. 491–514.

Francki, R. I. B., Kitajima, E. W. and Peters, D. (1981). Rhabdoviruses. In *Handbook of plant virus infections* (ed. Kurstak, E.), Elsevier/North-Holland: London. pp. 455–90.

Fulton, R. W. (1981). Ilarviruses. In *Handbook of plant virus infections* (ed. Kurstak, E.), Elsevier/North Holland: London. pp. 377–414.

Galindo, J. A., Smith, D. R., Diener, T. O. (1982). Etiology of planta macho, a viroid disease of tomato. *Phyt* **72**, 49–54.

Ghabrial, S. A. (1980). Effects of fungal viruses on their hosts. *Ann Rev Phyto* **18**, 441–61.

Gibbs, A. J. and Harrison, B. D. (1968). Realistic approach to virus classification and nomenclature. *Nature* **218**, 927–9.

Gibbs, A. J. and Harrison, B. D. (1976). *Plant virology: The principles*. Edward Arnold: London.

Gibbs, A. J., Harrison B. D., Watson, D. H. and Wildy, P. (1966). 'What's in a virus name?' *Nature* **209**, 450–4.

Gibbs, A. J., Skotnicki, A. H., Gardner, J. E., Walker, E. S. and Hollings, M. (1975). A tobamovirus of a green algae. *Virol* **64**, 571–4.

Gierer, A. and Mundry, K. W. (1958). Production of mutants of tobacco mosaic virus by chemical alteration of its ribonucleic acid *in vitro*. *Nature* **182**, 1457–8.

Gingery, R. E., Gordon, D. T., Nault, L. R. and Bradfute, O. E. (1981). Maize chlorotic dwarf virus. In *Handbook of plant virus infections* (ed. Kurstak, E.), Elsevier/North Holland: London. pp. 19–32.

Goheen, A. C., Nyland, G. and Lowe, S. K. (1973). Association of a rickettsia-like organism with Pierce's disease of grapevine and alfalfa dwarf, and heat therapy of the disease in grapevines. *Phyt* **63**, 341–5.

Goodman, R. M. (1981). Geminiviruses. In *Handbook of plant virus infections* (ed. Kurstak, E.), Elsevier/North Holland: London. pp. 879–912.

Harrison, B. D. (1964). Infection of gymnosperms with nematode-transmitted viruses of flowering plants. *Virol* **24**, 228–9.

Harrison, B. D. and Robinson, D. J. (1981). Tobraviruses. In *Handbook of plant virus infections* (ed. Kurstak, E.), Elsevier/North Holland: London. pp. 515–40.

Herring, A. J. and Bevan, E. A. (1974). Virus-like particles associated with double stranded RNA species found in killer and sensitive strains of yeast *Saccharomyces cerevisiae*. *J Gen Virol* **22**, 387–94.

Hewitt, W. B., Frazier, N. W. and Houston, B. R. (1942). Transmission of Pierce's disease of grapevine with a leafhopper. *Phyt* **32**, 8.

Hollings, M. (1962). Viruses associated with a dieback disease of cultivated mushrooms. *Nature* **196**, 962–5.

Hollings, M. (1978). Mycoviruses: viruses that infect fungi. *Adv Virus Res* **22**, 2–54.

Hollings, M. (1982). Mycoviruses and plant pathology. *Plant Dis R* **66**, 1106–12.

Hollings, M. and Brunt, A. A. (1981). Potyviruses. In *Handbook of plant virus infections* (ed. Kurstak, E.), Elsevier/North Holland; London. pp. 731–808.

Holmes, F. O. (1939). *Handbook of phytopathogenic viruses*. Burgess, Minneapolis.

Holmes, F. O. (1948). In *Manual of determinative bacteriology* (ed. Breed, R. S., Murray, E. G. D. and Hitchens, A. P.), Baillierè, Tindall and Cox.

Hull, R. (1968). A virus disease of hart's-tongue fern. *Virol* **35**, 333–5.

Hull, R. (1969). Alfalfa mosaic virus. *Adv Virus Res* **15**, 365–433.

Hull, R. (1981). Pea enation mosaic virus. In *Handbook of plant virus infections* (ed. Kurstak, E.), Elsevier/North Holland; London. pp. 239–56.

Ishiie, T., Doi, Y., Yora, K. and Asuyama, H. (1967). Suppressive effects of antibiotics of tetracycline group on symptom development of mulberry dwarf disease. *Ann Phytopathol Soc Jpn* **33**, 267–75.

Jackson, A. O. and Lane, L. C. (1981). Hordeiviruses. In *Handbook of plant virus infections* (ed. Kurstak, E.), Elsevier/North Holland: London. pp. 565–626.

Jensen, D. D., Griggs, W. H., Gonzales, C. Q. and Schneider, H. (1964). Pear psylla proven carrier of pear decline virus. *Calif Agric* **18**, 2–3.

Johnson, J. (1927). The classification of plant viruses. *Wis Agric Exp Stn Res Bull* **76**, 1–16.

Kaper, J. M. and Waterworth, H. E. (1977). Cucumber mosaic virus – associated RNA 5. Causal agent for tomato necrosis. *Sci* **196**, 429–31.

Kaper, J. M. and Waterworth, H. E. (1981). Cucumoviruses. In *Handbook of plant virus infections* (ed. Kurstak, E.), Elsevier/North Holland: London. pp. 257–332.

Kazama, F. Y., and Schornstein, K. L. (1973). Ultrastructure of a fungus herpes-type virus. *Virol* **52**, 478–87.

Koenig, R. and Lesemann, D. E. (1981). Tymoviruses. In *Handbook of plant virus infection* (ed. Kurstak, E.), Elsevier/North Holland: London. pp. 33–60.

Kunkel, L. O. (1926). Studies on aster yellows. *Am J Bot* **13**, 646–705.

Kunkel, L. O. (1940). In *The genetics of pathogenic organisms* (ed. Moulton, F. R.), AAAS Publ. No. 12.

Kunkel, L. O, (1948) Studies on a new corn virus disease. *Arch Gesamte Virusforsch* **4**,24–46.

Kurstak, E. (1981). *Handbook of plant virus infections* (ed. Kurstak, E.), Elsevier/North Holland: London.

Lane, L. C. (1981). Bromoviruses. In *Handbook of plant virus infections* (ed. Kurstak, E.), Elsevier/North Holland: London. pp. 333–76.

Lapierre, H., Lemaire, J. M., Jouan, B. and Molin, G. (1970). Mise en evidence due particulars virales associees a une perte de pathogenicite chez le pietin-echandage des cereales, *Ophiobolus graminis*. Sacc. *C R Hebd Séances Acad Sci Ser D* **271**, 1833–6.

Lemke, P. A. (1976). Viruses of eucaryotic micro-organisms. *Ann Rev Microbiol* **30**, 105–45.

Lemke, P. A. (1979). *Viruses and plasmids in fungi*. Marcel Dekker: New York.

Lister, R. M. and Bar-Joseph, M. (1981). Closteroviruses. In *Handbook of plant virus infections* (ed. Kurstak, E.), Elsevier/North Holland: London. pp. 809–46.

Maramorosch, K. (1974). Mycoplasmas and rickettsiae in relation to plant diseases. *Ann Rev Microbiol* **28**, 301–24.

Maramorosch, K. (1981). Spiroplasmas: agents of animal and plant diseases. *Biosci* **31**, 374–80.

Maramorosch, K. (1982). Control of vector-borne mycoplasmas. In *Pathogens, vectors and plant diseases* (ed. Harris, K. F. and Maramorosch, K.), Academic Press: London. pp. 265–95.

Maramorosch, K., Granados, R. R. and Hirumi, H. (1970). Mycoplasma diseases of plants and insects. *Adv Virus Res* **16**, 136–93.

Martelli, G. P. (1981). Tombusviruses. In *Handbook of plant virus infections* (ed. Kurstak, E.), Elsevier/North Holland: London. pp. 61–90.

Matsumoto, T., Lee, C. S. and Teng, W. S. (1969). Studies on sugarcane white leaf disease of Taiwan, with special reference to the transmission by a leaf-hopper, *Epitettix hiroglyphicus* Mats. *Ann Phytopath Soc Jpn* **35**, 251–9.

Matthews, R. E. F. (1979). Classification and nomenclature of viruses. 3rd report of the International Committee on Taxonomy of Viruses. *Intervirolo* **12**, 131–296.

Matthews, R. E. F. (1981). *Plant virology*, Academic Press: London.

Matthews, R. E. F. (1982) Classification and nomenclature of viruses. 4th report of the International Committee on Taxonomy of viruses. *Intervirolo* **15**, pp. 64–179.

Molitoris, H. P., Hollings, M. and Wood, H. A. (1979). *Fungal viruses*, Springer-Verlag.

Moll, J. N. and Martin, M. M. (1973). Electron microscope evidence that citrus psylla (*Trioza erytreae*) is a vector of greening disease in South Africa. *Phytophylactica* **5**, 41–4.

Murant, A. F. (1970). Arabis mosaic virus. CMI/AAB. Descriptions of plant viruses No. 16.

Murant, A. F. (1981). Nepoviruses. In *Handbook of plant virus infections* (ed.

Kurstak, E.), Elsevier/North Holland: London. pp. 197–238.

Murant, A. F. and Roberts, I. M. (1971). Mycoplasma-like bodies associated with *Rubus* stunt disease. *Ann Appl Biol* **67**, 389–93.

Nasu, S., Sugiura, M., Watimoto, T. and Iida, T. T. (1967). On the pathogen of rice yellow dwarf virus. *Ann Phytopathol Soc Jpn* **3**, 343–4.

Nienhaus, F. and Sikora, R. A. (1979). Mycoplasmas, spiroplasmas, and rickettsia-like organisms as plant pathogens. *Ann Rev Phyto* **17**, 37–59.

Palukaitis, P., Hatta, T., Alexander, D. M. and Symons, R. H. (1979). Characterisation of a viroid associated with avocado sunblotch disease. *Virol* **99**, 145–51.

Purcifull, D. E. and Edwardson, J. R. (1981). Potexviruses. In *Handbook of plant virus infections* (ed. Kurstak, E.), Elsevier/North Holland: London. pp. 627–94.

Randles, J. W. (1975). Association of two ribonucleic acid species with cadang-cadang disease of coconut palm. *Phyt* **65**, 163–7.

Rast, A. T. B. (1972). M11–16, an artificial symptomless mutant of tobacco mosaic virus for seedling inoculation of tomato crops. *Neth J Plant Pathol* **78**, 110–12.

Rawlinson, C. J., Hornby, D., Pearson, V. and Carpenter, J. M. (1973). Virus-like particles in take-all fungus, *Gaeumannomyces graminis*. *Ann Appl Biol* **74**, 197–209.

Rochow, W. F. (1970). Barley yellow dwarf virus; phenotypic mixing and vector specificity. *Sci* **167**, 875–8.

Rochow, W. F. and Duffus, J. E. (1981). Luteoviruses and yellow diseases. In *Handbook of plant virus infections* (ed. Kurstak, E.), Elsevier/North Holland: London. pp. 147–70.

Romaine, C. P. and Horst, R. K. (1975). Suggested viroid etiology for chrysanthemum chlorotic mottle disease. *Virol* **64**, 86–95.

Sanderlin, R. S. and Ghabrial, S. A. (1978). Physicochemical properties of two distinct types of virus-like particles from *Helminthosporium victoriae*. *Virol* **87**, 142–51.

Sanger, H. L. (1972). An infectious and replicating RNA of low molecular weight: The agent of the exocortis disease of citrus. *Adv Biosci* **8**, 103–16.

Sasaki, M. and Shikata, E. (1977). Studies on the host range of hop stunt disease in Japan. *Proc Jpn Acad* **53B**, 103–8.

Sehgal, O. P. (1981). Southern bean mosaic virus group. In *Handbook of plant virus infections* (ed. Kurstak, E.), Elsevier/North Holland: London. pp. 91–1 22.

Shepherd, R. J. and Lawson, R. H. (1981). Caulimoviruses. In *Handbook of plant virus infections* (ed. Kurstak, E.), Elsevier/North Holland: London. pp. 847–78.

Shikata, E. (1981). Reoviruses. In *Handbook of plant virus infections* (ed. Kurstak, E.), Elsevier/North Holland: London. pp. 423–54.

Siegel, A. (1960). Studies on the induction of tobacco mosaic virus mutants with nitrous acid. *Virol* **11**, 156–67.

Singh, R. P., Finnie, R. E. and Bagnall, R. H. (1971). Losses due to potato spindle tuber virus. *Am Potato J* **48**, 262–7.

Sinha, R. C. and Paliwal, Y. C. (1970). Localization of a mycoplasma-like organism in tissues of a leaf-hopper vector carrying clover phyllody agent. *Virol* **40**, 665–72.

Smith, K. M. (1937). *A textbook of plant virus diseases*, Churchill Livingstone.

Stace-Smith, R. (1981). Comoviruses. In *Handbook of plant virus infections* (ed. Kurstak, E.), Elsevier/North Holland: London. pp. 171–96.

Stanley, W. M. (1935). Isolation of a crystalline protein possessing the properties of tobacco mosaic virus. *Sci* **81,** 644–5.

Story, G. E. and Halliwell, R. S. (1969). Association of a mycoplasma-like organism with the bunchy top disease of papaya. *Phyt* **59**, 1336–7.

Turner, W. F. and Pollard, H. N. (1959). Insect transmission of phony peach disease. *Tech Bull U.S.D.A.* 1193.

Ushiyama, R., Bullivant, S. and Matthews, R. E. F. (1969). A mycoplasma-like organism associated with *Phormium* yellow leaf disease. *NZ J Bot* **7**, 363–71.

Uyemoto, J. K. (1981). Tobacco necrosis and satellite viruses. In *Handbook of plant virus infections* (ed. Kurstak, E.). Elsevier/North Holland: London. pp. 123–46.

Van Dorst, H. J. M. and Peters, D. (1974). Some biological observations on pale fruit, a viroid-incited disease of cucumber. *Neth J Plant Path* **80**, 85–96.

Van Regenmortel, M. H. V. (1981). Tobamoviruses. In *Handbook of plant virus infections* (ed. Kurstak, E.), Elsevier/North Holland: London. pp. 541–64.

Van Regenmortel, M. H. V. and Pinck, L. (1981). Alfalfa mosaic virus. In *Handbook of plant virus infections* (ed. Kurstak, E.), Elsevier/North Holland: London. pp. 415–22.

Walkey, D. G. A. and Innes, N. L. (1979). Resistance to bean common mosaic virus in dwarf beans (*Phaseolus vulgaris* L.), *J Agric Sci* **92**, 101–8.

Walkey, D. G. A., Innes, N. L. and Miller, A. (1983). Resistance to bean yellow mosaic virus in *Phaseolus vulgaris*. *J Agric Sci* **100**, 643–50.

Walter, B. (1982). C R Hebd Séances Acad Sci **292**, 537–42.

Wetter, C. and Milne, R. G. (1981). Carlaviruses. In *Handbook of plant virus infections* (ed. Kurstak, E.), Elsevier/North Holland: London. pp. 695–730.

Yarwood, C. E. (1959). Virus increase in seedling roots. *Phyt* **49**, 220–3.

2.7 Further Selected Reading

Diener, T. O. (1982). Viroids and their interaction in host cells. *Ann Rev Microbiol* **36**, 239–58.

Ghabrial, S. A. (1980). Effects of fungal viruses on their hosts. *Ann Rev Phyto* **18**, 441–61.

Kurstak, E. (1981). *Handbook of plant virus infections*, Elsevier/North Holland: London. pp. 1–943.

Maramorosch, K. (1982). Control of vector-borne mycoplasmas. In *Pathogens, vectors and plant diseases* (ed. Harris, K. F. and Maramorosch, K.), Academic Press: London. pp. 265–95.

Matthews, R. E. F. (1979). Classification and nomenclature of viruses 3rd report of the International Committee on Taxonomy of Viruses. *Intervirolo* **12**, 131–296.

Nienhaus, F. and Sikora, R. A. (1979). Mycoplasmas, spiroplasmas and rickettsia-like organisms as plant pathogens. *Ann Rev Phyto* **17**, 37–59.

3 *Virus Symptoms*

3.1 Introduction

Symptoms are the observable effects that a virus has on the growth, development and metabolism of an infected host plant. In the early days of plant virology, symptoms were of major importance, for they were the main means by which a virus disease was diagnosed and named. Viruses are still named after the type of symptom they cause in the diseased plant (*see* Section 2.2), but many other techniques have now become available to assist in virus diagnosis (*see* Chapter 6). These techniques not only accelerate the process of virus identification, but they also enable us to avoid confusing virus induced symptoms with those caused by other disease agents such as viroids, mycoplasma and rickettsia-like organisms (*see* Section 2.5).

Host symptoms are still very important to the applied plant virologist. In the field symptoms give the first clue to a virus's identity, and in the laboratory, the symptoms produced in a range of test plants may often be of considerable diagnostic value. For the grower, symptoms are the most important aspect of virus infection. The nature and severeity of the disease symptoms will determine the economic importance of a particular virus, in terms of yield loss and reduced quality.

In a susceptible plant, following infection, the virus begins to replicate in the host cell. This process alters the cell's metabolism which results in biochemical and physiological changes. The virus symptoms that are described in this chapter, are the result of the abnormal metabolism that the virus causes in the host's tissues. These changes may be macroscopic and clearly visible on the external surfaces of the plant's organs, or they may be internal changes that may only be seen in tissues examined using a light or electron microscope. Alterations in the infected host plant start in the inoculated cells, spread to the surrounding tissues and finally, may affect the whole plant. Detailed information on the processes of plant virus infection and replication within the cell, has been described by Matthews (1981).

When considering virus symptoms, it should be remembered that a virus is unlikely to cause just one symptom in an infected plant. Usually, infection results in more than one type of symptom and frequently there may be a sequence of symptoms as the disease spreads within the plant. For example, stunted growth and dwarfing may be associated with necrotic symptoms, and in extreme cases, the necrosis may spread to the whole plant to cause eventual death.

The occurrence of more than one type of symptom in a diseased host is called a *syndrome* and if the host is affected simultaneously by more than one type of virus or pathogen, their effects may be greater than expected from their individual symptoms. These cumulative symptoms are referred to as *synergism* or a *synergistic* effect.

In this chapter the major external and internal symptoms caused by virus infection in the host plant are described, together with the main factors that may influence or govern the expression of these symptoms. Further information on plant virus symptoms may be obtained from Smith (1972); Bos (1978) and Matthews (1981).

3.2 Principal External Symptoms

The visible symptoms caused by virus infection may be considered under two headings, those resulting from *primary infection* in the inoculated cells of the host plant (*see* Plate 3.2), and those caused by *secondary* or *systematic infection* as the virus moves from the sites of primary inoculation into the remainder of the plant (*see* Plate 3.2).

3.2.1 Primary infection

Viruses, unlike fungal plant pathogens, are unable to get into the cells of the host plant, unless they can enter through a wound. In the laboratory, virus entry can be brought about by dusting the leaf surface with a fine abrasive and then rubbing the surface with virus infected sap (*see* Section 4.2). The virus enters the cells through broken epidermal hairs, or through small abrasions in the epidermal layer of cells (*see* Plate 4.1) In nature, the infection process may occur as a result of infected and healthy leaves rubbing together, but most frequently it occurs during vector feeding (*see* Chapter 7).

The primary symptoms that develop at the site of virus entry in the inoculated leaves are known as *local* symptoms. These often take the form of distinct areas of diseased cells referred to as *lesions* and these symptoms are commonly called *local lesions*.

Local lesions vary in size from small pin-points to large patches and they may be chlorotic due to loss of chlorophyll, or necrotic, if, as often happens, the cells die (*see* Plate 3.1). They occur most frequently after

Plate 3.1 Primary infection symptoms on inoculated leaves.
(*a*) Chlorotic ringspots on cotyledons of marrow caused by cucumber mosaic virus; (*b*) necrotic lesions and veinal necrosis caused by bean common mosaic virus on primary leaves of dwarf bean (*Phaseolus vulgaris*); (*c*) necrotic lesions caused by tobacco mosaic virus on *Nicotiana glutinosa*; (*d*) lesions caused by arabis mosaic virus on *Chenopodium quinoa*.

mechanical sap transmission of virus to a leaf surface. Local lesions are very useful to virologists since they provide a bioassay of virus infectivity (*see* Section 4.4.). Local lesions may also occur on a leaf surface at the site of feeding of a viruliferous insect, but the occurrence of local lesions following insect transmission is less common than after mechanical transmission.

In some virus/host infections, the virus is unable to spread in the plant beyond the site of primary infection and local lesions may be the only observed symptoms. The cells within these lesions die as the infection process ceases, and this type of restricted response is often referred to as a *hypersensitive* reaction. The phenomenon of *hypersensitivity* and hypersensitive reactions is covered in greater detail in Chapter 10.

In other primary infections, no local lesions are visible at the surface of the inoculated leaves, but if the leaf is cleared with ethyl alcohol and then stained with iodine, it is sometimes possible to see starch lesions (Holmes, 1931).

If a virus is not confined to the site of primary infection, it will spread from the initial inoculation site from cell to cell within the leaf mesophyll. Spread probably occurs through the plasmodesmata connections between the cells (Gunning and Roberts, 1976). Eventually the virus may reach the vascular system, and once there, it is likely to spread fairly rapidly through the entire plant. This causes the *secondary*, so-called *systemic infection*. Virus movement in the vascular system is usually in the phloem, along with other plant assimilates, but a few viruses, such as lettuce necrotic yellows and potato mop top viruses, have been shown to move in xylem vessels (Matthews, 1981).

From the vascular elements in the veins of systemically infected leaves, the virus moves from cell to cell in the leaf mesophyll to produce the systemic symptoms, the most important of which are described in the following section. Similarly, virus moves from the vascular systems to adjacent cells in roots, fruits and other organs.

3.2.2 Secondary, systemic symptoms

The secondary, systemic symptoms caused by viruses in diseased plants may be divided into visible, macroscopic changes and those which occur internally in the form of abnormal cell structures. The latter symptoms can usually only be observed with the light or electron microscope. In this section, the most important of the macroscopic symptoms are summarized and internal symptoms are described in Section 3.3.

In considering systemic symptoms, it must again be emphasized that there is frequently a sequential development of symptoms in the infected plant, and that one or more of those described may occur together or following one another in the same plant.

Stunting and dwarfing

Reduced plant size is a frequent symptom of most plant virus infections (*see* Plate 3.10) and is the one most likely to be found in combination with any of the symptoms described in the following pages. In some infections, unless uninfected plants are grown side by side with infected ones, it may be difficult to observe reduced growth, and even in symptomless, latent infections (*see* Section 3.4) some less obvious stunting is likely to occur.

Growth may be evenly reduced throughout the plant, or the stunting may be confined to specific parts or organs of the plant, as is the case with apical stunting of pea stunt disease, caused by red clover vein mosaic virus (Hagedorn and Walker, 1949).

Plate 3.2 *Chenopodium amaranticolor* plant infected with arabis mosaic virus showing necrotic lesions on the inoculated leaves, and secondary, systemic spread of the virus to cause necrosis and stunting of the apical shoot.

Plate 3.3 Examples of leaf mosaic symptoms.
(*a*) Bean yellow mosaic virus infected *Vicia faba* bean leaves; (*b*) mottle symptoms caused by turnip mosaic virus in cabbage; (*c*) cucumber mosaic virus symptoms in marrow; (*d*) streak symptoms caused by maize white line mosaic virus in maize (courtesy of M Conti); (*e*) sugar cane mosaic virus symptoms in maize (courtesy of M. Conti); (*f*) light and dark green banding symptoms caused by cauliflower mosaic virus in cauliflower.

Root growth, in common with the aerial parts of the infected plant, may be stunted, but tend to be overlooked when diseased plants are examined. Consequently, the number of reports of the effects of virus infection on root growth are limited.

Although symptoms of reduced size are not always as dramatic as some other systemic symptoms, in terms of reduced yield, they can be very important to the commercial grower.

Mosaic

The term mosaic refers to a number of leaf symptoms in which cells in some areas of the leaf or organ are infected and discoloured, and other cells in other areas are not. The infected areas are usually pale green or chlorotic due to the loss, or reduced production of chlorophyll, while the cells of the adjacent areas remain green in colour. The shape and pattern of this type of symptom varies considerably (*see* Plate 3.3). If the discoloured portions of the leaf are rounded, the symptoms are often referred to as a *mottle*, and *chlorotic flecking*, *spotting* and *blotching* may also occur. In some infections such as cauliflower mosaic virus in brassicas, regular light and dark green banding may occur (*see* Plate 3.3*f*).

In infected monocotyledonous plants, the mosaic symptom usually takes the form of light and dark green striping or streaking of the tissues (*see* Plate 3.3*d*, *e*). Mosaic symptoms may also occur on the stems, or fruits of some infected plants, as is the case with marrow infected with cucumber mosaic virus.

Chlorosis

In some virus infections, the whole leaf may become chlorotic due to decreased chlorophyll production and the breakdown of chloroplasts. Most chlorotic symptoms are linked with internal histological disorders, such as abnormal changes in the palisade cells and intracellular vacuoles (Esau, 1968). Chlorosis is the main symptom associated with the economically important 'yellowing' viruses, beet yellows and barley yellow dwarf.

Symptoms of chlorosis usually start as interveinal chlorosis and spread through the leaf. Sometimes the area adjacent to the vein of the leaf remains green in contrast to the remainder of the leaf, as in lettuce infected with beet western yellows virus (*see* Plate 3.4*a*). In other infections, such as in strawberries infected with strawberry yellow edge virus, the chlorosis may be restricted to certain areas of the leaf. In some instances the chlorosis may be confined to the area of the vein and is referred to as veinal chlorosis or *vein yellowing* (*see* Plate 3.4*c*), and a variation on this type of symptom is seen with *vein clearing*, in which the cells adjacent to the vein become translucent (*see* Plate 3.4*b*).

Plate 3.4 Examples of chlorotic leaf symptoms.
(*a*) A lettuce plant infected with beet western yellows virus showing chlorosis of its outer leaves (*right*) healthy plant (*left*); (*b*) vein clearing symptoms in lettuce infected with big-vein disease (courtesy of J. A. Tomlinson); (*c*) veinal chlorosis caused by turnip mosaic virus in mustard.

Plate 3.5 Examples of ringspotting symptoms.
(*a*) Necrotic ringspots caused by turnip mosaic virus in cabbage; (*b*) concentric ringspots induced by cherry leaf-roll virus in *Nicotiana tabacum* cv. White Burley (courtesy J. A. Tomlinson); (*c*) necrotic ringspots caused by turnip mosaic virus in *Nicotiana clevelandii*; (*d*) chlorotic ringspots and broken ringspots caused by celery mosaic virus in celery.

Plate 3.6 Necrotic symtoms induced by turnip mosaic virus in Dutch white storage cabbage.
(*a*) Necrosis on the outer leaves of plants in the field; (*b*) a cross-section of a cabbage from cold-storage showing internal necrosis.

Plate 3.7 (*a*) Surface view of internal necrotic lesions caused by turnip mosaic virus in stored white cabbage; (*b*) necrotic symptoms (the so-called 'black-root' symptoms) caused by bean common mosaic virus in dwarf bean (*Phaseolus vulgaris*). Large necrotic patches develop on the inoculated leaves, followed by apical necrosis and necrosis of the veins.

Plate 3.8 Necrotic symptoms on seeds and fruits.
(*a*) Healthy *Vicia faba* bean seed; (*b*) seeds from a plant infected with broad bean stain virus; (*c*) necrotic pitting on fruit from a pepper plant (cv. Quadrato) infected with tobacco mosaic virus (courtesy M. Conti); (*d*) necrotic ringspots on fruits of a pepper plant (cv. Quadrato giallo) infected with cucumber mosaic virus (courtesy M. Conti).

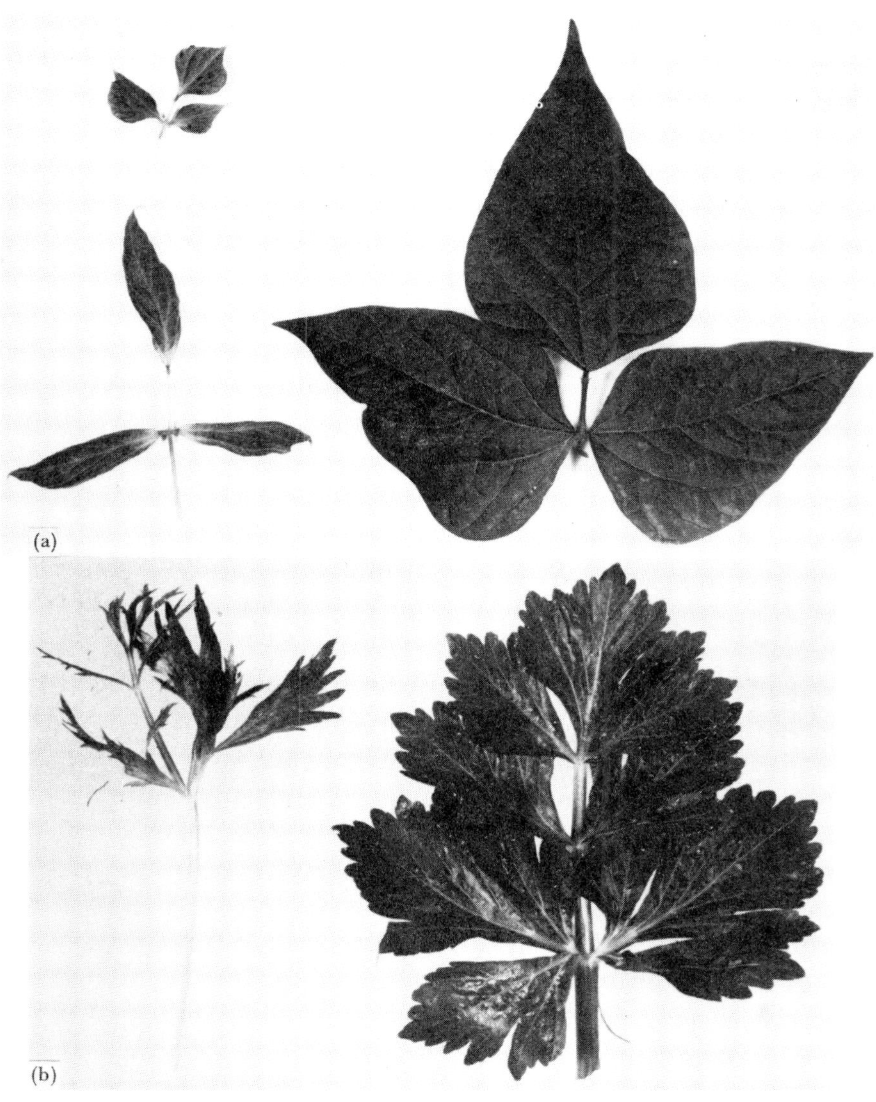

Plate 3.9 Leaf distortion symptoms.
(*a*) Distorted and stunted leaves of dwarf bean (*Phaseolus vulgaris*) infected with bean common mosaic virus, healthy leaf on right; (*b*) strap-like and stunted leaves of celery infected with strawberry latent ringspot virus, healthy leaf on right.

Ringspotting

The symptom of ringspotting is a common feature of some viruses. In these infections, the diseased area is restricted to a ring or broken ring of infected cells. The infected cells may be chlorotic or necrotic, and sometimes the rings may occur in concentric circles (*see* Plate 3.5). Ringspotting symptoms are most frequent on the leaves of infected plants, but may also occur on stems, fruits (*see* Plate 3.8*d*) and pods.

Necrosis

In addition to the primary necrotic local-lesion symptoms that have already been described, there are many types of virus-induced necrosis that occur in systemically infected tissues, that are caused by the death of cells.

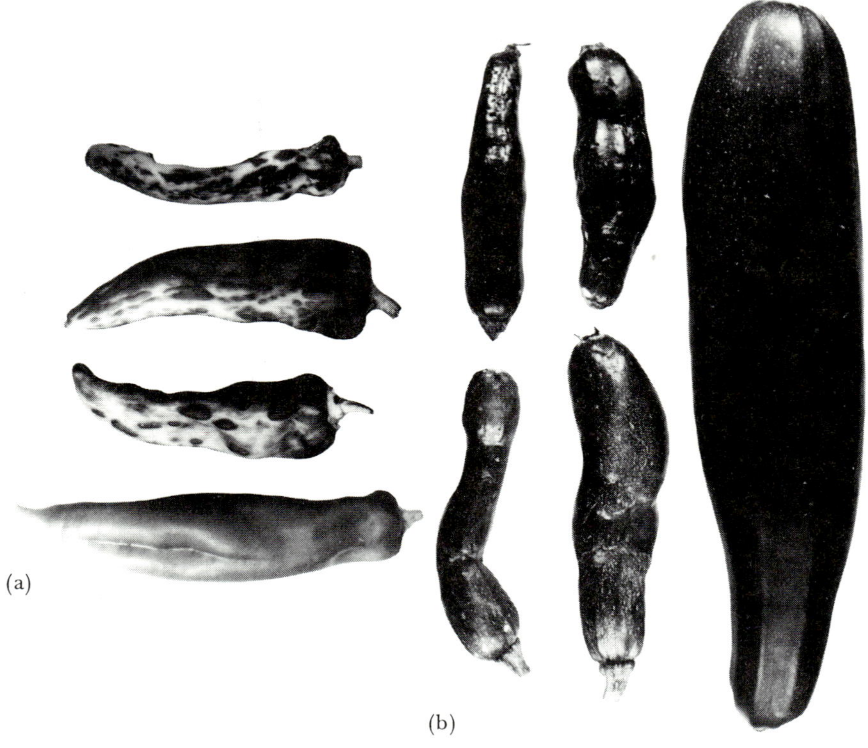

Plate 3.10 Distortion symptoms on fruit.
(*a*) Stunted and distorted fruits from pepper plant (cv. Corno di Toro) infected with tobacco mosaic virus, lower fruit is healthy (courtesy of M. Conti); (*b*) fruits from a marrow plant (cv. Brimmer) infected with cucumber mosaic virus, healthy fruit on right.

Systemic necrosis may take the form of small or large lesions, as for instance with those caused by turnip mosaic virus on the external and internal leaves of cabbage (*see* Plates 3.6 and 3.7), or veinal systems may become necrotic, and the necrosis may spread to the stem and root apices to eventually kill the plant. This happens in some cultivars of *Phaseolus vulgaris* beans infected with bean common mosaic virus (*see* Plate 3.7*b*). In other circumstances necrosis may occur on fruits or seeds (*see* Plate 3.8).

Leaf and stem distortion

In some infections the leaf lamina is affected and it may become irregularly distorted or strap-like. Virus infections such as bean common mosaic in *Phaseolus vulgaris* and strawberry latent ringspot in celery will cause this type of symptom (*see* Plate 3.9). Such abnormal growth is the result of hormonal imbalance within the leaf, and is similar to the type of leaf disorder that may occur as a result of hormonal spray damage. Other examples of distorted leaf symptoms are seen in tomatoes infected with tobacco mosaic virus and *Vicia faba* infected with bean leaf roll virus (Bos, 1978). Abnormal cell proliferation caused by a virus in the stem, is seen in the case of cacao infected with cacao swollen shoot disease.

In this type of malformation, an increase in the number of cells is referred to as *hyperplasia* and an abnormal size increase in an organ as *hypertrophy*. In contrast, a decrease in the number of cells is referred to as *hypoplasia* and a reduction in organ size is called *atrophy*. An example of hypoplasia is stem pitting virus infection in apple and citrus. This is caused by the failure of some cambial cells to differentiate normally, which results in a wedge of phloem being embedded in the developing xylem (Hilborn *et al.*, 1965). The pitting is visible as elongated pits or furrows on the surface of the stem when the bark is removed (*see* Plate 3.11). It is common in certain cultivars of apple and a similar condition appears to be caused by citrus tristeza virus in Mexican lime (*Citrus aurantifolia*) (Schneider, 1959).

Enations or tumours

Some virus infections are characterized by tumour-like outgrowths on the leaves or roots. The outgrowths on the leaves are commonly referred to as *enations* and these appear like 'warts' on the upper or lower surface of the leaf. They are common in pea plants infected with pea enation mosaic virus (*see* Plate 3.12*b*). Such growth is caused by abnormal cell proliferation which is probably due to virus induced changes in hormone concentration (Bos, 1978). Other examples of enations are those caused by sugar cane Fiji disease on the lower surface of sugar-cane leaves (Kunkel, 1924) and those produced in

Plate 3.11 Elongated furrows in the stem of Virginia Crab apple caused by apple stem pitting virus (courtesy of J. M. Thresh).

maize leaves by maize rough-dwarf disease. Both arise from abnormal proliferation of the underlying phloem tissues.

Wound tumour virus causes round, wart-like *tumours* on both the stems and roots of infected clover plants. The root tumours arise from cells in the pericycle that are wounded as the side shoots develop, and break through the root cortex (Lee, 1955) (*see* Plate 3.12*a*).

Plate 3.12 Tumour and enation symptoms.
(*a*) Tumours on clover roots caused by wound tumour virus (courtesy A. A. Brunt); (*b*) enations on the under-surface of pea leaves caused by pea enation mosaic virus (courtesy of D. J. Hagedorn); (*c*) enations on the under surface of *Melilotus alba* leaves caused by PEMV (courtesy of D. J. Hagedorn).

Petal or flower 'break'

Virus-induced colour 'break' symptoms in tulip petals caused by tulip mosaic virus (*see* Plate 1.1) was mentioned in Chapter 1 in relation to its importance in the history of plant viruses. Such colour 'break' symptoms are also common in flowers of other infected plants, including wallflowers (*Cheiranthus cheiri*) and stocks (*Matthiola incana*) infected with turnip mosaic virus, and gladiolus infected with bean yellow mosaic virus or cucumber mosaic virus. The 'break' symptoms may take the form of streaking, flecking or sectoring of the petal tissues with a colour different from the normal. Purple stocks may be flecked white, and red wallflowers yellow, since the break-colour usually results from the loss of anthocyanins which cause an underlying pigment to be exposed (*see* Plate 3.13).

In addition to colour-break symptoms, the flowers of many virus infected plants may be stunted and deformed, and frequently fewer flowers develop on infected than on healthy plants.

Plate 3.13 Petal-break symtoms in flowers of Virginia Stock infected with turnip mosaic virus.

Fruit, seed and pollen abnormalities

The occurrence of mosaic, ringspotting and necrotic symptoms on fruits and seeds have been mentioned earlier in this section, but virus infection may affect these structures in many other ways. Fruits from infected plants may be fewer, smaller, misshapen or of changed texture (*see* Plate 3.10). For example, fewer fruits are produced in plums infected with prune dwarf virus, and sugar beet curly top virus infection results in smaller cantaloupe fruits. Fruits of gherkin (*Cucumis sativa*) may be badly misshapen by cucumber mosaic virus infection (Tjallingii, 1952), and apples can have abnormal skin textures when infected with apple rough skin (Van Katwijk, 1956), or apple star cracking (Jenkins and Storey, 1955) viruses.

Virus infection of the mother plant may also have a drastic affect upon seed production. Some cultivars of lettuce infected with lettuce mosaic virus have greatly reduced seed production and in some instances the virus can cause complete abortion of the lettuce seed (Couch, 1955). In other cases, when the virus is transmitted in the seed of the infected mother plant, the germination and vigour of the infected seed is significantly impaired (Walkey *et al.*, 1983).

Pollen from an infected plant is frequently sterile, or its viability is impaired. For instance, the rate of germination of *Nicotiana rustica* pollen infected with cherry leaf roll virus is slower and the length of the pollen tube shorter, than healthy pollen (Cooper, 1976).

3.3 Principal Internal Symptoms

Some of the changes that occur within the tissues of virus infected plants have been mentioned in the previous section, since they may be directly responsible for the macroscopic symptoms that are seen externally on the diseased plant. The symptoms associated with mosaic and chlorosis are for instance, caused by the breakdown, or failure of the cell chloroplasts to produce chlorophyll, and the enations and tumours induced by certain viruses are due to abnormal cell divisions within the infected tissues. Besides such obvious symptoms, there are numerous other cytological and histological changes that occur in the infected tissues that are only visible in the light or electron microscope. In addition to abnormal cell structure, various virus-induced structures may be present in the infected cells. Such structures are called *inclusion bodies* and are sometimes characteristic of infection by specific viruses.

In this section the major types of cytological and histological changes that occur in the diseased cell are described, together with the most characteristic types of inclusion bodies.

3.3.1 Cytological and histological changes

Among the most important cytological effects of virus infection are changes in the cell nuclei. Various viruses have been observed in the nuclei, including southern bean mosaic (Weintraub and Ragetli, 1970) and tomato bushy stunt (Russo and Martelli, 1972). Pea enation mosaic virus has been observed to cause the breakdown of the nucleolus as the virus multiplies in infected pea leaves (Shikata and Maramorosch, 1966), and the nucleolus of bean plants infected with bean golden mosaic virus has been reported to increase in size to fill three-quarters of the nuclear space (Kim *et al.*, 1978).

Many virus infections cause changes in the chloroplasts, and most of these changes result in structural and biochemical degeneration. The chloroplasts may become colourless as the chlorophyll is lost, after which they may become misshapen, fragmented or grouped into abnormal clumps within the cell wall (Matthews, 1981).

The mitochondria of cells may become associated with virus particles which could suggest that they play a part in virus replication (Harrison and Roberts, 1968). In some infections the mitochondria become aggregated (Kitajima and Lovisolo, 1972), and in others abnormal membrane systems develop within the mitochondria. Various changes may also occur in the cell wall as a result of infection, and these and other effects are described in detail by Matthews (1981).

The major histological effects of virus infection that occur internally in diseased plants, are frequently associated with externally visible symptoms. These symptoms may involve either a reduction or an increase in cell numbers, or internal cell necrosis. A reduction in cambial cell numbers was described in Section 3.2 as the cause of stem pitting disease in apples, and another example of reduced cell formation is seen with apple stem grooving virus. Here the grooves are caused by the replacement of the normal phloem and xylem cells with parenchyma cells (Pleše *et al.*, 1975). An abnormal increase in the division of cambial cells may occur to induce increased amounts of xylem, as is the case with swollen shoot disease of cacao (Posnette, 1947). In other cases, abnormal numbers of sieve elements may be produced as with sugar-beet infected with sugarbeet curly top virus (Esau and Hoefert, 1978).

Other examples of virus-induced histological changes, include the lignification of xylem elements in grapevines infected with grapevine fanleaf virus, and the callosing of sieve plates and degeneration and death of phloem cells in barley infected with barley yellow dwarf virus.

Plate 3.14 Crystalline inclusion bodies.
(*a*) An electron micrograph of a section through a cytoplasmic crystal of artichoke mottled crinkle virus, magnification bar = 250 nm (courtesy of M. Russo); (*b*) section through a crystal of tobacco mosaic virus showing the particles arranged in rows, end to end, magnification bar = 100 nm (courtesy of G. J. Hills).

3.3.2 Plant virus inclusion bodies

Inclusion bodies may occur in the nucleus, but are most common in the cytoplasm. They have been observed in infections involving most groups of plant viruses, and vary greatly in their composition, size, shape and location (Edwardson and Christie, 1978).

Cytoplasmic inclusion bodies

When viruses multiply, they may accumulate in large numbers within the cell to form inclusion bodies composed almost entirely of virus particles. The particles may be regularly arranged side by side, end to end, arranged in a three-dimensional lattice, or aggregated at random. According to the arrangement (of the virus particles) the inclusion bodies may be fibrous, paracrystalline or crystalline (Martelli and Russo, 1977). Fibrous and paracrystalline bodies are formed with rod-shaped viruses, such as those of the potexvirus group (Esau, 1968; Kitajima and Galves, 1973), and consist of bundles of particles arranged in a two dimensional array. Often the inclusion body appears 'banded', as with those formed by beet yellows virus (Esau, 1968) and tobacco mosaic virus, because of the periodicity in the regular arrangement of the particles (*see* Plate 3.14).

Crystalline inclusion bodies may be formed by both isometric and rod-shaped viruses (*see* Plate 3.14). The regular arrangement of the particles results in crystals that may be hexagonal or rounded in shape. The crystals of tobacco mosaic virus have been extensively studied and shown to consist of closely stacked layers of particles arranged in parallel (Martelli and Russo, 1977) (*see* Figure 3.1). Crystals are produced by viruses belonging to many different virus groups including the cucumoviruses (Russo and Martelli, 1973), the nepoviruses (Roberts *et al.*, 1970) and the comoviruses (Kim and Fulton, 1972). The crystals may vary in size ranging from minute bodies that can only be

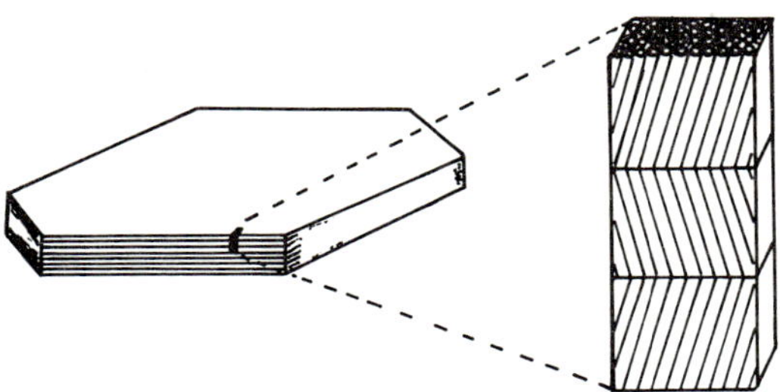

Fig. 3.1 Diagram of a portion of a tobacco mosaic virus inclusion crystal, showing rows of rod-shaped particles stacked in the component layers of the crystal (based on Steere, 1957).

seen in the electron microscope, to structures 15 to 20 μm in length that are visible in the light microscope.

Proteinaceous inclusion bodies

Several types of inclusion body associated with virus infection consist of proteins which are not identical to those of the virus particles. They

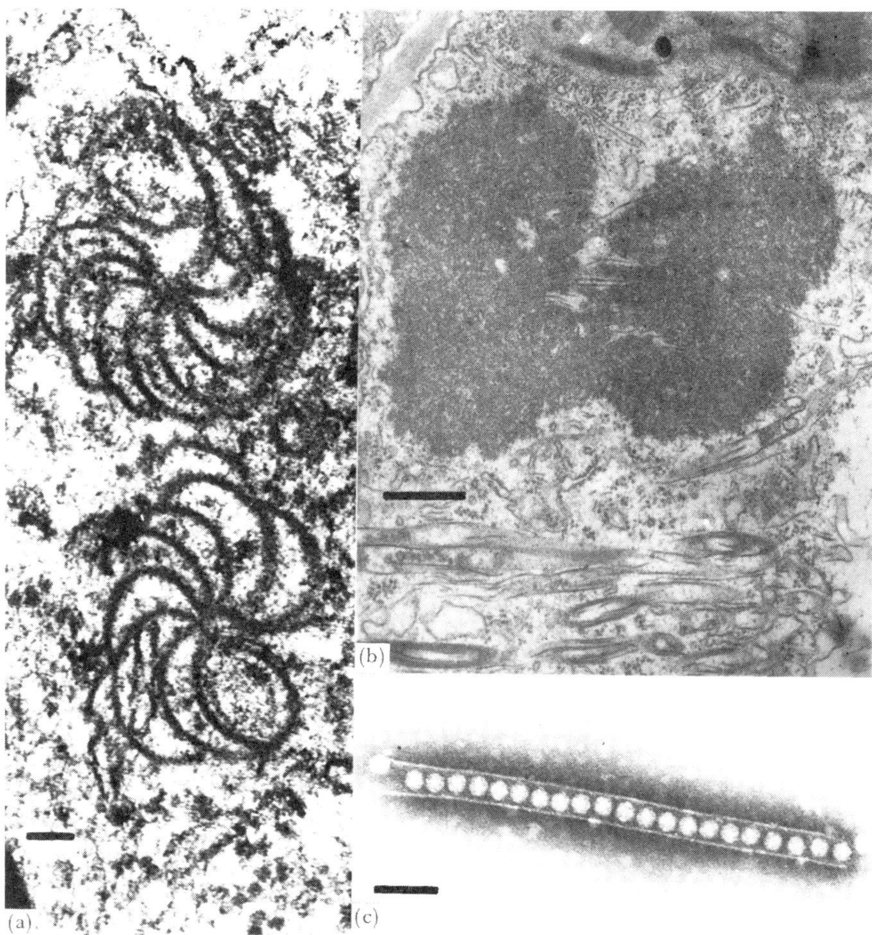

Plate 3.15 Virus associated inclusion bodies.
(*a*) An electron micrograph of a pin-wheel inclusion associated with white bryony mosaic virus (potyvirus group) infection of white bryony, magnification bar = 100 nm (courtesy of M. J. W. Webb); (*b*) a section through an amorphous cytoplasmic inclusion in a squash leaf infected with watermelon mosaic virus, magnification bar = 250 nm (courtesy of M. Russo); (*c*) a tubule associated with cherry leaf roll virus particles in a *Nicotiana rustica* plant, magnification bar = 100 nm (courtesy of M. J. W. Webb).

may be amorphous, or rounded granular bodies consisting of electron opaque material, like those formed by clover yellow and water melon mosaic viruses (Martelli and Russo, 1976) (*see* Plate 3.15*b*). It has been suggested that these bodies consist of a pool of excess viral protein, that has not yet been used to produce the complete virus particle (Schlegel and DeLisle, 1971). In the past, amorphous inclusion bodies have often been referred to as *X-bodies* (Esau and Hoefert, 1971).

Another type of proteinaceous inclusion body are the *cylindrical* bodies produced by the potyvirus group. These are so diagnostic of various members of the group, that Edwardson (1974) has used them as a means of classifying the group. They may also be found in plants infected with rod-shaped viruses of other groups (Weintraub and Ragetli, 1971; Hooper and Wiese, 1972).

Cylindrical inclusion bodies appear in electron microscope sections as scrolls, laminated plates or pinwheels when viewed in cross-section (*see* Plate 3.15*a*). Pinwheels are composed of curved plates radiating from a central axis of a cylinder with a central core (*see* Figure 3.2) and scrolls are cylinders which consist of a rolled-up plate. These bodies have been shown to consist of proteins that are not serologically related to the parent virus, but it is thought that the parent virus carries the genetical information for their production (Shepard *et al.*, 1974).

Fig. 3.2 Model of a cylindrical pinwheel inclusion body (based on Hollings and Brunt, 1981).

A further type of proteinaceous inclusion body is the *tubular* body frequently associated with infections by viruses of the nepo- and comovirus groups. These tubules contain single rows of virus particles and are readily seen in 'squash homogenates' observed in the electron microscope (Walkey and Webb, 1968; 1970) (*see* Plate 3.15*c*). The nature and function of these tubular bodies is not known, but Kim and Fulton (1975) have suggested that they are sites of virus assembly, following migration of virus protein and nucleic acid into the tubule.

Alternatively, they may help in the translocation of viruses between cells, possibly via the plasmodesmata. The tubules of strawberry latent ringspot and cherry leaf roll viruses have been shown to undulate in bundles of sufficient length to pass between adjacent cells (Walkey and Webb, 1970).

Further details of the various types of cytoplasmic inclusion bodies may be obtained from the review by Martelli and Russo (1977).

Nuclear inclusion bodies

Inclusion bodies are sometimes observed within nuclear tissues and may be amorphous or crystalline. Dense, electron-opaque, amorphous bodies have been reported associated with the nucleolus of cells in plants infected with beet mosaic virus (Martelli and Russo, 1969). Crystalline inclusion bodies are common in plants infected with bean yellow mosaic virus and may be cubic or octahedral in shape. Crystalline bodies have also been found in the nucleolus of plants infected with potato virus Y (Kitajima *et al.*, 1968) and alfalfa mosaic virus (Hull *et al.*, 1970).

Other nuclear-associated inclusion bodies may be caused by crystals of virus particles, similar to those that occur in the cytoplasm. Such virus crystals have been observed in the nucleus of plants infected with southern bean mosaic (Weintraub and Ragetli, 1970), tomato bushy stunt (Russo and Martelli, 1972) and eggplant mosaic (Hatta, 1976) viruses. Crystals composed of empty capsids have been seen in the nuclei of plants infected with turnip yellow mosaic and other tymoviruses (Hatta, 1976).

3.4 Latent Infection

Viruses may infect some hosts and multiply in them, but induce no visible symptoms. Such infection is called *latent infection* and the phenomenon is known as *latency*. Latent infection is quite frequent in naturally infected wild plants, such as in common agricultural weeds (Tomlinson *et al.*, 1970). Cucumber mosaic virus quite often infects chickweed (*Stellaria media*) without causing symptoms. The latent virus may be detected by back inoculation to a susceptible host plant or by serological assay.

Latent virus infection may result from a high level of host tolerance (*see* Chapter 10). Presumably in wild plants, latency has evolved as a result of natural selection over a long period of time. Plants which showed high levels of susceptibility and developed severe symptoms would be selected against, in favour of individuals in which a virus causes no symptoms.

In agriculture, man has often unwittingly selected for latent viruses in cultivated crops, especially those that are vegetatively propagated. In the rhubarb (*Rheum rhaponticum*) crop for instance, some infected commerical clones produce higher yields than virus-free clones (Walkey *et al.*, 1982).

Although latent infection may cause no obvious symptoms in some virus/host infections, plant virologists should be aware of the possibility of reduced yields in plants with latent infections. Such losses may be easily overlooked and can only be detected if yield comparisons are made with plants known to be virus-free.

Some plants will show severe symptoms following virus inoculation and pass through an *acute* stage of symptom expression. A period of *recovery* may follow and during the recovery phase the young developing leaves may be symptomless or show only mild symptoms, even though the virus is still present and recoverable in these symptomless leaves. Such recovery is common in plants infected with viruses of the nepovirus group.

The recovery phase in some virus infections may be associated with *acquired resistance*, a phenomenon that has been studied in detail by Ross and other workers (1961) (*see* Chapter 10). In other infections the recovery phase may be followed by a further acute phase of disease, and sometimes a cyclic occurrence of acute and recovery phases may occur (Cheo and Pound, 1952; Paul and Quantz, 1959).

Latent infection may often occur in seedlings that have been grown from infected seed. This type of latency is particularly common with seed-borne nepoviruses such as arabis mosaic and cherry leaf roll viruses.

3.5 Factors Influencing Symptom Expression

Various factors may influence the development and severity of symptoms in virus infected plants. Among the most important are those relating to the genetical composition of the host plant and the virus, the age of the host and existing environmental conditions prior to, and after infection.

3.5.1 Genetical and host factors

The genetical composition of either the virus or the host may control whether infection actually occurs, and the nature of the symptoms produced. The occurrence of different virus strains, possessing different virulence genes that will control infection and symptoms, was discussed in Section 2.3. The symptoms produced by different strains of bean common mosaic virus in *Phaseolus vulgaris* beans, clearly illustrates this point (Drijfhout *et al.*, 1978). Similarly, some host cultivars may be susceptible to a particular virus or virus strain, while others may carry a resistance gene which may prevent infection altogether, or modify the nature of the symptoms produced. The studies of Drijfhout *et al.* (1978) with bean common mosaic virus in various cultivars of *P.vulgaris* beans, also illustrate the importance of genetical variation in the host plant.

In addition to its genetical composition, the age of the host plant at the time of infection can be a critical factor in determining symptom expression. In general, the younger the plant, the more susceptible it is to virus infection, and very old leaves, or old plants, are usually relatively resistant to infection. The probable reason for this is that the virus is completely dependent upon the host cells for its multiplication, and in the older leaves the transport of assimilates and metabolism, is slower than in younger leaves. An example of this age effect is seen when marrow (*Cucurbita pepo*) plants are inoculated with cucumber mosaic virus. If the seedlings are infected at the cotyledon stage, they are more susceptible, and the symptoms more severe than if they are inoculated 14 to 21 days later (Walkey and Pink, 1984). Similarly, in potatoes infected with potato virus Y, all the progeny tubers were infected from plants inoculated with virus when they were 8 weeks old, but only 25% were infected if the plants were 13 weeks old when they were inoculated (Beemster, 1966).

The age of a crop at the time of infection may also determine the loss of yield that the virus causes. This has been demonstrated in cereals infected with barley yellow dwarf virus, where early infection causes considerably greater yield loss than later infections (Smith, 1967).

3.5.2 Environmental factors

The effect of the environment upon symptom development and expression, is seen most clearly in plants used in the laboratory and glasshouse, where environmental conditions can be easily controlled. The conditions the host plant is grown under, both before and after virus inoculation, can greatly influence symptom expression. In general, for test purposes, glasshouse plants should be grown at temperatures between 18° and 25°C, under low to moderate light intensities, and should be well watered to produce soft, lush growth.

Temperature

Often, high temperatures will reduce virus symptoms. This is particularly noticeable in glasshouses when temperatures rise repeatedly above 26°C, a factor that makes refrigeration-cooling essential for satisfactory glasshouse experimentation with viruses in tropical climates, or temperate climates in summer. Experiments with cucumber mosaic virus infected marrow seedlings clearly demonstrate the effects of temperature upon symptom development (Pink and Walkey, 1984). When grown at 25°C, CMV infected seedlings showed few symptoms, but the symptoms greatly increased when the plants were grown at 20° or 15°C. (*see* Table 3.1). Other workers have demonstrated, that incubation of the host plant at a high temperature (36°C) prior to virus inoculation, will increase the plant's susceptibility, but high temperature treatment after inoculation may cause variable reactions, depending on the virus and host plant concerned (Kassanis, 1952; Helms, 1965).

Table 3.1 Effect of temperature and light intensity upon symptoms caused by cucumber mosaic virus in marrow (*Cucurbita pepo*)

Cultivar	Symptom severity† 15°C	20°C	25°C	Symptom severity 13 Wm^{-2}*	40 Wm^{-2}	120 Wm^{-2}
Gobham Bush Green	3.7	1.3	0.3	4.9	4.0	1.4

*Watts per square meter.
†Symptom severity based on a 0 (no symptoms) to 5 (severe symptoms) scale. After Pink and Walkey (1984).

High temperatures may also affect a host plants' ability to resist virus infection. The existence of a temperature-sensitive resistance gene, the *I* gene, has been demonstrated in certain cultivars of *Phaseolus vulgaris* beans. Below 30°C, the gene confers resistance to certain strains of bean common mosaic virus, but above this temperature the resistance is ineffective, and inoculated beans develop systemic necrotic symptoms (Grogan and Walker, 1948). Another example of high temperature breakdown of the host plant's resistance, has been demonstrated in tissues of tobacco infected with cucumber mosaic or alfalfa mosaic viruses (Walkey, 1976). Treatment of tissues infected with these viruses at 32° or 40°C, had two effects. First, the high temperature arrested virus replication, and secondly, it inactivated a reversible resistance mechanism in the host cells. Consequently, when the restraint of high temperature was removed, the virus still present in the tissues was able to multiply in the absence of the resistance mechanism, to abnormally high concentrations (*see* Section 11.2.2. and Figure 11.3).

Light

Usually, high light intensities produce 'hard' plants, which are less susceptible to virus infection than plants grown under low light intensities (Costa and Bennett, 1955; Kimmins, 1967). Many plant virologists, therefore, shade their glasshouse test plants for 24 to 48 hours prior to virus inoculation.

High light intensities after inoculation also tend to reduce symptoms.

The symptoms produced by cucumber mosaic virus in marrow seedlings for example, are much less severe under high than under low light intensities (Pink and Walkey, 1984, *see* Table 3.1).

The combined effect of high light intensity and high-temperatures on plant virus symptoms, are very pronounced in temperate regions. In summer months, especially under glasshouse conditions, host plant growth is frequently rapid and virus susceptibility low. In contrast, during winter months, low light intensity and low temperatures, result in slower plant growth and increased virus susceptibility.

Nutrition

The effect of host plant nutrition upon virus symptoms may be quite variable, but in general, nutritional conditions that favour plant growth, also favour increased host susceptibility to virus infection (Bawden and Kassanis, 1950). High nitrogen levels for instance, have been reported to increase the susceptibility of marrow seedlings to infection by cucumber mosaic virus (Martin, 1959).

3.6 References

Bawden, F. C. and Kassanis, B. (1950). Some effects of host nutrition on the susceptibility of plants to infection by certain viruses. *Ann Appl Biol* **37**, 46–57.

Beemster, A. B. R. (1966). The rate of infection of potato tubers with potato virus Y in relation to position of the inoculated leaf. In *Viruses of Plants* (ed. Beemster, A. B. R. and Dijkstra, J.), North-Holland: London. pp. 44–7.

Bos, L. (1978). Symptoms of virus diseases in plants. *Cen Agric Pub Doc Wageningen.*

Cheo, P. C. and Pound, G. S. (1952). Relation of air temperature, soil temperature, photoperiod and light intensity to the concentration of cucumber virus 1 in Spinach. *Phyt* **42**, 306–10.

Cooper, V. C. (1976). Studies on the seed transmission of cherry leaf roll virus. PhD thesis, Univ. of Birmingham.

Costa, A. S. and Bennett, C. W. (1955). Studies on the mechanical transmission of the yellows virus of sugar beet. *Phyt* **45**, 233–8.

Couch, H. B. (1955). Studies on the seed transmission of lettuce mosaic virus. *Phyt* **45**, 63–70.

Drijfhout, E., Silbernagel, M. J. and Burke, D. W. (1978). Differentiation of strains of bean common mosaic virus. *Neth J Plant Pathol* **84**, 13–26.

Edwardson, J. R. (1974). *Some properties of the potato virus Y group.* Fla Agric Exp Stn Monogr. Series 4.

Edwardson, J. R. and Christie, R. G. (1978). Use of virus-induced inclusions in classification and diagnosis. *Ann Rev Phyto* **16**, 31–55.

Esau, K. (1968). *Viruses in plant hosts.* Univ Wisconsin Press.

Esau, K. and Hoefert, L. L. (1971). Cytology of beet yellows virus infection in *Tetragonia*, 111. Conformations of virus in infected cells. *Protoplasma* **73**, 51–65.

Esau, K. and Hoefert, L. L. (1978). Hyperplastic phloem in sugarbeet leaves infected with the beet curly top virus. *Am J Bot* **65**, 772–83.

Grogan, R. G. and Walker, J. C. (1948). The relation of common mosaic to black root of bean. *J Ag Res* **77**, 315–31.

Gunning, B. E. S. and Roberts, A. W. (1976). *Intercellular communications in plants, studies on plasmodesmata*. Springer-Verlag: Berlin and New York.

Hagedorn, D. J. and Walker, J. C. (1949). Wisconsin pea stunt, a newly described disease. *J Ag Res* **78**, 617–26.

Harrison, B. D. and Roberts, I. M. (1968). Association of tobacco rattle virus with mitochondria. *J Gen Virol* **3**, 121–4.

Hatta, T. (1976). Recognition and measurement of small isometric virus particles in thin sections. *Virol* **69**, 237–45.

Helms, K. (1965). Role of temperature and light in lesion development of tobacco mosaic virus. *Nature* **205**, 421–2.

Hillborn, M. T., Hyland, F. and McCrum, R. C. (1965). Pathological anatomy of apple trees affected by the stem-pitting virus. *Phyt* **55**, 34–9.

Hollings, M. and Brunt, A. A. (1981). Potyviruses. In *Handbook of plant virus infections*, Elseveir/North Holland; London. pp. 731–806.

Holmes, F. O. (1931). Local lesions of mosaic in *Nicotiana tabacum L. Contr Boyce Thompson Inst Pl Res* **3**, 163–72.

Hooper, G. R. and Wiese, M. W. (1972). Electron microscopy of leaf enations on Chinese white winter radish infected with radish mosaic virus. *Virol* **47**, 833–7.

Hull, R., Hills, G. J. and Plaskitt, A. (1970). The in vivo behaviour of twenty-four strains of alfalfa mosaic virus. *Virol* **42**, 753–72.

Jenkins, J. E. E. and Storey, I. F. (1955). Star cracking of apples in East Anglia. *Plant Path* **4**, 50–2.

Kassanis, B. (1952). Some effects of high temperature on the susceptibility of plants to infection with viruses. *Ann Appl Biol* **39**, 358–69.

Kim, K. S. and Fulton, J. P. (1972). Fine structure of plant cells infected with bean pod mottle virus. *Virol* **49**, 112–21.

Kim, K. S. and Fulton, J. P. (1975). An association of plant cell microtubules and virus particles. *Virol* **64**, 560–5.

Kim, K. S., Shock, T. L. and Goodman, R. M. (1978). Infection of *Phaseolus vulgaris* by bean golden mosaic virus. Ultrastructural aspects. *Virol* **89**, 22–33.

Kimmins, W. C. (1967). The effect of darkening on the susceptibility of French bean to tobacco necrosis virus. *Can J Bot* **45**, 543–53.

Kitajima, E. W. and Galves, G. E. (1973). *Cienc Cult (Sao Paulo)* **25**, 979.

Kitajima, E. W. and Lovisolo, O. (1972). Mitochondrial aggregates in *Datura* leaf cells infected with henbane mosaic virus. *J Gen Virol* **16**, 265–71.

Kitajima, E. W., Camargo, I. J. B., and Costa, A. S. (1968). *J Elec Micr* **17**, 144.

Kunkel, L. O. (1924). Histological and cytological studies on the Fiji disease of sugar cane. *Bull Hawaiian Sug Plrs Ass Exp Stn, Bot Ser* **3**, 99–107.

Lee, C. L. (1955). Virus tumour development in relation to lateral-root and bacterial-nodule formation in *Melilotus alba*. *Virol* **1**, 152–64.

Martelli, G. P. and Russo, M. (1969). Nuclear changes in mesophyll cells of

Gomphrena globosa L. associated with infection by beet mosaic virus. *Virol* **38**, 297–308.

Martelli, G. P. and Russo, M. (1976). Unusual cytoplasmic inclusions induced by watermelon mosaic virus. *Virol* **72**, 352–62.

Martelli, G. P. and Russo, M. (1977). Plant virus inclusion bodies. *Adv Virus Res* **21**, 175–266.

Martin, M. W. (1959). Inheritance and nature of cucumber mosaic virus resistance in squash. PhD thesis, Univ. Cornell.

Matthews, R. E. F. (1981). *Plant virology*. Academic Press: London.

Paul, H. L. and Quantz, L. (1959). Uber den Wechsel der Konzentration des echten Ackerbohnenmosaikvirus in Ackerbohnen. *Arch Mikrobiol* **32**, 312–18.

Pink, D. A. C. and Walkey, D. G. A. (1984). Unpublished results.

Pleše, N., Hoxha, E. and Miličić, D. (1975). Pathological anatomy of trees affected with apple stem grooving virus. *Phyt Z.* **82**, 315–25.

Posnette, A. F. (1947). Virus diseases of cacao in West Africa. 1. Cacao virus 1A, 1B, 1C and 1D. *Ann Appl Biol* **34**, 388–402.

Roberts, D. A., Christie, R. G. and Archer, M. C. (1970). Infection of apical initials in tobacco shoot meristems by tobacco ringspot virus. *Virol* **42**, 217–20.

Ross, A. F. (1961). Systemic acquired resistance induced by localised virus infections in plants. *Virol* **14**, 340–58.

Russo, M. and Martelli, G. P. (1972). Ultrastructural observations on tomato bushy stunt virus in plant cells. *Virol* **49**, 122–9.

Russo, M. and Martelli, G. P. (1973). The fine structure of *Nicotiana glutinosa* L. cells infected with two viruses. *Phytopathol Mediterr* **12**, 54–60.

Schlegel, D. E. and DeLisle, D. E. (1971). Viral protein in early stages of clover yellow mosaic virus infection of *Vicia faba*. *Virol* **45**, 747–54.

Schneider, H. (1959). The anatomy of tristeza-virus infected citrus. *Proc Conf Citrus Virus Dis Riverside, Calif* 1957, 73–84.

Shepard, J. F., Gaard, G. and Purcifull, D. E. (1974). A study of tobacco etch virus-induced inclusions using indirect immunoferritin procedures. *Phyt* **64**, 418–25.

Shikata, E. and Maramorosch, K. (1966). Electron microscopy of pea enation mosaic virus in plant cell nuclei. *Virol* **30**, 439–54.

Smith, H. C. (1967). The effect of aphid numbers and stage of plant growth in determining tolerance to barley yellow dwarf virus in cereals. *NZ J Agric Res* **10**, 445–66.

Smith, K. M. (1972). *A textbook of plant virus diseases*. Longman: Harlow.

Steere, R. L. (1957). Electron microscopy of structual detail in frozen biological specimens. *J Biophys Biochem Cytol* **3**, 45–60.

Tjallingii, F. (1952). Investigations on the mosaic disease of gherkin, *Cucumis sativus* L. Meded. *Inst plziektenk Onderz* Wageningen 47.

Tomlinson, J. A., Carter, A. L., Dale, W. T and Simpson, C. J. (1970) Weed plants as sources of cucumber mosaic virus. *Ann Appl Biol* **66**, 11–16.

Van Katwijk, W. (1956). Rough-skin of apples. *Tijdschr Plantenziekten* **62**, 46–9.

Walkey, D. G. A. (1976). High temperature inactivation of cucumber and alfalfa mosaic viruses in *Nicotiana rustica* cultures. *Ann Appl Biol* **84**, 183–92.

Walkey, D. G. A. and Pink, D. A. C. (1984). Resistance in vegetable marrow and other *Cucurbita* spp. to two British strains of cucumber mosaic virus. *J Agric Sci* **102**, 197–205.
Walkey, D. G. A. and Webb, M. J. W. (1968). Virus in plant apical meristems. *J Gen Virol* **3**, 311–13.
Walkey, D. G. A. and Webb, M. J. W. (1970). Tubular inclusion bodies in plants infected with viruses of the NEPO type. *J Gen Virol* **7**, 159–66.
Walkey, D. G. A., Brocklehurst, P. and Parker, J. (1983). Seed transmission of viruses. *Nat Veg Res Stn, Rept 1982*, pp. 82–3.
Walkey, D. G. A., Creed, C., Delaney, H. and Whitwell, J. D. (1982) Studies on the reinfection and yield of virus-tested and commercial stocks of rhubarb cv. Timperley Early. *Plant Path* **31**, 253–61.
Weintraub, M. and Ragetli, H. W. J. (1970). Electron microscopy of the bean and cowpea strains of southern bean mosaic virus within leaf cells. *J Ultrastruct Res* **32**, 167–89.
Weintraub, M. and Ragetli, H. W. J. (1971). A mitochondrial disease of leaf cells infected with an apple virus. *J Ultrastruct Res* **36**, 669–83.

3.7 Further Selected Reading

Bos, L. (1978). *Symptoms of virus diseases in plants. Cen Agric Pub Doc*, Wageningen.
Esau, K. (1968). *Viruses in plant hosts*. Univ. Wisconsin Press: Wisconsin.
Edwardson, J. R. and Christie, R. G. (1978). Use of virus-induced inclusions in classification and diagnosis. *Ann Rev Phyto* **16**, 31–55.
Martelli, G. P. and Russo, M. (1977). Plant virus inclusion bodies. *Adv Virus Res* **21**, 175–266.
Matthews, R. E. F. (1981). *Plant virology*. Academic Press: London.
Smith, K. M. (1972). *A textbook of plant virus diseases*. Longman: Harlow.

4 *Mechanical Transmission and Virus Isolation*

4.1 Introduction

The transmission of a virus from infected to healthy tissues is a procedure fundamental to the study of virus diseases. In the laboratory this is usually accomplished by grinding the leaf of a diseased plant, and rubbing the infectious sap on to the leaf of a healthy plant. The procedure is referred to as *mechanical* or *sap transmission*. It is used in the laboratory to isolate viruses from diseased field plants; to transmit them to test hosts; to sub-culture viruses; to study virus symptoms in a range of host species; and to assay for virus infectivity.

Unaided, plant viruses are unable to pass through the cuticle of the host plant and enter the cells beneath. For infection to occur, the virus must enter the tissues of the host through a sub-lethal wound. This is normally accomplished in experimental mechanical transmission, by the use of mild abrasives which damage the cuticle and epidermis of the plant, when the infectious sap is rubbed on to the host's surface. The virus then enters cells through these wounds (*see* Plate 4.1).

In nature, the entry of viruses into the host tissues is achieved in a number of different ways, which are described in Chapter 7. Although mechanical sap transmission is the method most frequently used to transmit viruses in the laboratory, any of the other methods described in Chapter 7 may also be used experimentally.

In this chapter the various factors associated with the use of mechanical transmission for experimental purposes are described.

4.2 The Transmission Process

4.2.1 Source and preparation of inoculum

The virus inoculum for sap transmission may be obtained from various parts of the infected donor plant, but usually younger leaf material contains a higher concentration of virus than older woody

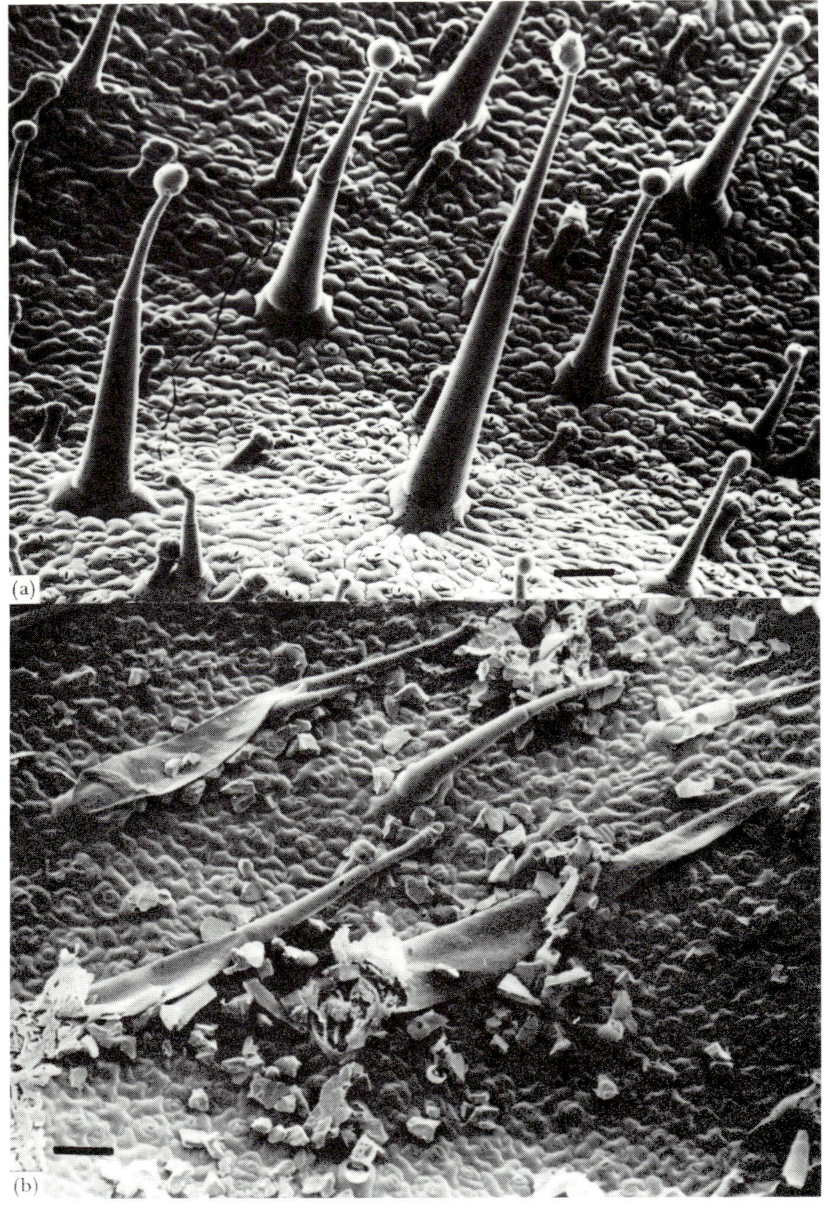

Plate 4.1 Scanning electron micrographs of the surface of *Nicotiana glutinosa* leaves before and after sap inoculation.
(*a*) An untreated leaf showing intact leaf hairs and epidermis; (*b*) the broken hairs following sap inoculation. The particles of the carborundum abrasive may be clearly seen, magnfication bar = 0.1 nm (courtesy of M. J. W. Webb).

tissues. Leaves showing virus symptoms are usually chosen and the actual area of the leaf taken can sometimes be important. In Chinese cabbage leaves infected with turnip mosaic virus for example, it has been shown that the chlorotic areas of the infected leaf contain far more virus than the dark green areas (Reid and Matthews, 1966).

In the case of a few viruses such as the soil-borne tobacco necrosis virus, the roots of the infected plants may be a better source of inoculum than the leaves (Smith, 1937), and occasionally the flower petals may be used. Sill and Walker (1952) for example, found that cucumber flowers contained less virus inhibitors than leaf material, which enabled them to transmit cucumber mosaic virus more efficiently.

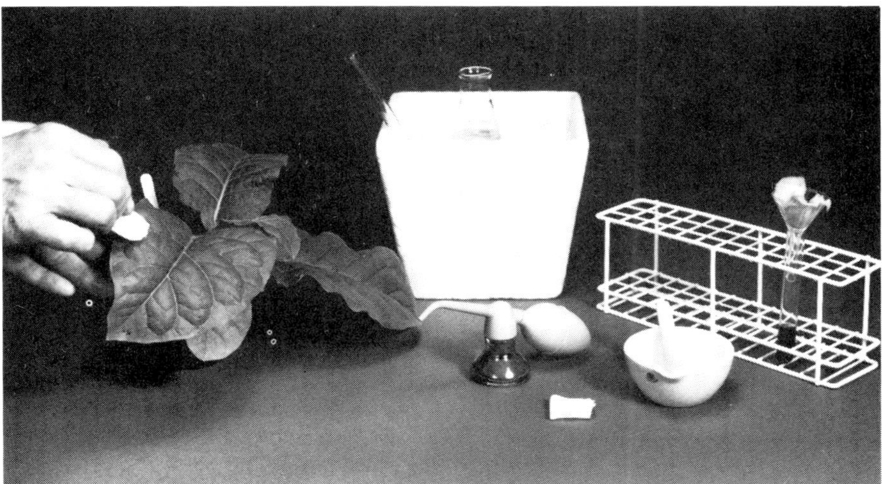

Plate 4.2 The procedure and equipment required for experimental transmission by sap inoculation.

After selection of a suitable piece of infected leaf or other tissue, the inoculum is prepared by thorough homogenization (grinding) of the tissue in a chilled mortar and pestle (*see* Plate 4.2) or by using some form of power-driven homogenizer. If large quantities of inoculum are required the infected material may be homogenized in an ordinary kitchen liquidizer. Whatever the means of grinding the tissue, however, various host metabolites and cellular debris will be released together with the virus. Some of these compounds may cause virus inactivation or inhibit infectivity. Consequently, it is usual to grind the leaf in a suitable buffer or other solution at a low temperature (0°C) to minimize loss of virus infectivity.

Research over many years has shown that if the leaf is homogenized in the presence of phosphate buffers, inoculum infectivity is greatly enhanced (Yarwood, 1952; Fulton, 1964). Infected tissues are often ground in 0.1 M potassium phosphate buffer at a pH of 7.0–7.5. If the sap of the donor plant is particularly acidic the buffer should be used at a pH of 8.0 to 8.5, but excessively alkaline conditions may also cause breakdown of the virus. The addition of a reducing agent to the phosphate solution may help to prevent oxidation and hence loss of virus infectivity. A solution of 1% di-potassium hydrogen phosphate (K_2HPO_4), containing 0.1% sodium sulphite is excellent for isolating and transmitting many viruses. To isolate viruses from some hosts, strong reducing agents such as 0.1% thioglycollic acid (syn. mercaptoacetic acid, $CH_2(SH)COOH$) may be added to the phosphate solution, whilst chelating agents such as 1–2% sodium EDTA (ethylenediaminetetra-acetic acid) may need to be added to prevent oxidation of polyphenols. Another additive that is sometimes used to reduce the activity of ribonuclease, is a powdered clay called bentonite (Fraenkel-Conrat *et al.*, 1961).

Once the infected tissues have been homogenized the sap-inoculum may be used directly, or filtered through a piece of cotton-gauze to remove the larger cellular debris. If the inoculum is not used immediately, or if a large number of test plants are to be inoculated, the inoculum should be kept cool in an ice-bucket until it is applied to the leaf.

The species used as the source for the virus inoculum, is often critical for successful isolation and sap transmission. When isolating viruses from naturally infected hosts the virologist has no choice, but in the

Table 4.1 Examples of susceptible laboratory host plants commonly used for virus transmission studies

Host family	*Host species*	
Solanaceae	*Nicotiana tabacum*	cv. White Burley
		cv. Xanthi.
	N. clevelandii	
	N. glutinosa	
	Petunia hybrida	
Chenopodiaceae	*Chenopodium quinoa*	
	C. amaranticolor	
Leguminosae	*Phaseolus vulgaris*	
	Vicia faba	
	Vigna sinensis	
Cruciferae	*Brassica perviridis*	cv. Tendergreen Mustard
Cucurbitaceae	*Cucumis sativus*	
	Cucurbita pepo	

laboratory a number of species have been shown to be particularly suitable donors for virus studies (*see* Table 4.1). Such hosts as various *Nicotiana spp.* (tobacco) *Chenopodium spp.* and *Phaseolus vulgaris*, have been found to be highly susceptible to infection by a wide range of plant viruses. They provide good hosts for the maintenance of virus cultures, virus multiplication and are often suitable for infectivity assay. Consequently, when a virus is first isolated from the field, it is often advantageous, and sometimes essential, to transmit it from its natural host, to one of the more suitable laboratory test species for further study.

If a virus occurs in very low concentrations in the donor host, it may not be possible to transmit it directly. This problem may be overcome by high speed centrifugation of the sap to concentrate the virus (*see* Chapter 5).

4.2.2 Inoculation

It has already been mentioned in the introduction to this chapter, that during sap transmission virus can only enter a plant's tissues through a surface wound. This wound must be minor and not so severe as to cause the death of the injured cells. Fine abrasive powders, such as 300–600 mesh carborundum (silicon carbide) or celite (crushed diatomaceous earth), are commonly used to produce the entry wounds. These powders are usually blown on to the leaf surface of the test plant using a throat-powder spray prior to inoculation, so that the leaf is covered with a light layer of the abrasive. Alternatively, the abrasive may be mixed with the sap inoculum, but the disadvantage of this procedure is that the abrasive tends to settle out of the suspension fairly quickly and the inoculum must be constantly shaken during use. Celite powder is lighter than carborundum and tends to blow around the laboratory when the leaf-dusting method is used, so many workers prefer to use the heavier carborundum powder.

Methods of applying the inoculum to the leaf tend to vary from laboratory to laboratory. Some workers apply the inoculum with their fingers, gently rubbing a wet finger over the whole upper surface of the test leaf. If a finger is used, however, the hands must be thoroughly decontaminated between different inoculations. Other workers prefer to use pads of cotton gauze or sponge (*see* Plate 4.2), a spatula or often the pestle used to grind the inoculum. The latter method is useful if many separate inoculations have to be made, for the fingers can avoid contact with the inoculum. Application of the inoculum by spraying will normally not result in virus infection, but has been successfully used by adding an abrasive to the inoculum before spraying (Mackenzie *et al.*, 1966).

The virus enters the plant tissues through broken leaf-hair cells or other wounds in the epidermis (*see* Plate 4.1). Although most viruses can be mechanically inoculated in sap by the methods described, infection will only occur if the virus is able to multiply within the epidermal cells of the recipient host plant. If the infection process is successful, the virus will multiply and spread into other cells. Local lesions may form and the virus may then spread into the plant's vascular system to move systemically throughout the plant (*see* Chapter 3).

In other cases, a virus may only be able to multiply within phloem or xylem cells and in these examples, the virus cannot usually be mechanically transmitted, even if the correct, susceptible host species is used. When mechanical transmission is impossible, experimental transmission and isolation is usually accomplished using an insect vector which feeds by probing deeply into the vascular tissues (*see* Chapter 7). A few viruses of this type, such as beet curly top, have been mechanically transmitted by using a pin to prick deeply into the vascular tissues through a pool of infectious sap (Gibbs and Harrison, 1976).

4.3 Factors Affecting Mechanical Transmission

4.3.1 Inhibitors and inactivators

Loss of inoculum infectivity, caused by chemicals released from the donor plant during homogenization, is a major problem in the mechanical transmission from some donors. In considering loss of virus infectivity in sap homogenates, it is important to distinguish between chemicals that inhibit and those that actually inactivate the virus.

Some donor species contain powerful inhibitors which make it difficult, or even impossible to sap transmit viruses to other species, even when the recipient host is known to be susceptible to the particular virus concerned. Such inhibitors include enzymes and polysaccharides (Bawden, 1954), and are common in the leaf sap of such plants as *Chenopodium spp.*, *Phytolacca spp.* and *Dianthus spp.*

Some inhibitors may bind to the virus particle, but the mechanism of their action is not fully understood. The inhibitory protein of *Phytolacca* is thought to block the translation of messenger RNA by ribosomes of hosts such as wheat, but not affect translation by ribosomes of *Phytolacca spp.* (Owens *et al.*, 1973). In general, it is thought that inhibitors act on the recipient host plant rather than against the virus itself.

The action of some inhibitors can be overcome by diluting a sap sample to a level at which the inhibitor becomes ineffective, but at which the virus concentration remains high enough for infection. The action of other inhibitors may be overcome by the addition of bentonite to the grinding buffer (Yarwood, 1966). Added at a concentration of 5 mg per ml, bentonite has been shown to be particularly effective in assisting the transmission of several viruses from spinach (*Spinacea oleracea*) to non-Chenopodiaceous hosts (Bailiss and Okonkwo, 1979). It is thought that the bentonite particles bind with the protein inhibitors in the spinach sap, to prevent their action on the recipient host species to which the virus inoculum is transmitted.

In contrast, inactivators cause permanent loss of virus infectivity by acting upon the virus nucleic acid. Treatment of virus preparations with formaldehyde or nitrous acid, for example, can cause complete loss of virus infectivity. In nature many woody plants contain powerful virus inactivators such as tanning and Oxidases. These may be rendered ineffective by the addition of high pH (8 to 8.5) buffers, reducing agents, chelating agents or protein-binders to the grinding medium (Gibbs and Harrison, 1976). Alternatively, workers may try to avoid donor hosts known to contain high levels of inactivators, and sometimes it is possible to use the young leaves of a woody host, which have a lower tannin content than the older parts of the plant (Fulton, 1964).

The effect of virus inhibitors and inactivators upon inoculum infectivity has been the subject of extensive reviews of Bawden (1954), Fulton (1964) and Loebenstein (1972).

4.3.2 Host plant

For most hosts the leaves are the most susceptible and convenient part of the test plant to inoculate with virus. In some species such as *Phaseolus vulgaris* beans, the primary leaves are usually more susceptible than the later developing trifoliate leaves. In species of *Cucurbitaceae*, the seedling cotyledons are more susceptible than the leaves (Alconero, 1973; Walkey and Pink, 1984), and occasionally the roots of the test plant may be the most suitable parts for inoculation (Yarwood, 1966; Teakle, 1973; Moline and Ford, 1974).

The choice of the recipient host plant to be used for mechanical transmission will depend very much on the virus concerned, and the experimental objectives. Frequently, different species of host plant are required for different aspects of study with the same virus (*see* Table 4.2). One host may be required for the long-term propagation of a virus, another for its rapid multiplication for purification (*see* Section 5.2), a further host for local lesion infectivity assay and a range of hosts for studying chacteristic symptoms for diagnostic purposes.

Frequently, the common laboratory host plants such as *Nicotiana spp*,

Table 4.2 Examples of the use of laboratory host species for different transmission purposes

Virus	*Propagation host*	*Multiplication host for purification*	*Local-lesion host*
Alfalfa mosaic	*Nicotiana glutinosa*	*N. tabacum*	*C. quinoa*
Arabis mosaic	*Petunia hybrida*	*N. clevelandii*	*C. amaranticolor*
Bean yellow mosaic	*Phaseolus vulgaris*	**Vicia faba*	**C. amaranticolor*
	Vicia faba	*N. clevelandii*	*C. quinoa*
Cucumber mosaic	*N. tabacum*	*N. clevelandii*	*C. quinoa*
Lettuce mosaic	*Lactuca sativa*	*Chenopodium quinoa*	*C. quinoa*
		L. sativa	*C. amaranticolor*
Potato virus Y	*N. glutinosa*	*N. tabacum*	*C. quinoa*
			C. amaranticolor
Tobacco ringspot	*N. tabacum*	*Cucumis sativus*	*Vigna sinensis*
	Cucumis sativus		

*Host used may depend on virus strain.

Chenopodium spp. or *Phaseolus vulgaris* beans (*see* Table 4.1), may be suitable for one or more of these purposes, and often all may be superior to the virus' natural host. In other instances, however, a virus may have a very narrow host range and infection may be restricted to a family or genus of plants (*see* Table 4.3). Celery mosaic virus, for example, only infects umbelliferous plants and bean common mosaic virus can only be effectively transmitted to members of the genus *Phaseolus*. In these circumstances the same host may have to be used to maintain, propagate and assay the virus and there may be no suitable host for local lesion assay. Consequently studies with such viruses may be severely restricted.

The investigator must also be particulary careful in the selection of specific host cultivars for test purposes. Some cultivars may be highly susceptible to a particular virus and other cultivars of the same species, relatively, or completely resistant. Often it is necessary to carry out

Table 4.3 Examples of viruses that have a restricted host range and no satisfactory local-lesion host

Virus	*Propagation, multiplication and infectivity assay host**
Barley yellow dwarf	*Avena byzantina* (oat)
Beet western yellows	*Claytonia perfoliata*
Celery mosaic	*Apium graveolens* (celery)
Maize rough dwarf	*Zea mays* (maize)
Sugarcane Fiji disease	*Saccharum officinarum* (sugarcane)
Wheat streak mosaic	*Triticum aestivum* (wheat)

*May be assayed by systemic symptoms only.

extensive susceptibility tests with a range of cultivars before a convenient one is selected. Cultivar selection is particularly important with viruses such as bean common mosaic (BCMV) which has many distinct strains (Drijfhout, 1978). In this case it was essential to select a bean cultivar which was susceptible, and a good propagation host for most of the BCMV strains (Walkey and Innes, 1979).

Even when a plant is susceptible to a particular virus, the degree and extent of infection will be governed by a number of other factors. The effects of high temperature, light intensity, nutrition and host age upon symptom development and expression have already been discussed in Chapter 3, but all these factors also affect mchanical transmission of the virus.

Amongst the most important factors influencing mechanical transmission, is the physiological state of the plant at the time of inoculation. Plants that have been kept in a shaded box for 24 to 48 h. before inoculation are more susceptible than 'harder' plants grown in high light intensities (Bawden and Roberts 1948). Well-watered plants are also generally more susceptible, and Tinsley (1953) demonstrated that as many as ten times more tobacco mosaic virus local lesions were produced on well-watered, than on poorly-watered plants. The well-watered plants had thinner cuticles, which allowed the abrasive to be more effective during the inoculation process.

It has also been shown that a high temperature treatment of 36°C immediately before inoculation may increase susceptibility, whilst the same temperature treatment after inoculation will decrease susceptibility (Kassanis, 1952). Moderate wilting before inoculation has also been shown to increase susceptibility (Matthews, 1981), as has dipping the leaves in hot water at 50°C (Gibbs and Harrison, 1976). In some species, a diurnal periodicity in susceptibility has been reported, and plants may be more susceptible in the late afternoon, than they are at the end of a night period (Matthews, 1953).

Post-inoculation treatment of the host plant may also increase virus infectivity. In many laboratories the inoculated leaves are immediately washed under a cold-water tap as a routine procedure. The effectiveness of this treatment has been confirmed for several, but not all viruses (Holmes, 1929; Yarwood, 1973). Yarwood (1955) found that washing for up to ten seconds after inoculation increased susceptibility, but infection was reduced if the leaves were washed for longer periods.

In contrast to washing leaves, it has also been shown that if leaves are dried quickly in a jet of air immediately after inoculation, infectivity may be increased up to almost 100-fold, compared with untreated leaves (Yarwood, 1973). In practice, it seems probable that washing or the quick-drying of inoculated leaves may be advantageous in increasing infectivity, and both treatments should be tried when optimal transmission conditions are being determined for the study of a newly isolated virus.

Other workers have demonstrated that *leaf water potential* in the epidermal cells following virus inoculation has an important effect upon virus transmission. A significant increase in the number of tobacco necrosis virus local lesions was induced in inoculated leaves of *Phaseolus vulgaris*, if the water potential in the leaves was rapidly reduced within three hours of inoculation (Bailiss and Plaza-Morales, 1980). These workers suggested that the normal gradient of water movement from the leaves to the atmosphere is reversed when the water potential of the leaves is lowered. This causes the virus to be carried with the reversed flow, from the leaf surface into the wounded cells.

The effects on mechanical transmission of various post-inoculation treatments have been the subject of a review by Yarwood and Fulton (1967).

4.4 Infectivity Assay

Mechanical transmission to suitable host plants is used extensively in plant virology as a quantitative bioassay. Usually, virus infectivity is measured by the number of local lesions induced by the inoculum on inoculated leaves. Considerable efforts are made by virologists, when studying a new virus, to find a host plant that will produce discrete, countable local lesions for assay purposes. If no suitable local lesion assay host can be found, a virus may still be assayed by inoculating dilution series of its inoculum to a number of susceptible plants and recording systemic symptoms (*see* Chapter 3).

It must be emphasized, however, that virus concentration values obtained by infectivity assays are not absolute, and are only relative to the total number of virus particles present in the inoculum. This is because not all particles present in the inoculum are infectious, and not all the cells innoculated become infected. In fact, it has been estimated that as many as 10^5 or more particles must be inoculated for infection of a cell to occur, and this figure may be even higher with multi-component viruses (*see* Section 1.4.2) such as cowpea mosaic (10^7, Van Kammen, 1968) and alfalfa mosaic (10^9, Bol and Van Vloten-Doting, 1973). This would suggest that the procedures of rubbing and wounding, used in mechanical inoculation, are not particularly efficient. This view is supported by the fact, that in the case of cell protoplasts, where the cell wall has been removed, only 400 particles of cowpea chlorotic mottle virus (a multicomponent virus with a tripartite genome) are required per protoplast to infect 50% of those treated (Motoyoshi *et al.*, 1973).

A further limitation upon local lesion assay procedures, is that the relationship between inoculum dilution and local lesion numbers is not necessarily linear, and is frequently variable from virus to virus (Kleczkowski, 1950). For many viruses the infectivity/dilution curves are sigmoidal in shape, although for some viruses they may be more linear (*see* Figure 4.1). In the case of viruses with sigmoidal dilution

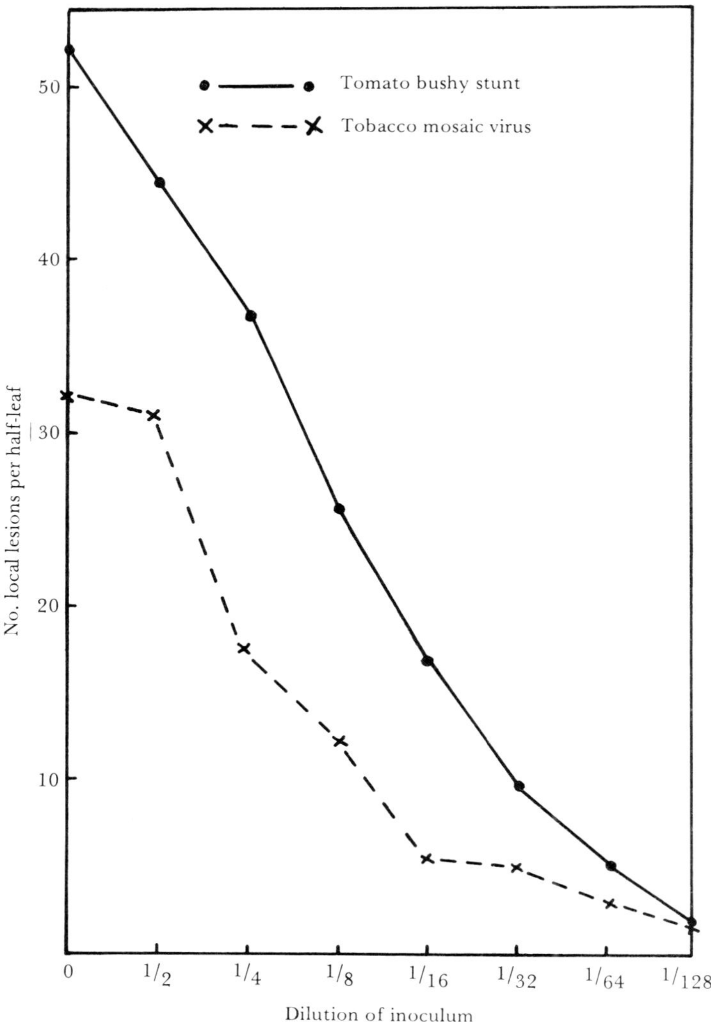

Fig. 4.1 The relationship between dilution of virus inoculum and local lesion numbers for tomato bushy stunt and tobacco mosaic viruses assayed on *Nicotiana glutinosa* (based on Kleczkowski, 1950).

curves, at high concentrations, dilution of the virus causes little change in lesion numbers, at medium concentrations the change is linear, and at low concentration there is again little change in lesion numbers. It is, therefore, important to consider the dilution of the inoculum when comparing the virus concentration in different preparations by local lesion assay. It is usually advisable to compare more than one dilution of the samples to obtain accurate comparisons of relative infectivity.

Despite these limitations, however, infectivity assay is widely used by plant virologists for the comparison of different virus preparations. Compared with other methods of virus assay, which depend on physical, chemical or serological procedures, it has the advantage of quantifying the relative amounts of infectious virus, rather than the total amount of nucleoprotein, all of which may not be infectious.

4.4.1 Experimental designs for infectivity assays

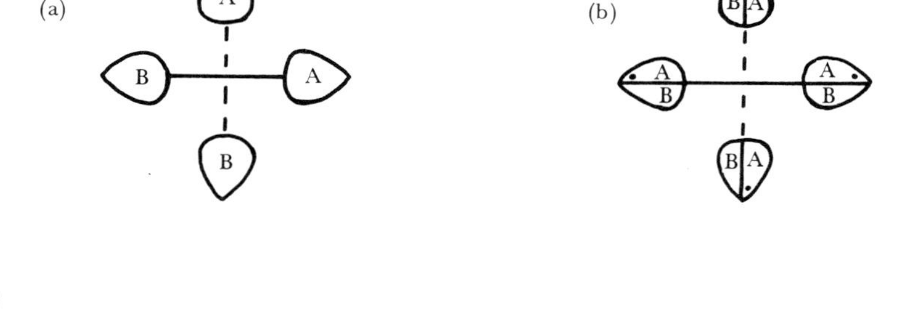

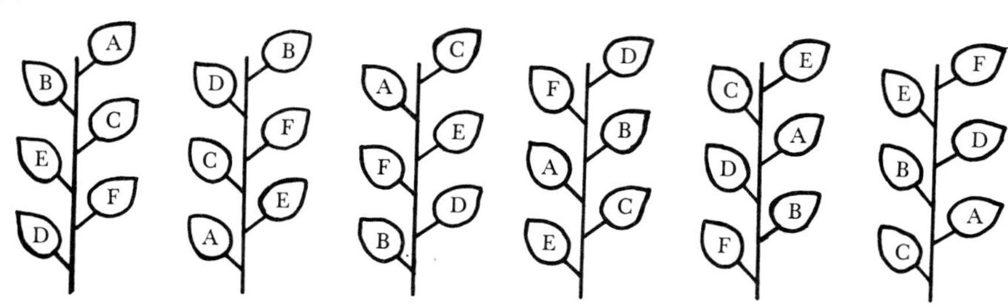

Fig. 4.2 Various assay designs for comparing the relative virus concentrations in different preparations. (*a*) Two virus samples inoculated to whole, opposite leaves at the same level on the stem; (*b*) two samples compared on opposite halves of the same leaf; (*c*) a Latin square design for comparing six different samples on six leaves of six plants.

Local lesion assay

When considering experimental designs for infectivity assay, it should be remembered that the host plants and leaves to be used for assay are usually variable to some degree, and every precaution must be taken to minimize this variation if accurate assays are required. In all infectivity assays the plants used should be as uniform as possible, and if some of the plants do vary in size, it must be ensured that equal numbers of plants of each size group are included in each sample replicate.

Assays may be designed so that the virus samples for comparison are inoculated to single whole leaves on the same plant, in which case the leaves selected for each sample should be as similar as possible. With plants that have their leaves arranged opposite to one-another (such as in *Chenopodium spp.*), this may be accomplished by inoculating the two virus samples to be compared, to opposite leaves at the same level on the stem (*see* Figure 4.2*a*), and the assay may be replicated by inoculating other pairs of leaves on the same plant, or on other plants (*see* Plate 4.3*a*). The actual number of replicates used will depend on the number of plants available and the number of leaves on each plant, but the greater the number of replicates used, the greater will be the accuracy of the results.

Plate 4.3 Local lesion assay.
(*a*) Lesions induced by turnip mosaic virus on an inoculated whole-leaf of *Chenopodium quinoa*; (*b*) cucumber mosaic virus lesions on the inoculated half of a *C. quinoa* leaf.

A more accurate method of assay when comparing two different virus samples, is to inoculate each sample to opposite halves of the same leaf, so minimizing leaf to leaf variation (*see* Figure 4.2*b*). This is of course only practical, if a leaf is divided by a midrib into two equal halves (*see* Plate 4.3*b*). This procedure allows more economical use of assay plants and it is possible to compare four or more replicates on one plant. If this design is used, then it is important to inoculate an equal number of right and left half-leaves with each test sample, to eliminate any personal bias caused by the inoculator's right or left-handedness. If a sample comparison between two virus samples is to be made by the half-leaf method, probably six to eight replicates would give sufficient accuracy.

If more than two samples are to be compared, then more complicated assay designs are used. One good design is to compare each individual virus inoculum with every other sample on opposite half-leaves an equal number of times. For example, if six comparisons are to be made, the designs might be:

$$\frac{A}{B}\quad\frac{A}{C}\quad\frac{A}{D}\quad\frac{A}{E}\quad\frac{A}{F}\quad\frac{B}{C}\quad\frac{B}{D}\quad\frac{B}{E}\quad\frac{B}{F}\quad\frac{C}{D}\quad\frac{C}{E}\quad\frac{C}{F}\quad\frac{D}{E}\quad\frac{D}{F}\quad\frac{E}{F}$$

Such a design is particularly suitable when using an assay host such as *Phaseolus vulgaris* bean or cowpea (*Vigna sinensis*), in which only the two opposite primary leaves are inoculated on each plant. Increased replication can easily be achieved by duplicating the complete design with a second batch of plants.

If whole leaves are used instead of half-leaves, for multiple comparisons, and many leaves are inoculated on each assay plant, a Latin square design is often appropriate. Such a design to compare six different virus samples, might be achieved by inoculating each sample to six different leaves on six test plants, ensuring that each sample occurs in a different leaf position on each plant (*see* Figure 4.2*c*). This design overcomes any problems arising from differences in leaf susceptibility at different positions on the plant.

In the case of a few viruses (such as tobacco necrosis virus inoculated into *P. vulgaris* leaves (Bailiss and Plaza-Morales, 1980) local lesions develop very quickly. It is therefore possible, to carry out infectivity assays on detached leaves that are kept moist in sealed containers until the lesions develop.

When no suitable local lesion assay host is available, the virus is assayed by mechanically inoculating all the leaves of a susceptible host and recording systemic symptoms. Using this procedure, it is essential to have a uniform group of assay plants, and each virus inoculum to be tested is usually diluted in a ten-fold dilution series e.g. 0, 10^{-1}, 10^{-2}, 10^{-3}, 10^{-4}, 10^{-5}, etc. Two or more test plants are usually used for each dilution of each inoculum tested.

Further details of infectivity assay designs, and the requirement to transform lesion number data before testing for statistical significance of differences, may be obtained from Kleczkowski (1950), Fry and Taylor (1954), Preece (1967), Gibbs and Harrison (1976). It should be emphasized, however, that it is generally advisable to avoid complicated designs, as simple errors can easily occur when numerous virus samples have to be inoculated.

4.5 Storage of Virus Isolates

For short-term storage (minutes or hours) during experiments in the laboratory or glasshouse, it is advisable that virus inocula are kept at, or close to 0°C. This is usually accomplished by plunging the tube containing the inoculum in an ice-bucket.

Long-term virus storage, however, is a continuing problem for all workers involved in plant virus studies. If the virus isolates are continually sub-cultured in laboratory host plants there are several problems that may occur. First, the virus may become contaminated by a second virus. This may occur in the glasshouse by insect transmission or even by plants touching and rubbing together (*see* Chapter 7). Secondly, mutation or attenuation of the culture may occur, with the progressive selection of an atypical strain during sub-culture (*see* Section 2.3). Thirdly, the culture may be lost through death of the host plant, and finally, but not least, the propagation host may occupy valuable glasshouse space over long periods of time. Various methods are therefore used for the long-term storage of virus isolates, to maintain them in their original, uncontaminated condition.

Most methods involve storing material from which the water content has been removed. One such method is simply to dry the leaves rapidly over calcium chloride ($CaCl_2$) under a vacuum pressure. The dried material may then be ground to a powder and stored (McKinney and Silber, 1968; Bos, 1969). This procedure is quite effective for some, but not all viruses. A more efficient method for the long-term storage of most plant viruses, is to freeze-dry (a process referred to as *lyophilization*) infected sap in the presence of glucose and peptone. A suitable method is to add 0.7% (W/V) of d-glucose and peptone to filtered sap in a glass ampoule (Hollings and Stone, 1970). After lyophilization, the ampoule is sealed and may be stored at room temperature. Using this procedure some viruses have been stored for ten years or more (Hollings and Stone, 1970). Purified virus preparations (*see* Chapter 5) may also be lyophilized in this way. Another satisfactory procedure, frequently used by the author, is to lyophilize small pieces of infected leaf in an ampoule without grinding or the addition of other chemicals.

Other workers have found that deep-freezing infectious sap is satisfactory for storing some viruses (De Wijs and Suda-Bachmann, 1979), but unsatisfactory for others (Marcinka and Musil, 1977). In the author's laboratory ampoules of sap inoculum or purified virus are frozen and stored in liquid nitrogen. Using this procedure, infectivity of many viruses has been maintained for ten years or more. Sometimes, if a virus is seed-borne, it is possible to store the virus in the infected seed, such viruses usually remain infectious provided the seed remains viable.

In practice, it is probably advisable to store a virus isolate in several different ways at the same time, to increase the probability of at least one method being successful. Every few years the stored isolate should be reactivated and its infectivity tested. In the case of dried or lyophilized virus isolates this is simply accomplished by resuspending the virus in a minimal amount of phosphate buffer and inoculating the mixture to a test plant. The reactivated isolate can then if necessary, be stored again from the fresh leaf material.

More detailed descriptions of the methods used for virus storage have been given by McKinney and Silber (1968), Bos (1969) and Hollings and Stone (1970).

4.6 References

Alconero, R. (1973). Mechanical transmission of viruses from sweet potato. *Phyt* **63**, 377–80.

Bawden, F. C. (1954). Inhibitors and plant viruses. *Adv Virus Res* **2**, 31–57.

Bawden, F. C. and Roberts, F. M. (1948). Photosynthesis and predisposition of plants to infection with certain viruses. *Ann Appl Biol* **35**, 418–28.

Bailiss, K. W. and Okonkwo, V. N. (1979). Isolation of sap-transmissible viruses from spinach (*Spinacea oleracea L.*). *Phytopathol Z* **96**, 146–55.

Bailiss, K. W. and Plaza-Morales, G. (1980). Effects of postinoculation leaf water status on infection of French bean by tobacco necrosis virus. *Physiol Plant Patho* **17**, 357–67.

Bol, J. F. and Van Vloten-Doting, L. (1973). Function of top component a RNA in the initiation of infection by alfalfa mosaic virus. *Virol* **51**, 102–8.

Bos, L. (1969). Some experiences with a collection of plant viruses in leaf material dried and stored over calcium chloride and a discussion of literature on virus preservation. *Meded Rijksfac Landbouwwet Gent* **34**, 875–87.

De Wijs, J. J. and Suda-Bachmann, F. (1979). The long term preservation of potato virus Y and water-melon mosaic virus in liquid nitrogen in comparison to other preservation methods. *Neth J Plant Path* **85**, 23–9.

Drijfhout, E. (1978). Genetic interaction between *Phaseolus vulgaris* and bean common mosaic virus. *Cen Agric Pub Doc* Wageningen.

Fraenkel-Conrat, H., Singer, B., and Tsugita, A. (1961). Purification of viral RNA by means of bentonite. *Virol* **14**, 54–8.

Fry, P. R. and Taylor, W. B. (1954). Analysis of virus local lesions experiments. *Ann Appl Biol* **41**, 664–74.

Fulton, R. W. (1964). Transmission of plant viruses by grafting, dodder, seed and mechanical inoculation. In *Plant Virology* (ed. Corbett, M. K. and Sisler, H. D.). Univ. Florida Press: Florida. pp. 39–67.

Gibbs, A. J. and Harrison, B. D. (1976). *Plant virology, the principles*. Edward Arnold: London.

Hollings, M. and Stone, O. M. (1970). The long-term survival of some plant viruses preserved by lyophilization. *Ann Appl Biol* **65**, 411–18.

Holmes, F. O. (1929). Local lesions in tobacco mosaic. *Bot Gaz* **87**, 39–55.

Kassanis, B. (1952). Some effects of high temperature on the susceptibility of plants to infection with viruses. *Ann Appl Biol* **39**, 358–69.

Kleczkowski, A. (1950). Interpreting relationships between concentrations of plant viruses and numbers of local lesions. *J Gen Micro* **4**, 53–69.

Loebenstein, G. (1972). In *Principles and techniques in plant virology* (ed. Kado, C. I. and Agrawal, H. O.), Van Nostrand Reinhold: New York. pp. 32–61.

McKinney, H. H., and Silber, G. (1968). In *Methods in virology*, Vol. 4 (ed. Maramorosch, K. and Koprowski, H.), Academic Press: New York.

Mackenzie, D. R., Anderson, P. M. and Wernham, C. C. (1966). A mobile air blast inoculator for plot experiments with maize dwarf mosaic virus. *Plant Dis R* **50**, 363–7.

Marcinka, K. and Musil, M. (1977). Disintegration of red clover mottle virus virions under different conditions of storage invitro. *Acta Virol* **21**, 71–8.

Matthews, R. E. F. (1953). Factors affecting the production of local lesions by plant viruses. I. Effect of time of day on inoculation. *Ann Appl Biol* **40**, 377–83.

Matthews, R. E. F. (1981). *Plant Virology*. Academic Press.

Moline, H. E. and Ford, R. E. (1974). Clover yellow mosaic virus infection of seedling roots of *Pisum sativum*. *Physiol Plant Pathol* **4**, 219–28.

Motoyoshi, F., Bancroft, J. D. and Watts, J. W. (1973). The infection of tobacco protoplasts with cowpea chlorotic mottle virus and its RNA. *J Gen Virol* **20**, 177–93.

Owens, R. A., Bruening, G. and Shepherd, R. J. (1973). A possible mechanism for the inhibition of plant viruses by a peptide from *Phytolacca americana*. *Virol* **56**, 390–3.

Preece, D. A. (1967). Nested balanced incomplete block designs. *Biom* **54**, 479–86.

Reid, M. S. and Matthews, R. E. F. (1966). On the origin of the mosaic induced by turnip yellow mosaic virus. *Virol* **28**, 563–70.

Sill, W. H. and Walker, J. C. (1952). A virus inhibitor in cucumber in relation to mosaic resistance. *Phyt* **42**, 349–52.

Smith, K. M. (1937). *A textbook of plant virus diseases*. Churchill: London.

Teakle, D. S. (1973). Use of the local lesion method to study the effect of celite and inhibitors on virus infection of roots. *Phytopathol Z* **77**, 209–15.

Tinsley, T. W. (1953). The effects of varying the water supply of plants on their susceptibility to infection with viruses. *Ann Appl Biol* **40**, 750–60.

van Kammen, A. (1968). The relationship between the components of cowpea mosaic virus. I. Two ribonucleoprotein particles necessary for the infectivity of CPMV. *Virol* **34**, 312–18.

Walkey, D. G. A. and Innes, N. L. (1979). Resistance to bean common mosaic virus in dwarf beans (*Phaseolus vulgaris* L.). *J Agric Sci* **92**, 101–8.

Walkey, D. G. A. and Pink D. A. C. (1984). Resistance in vegetable marrow and other *Cucurbita* spp. to two British strains of cucumber mosaic virus. *J Agric Sci* **102**, 197–205.

Yarwood, C. E. (1952). The phosphate effect in plant virus inoculations. *Phyt* **42**, 137–43.

Yarwood, C. E. (1955). Deleterious effects of water in plant virus inoculations. *Virol* **1**, 268–85.

Yarwood, C. E. (1966). Bentonite aids virus transmission. *Virol* **28**, 459–62.

Yarwood, C. E. (1973). Quick drying versus washing in virus inoculations. *Phyt* **63**, 72–6.

Yarwood, C. E. and Fulton, R. W. (1967). Mechanical transmission of plant viruses. In *Methods in virology I* (ed. Maramorosch, K. and Koprowski, H.), Academic Press: New York.

4.7 Further Selected Reading

Kleczkowski, A. (1950). Interpreting relationships between concentrations of plant viruses and numbers of local lesions. *J Gen Micro* **4**, 53–69.

Loebenstein, G. (1972). In *Principles and techniques in plant virology* (ed. Kado, C. I. and Agrawal, H. O.), Van Nostrand Reinhold: New York. pp. 32–61.

Yarwood, C. E. (1979). Host passage effects with plant viruses. *Adv Virus Res* **25**, 169–91.

Yarwood, C. E. and Fulton, R. W. (1967). Mechanical transmission of plant viruses. In *Methods in virology I* (ed. Maramorosch, K. and Koprowski, H.), Academic Press: New York.

5 *Virus Purification*

5.1 Introduction

The information that can be obtained about any plant virus is limited, if the investigation is confined to an '*in vivo*' study. To obtain information on many biochemical and physical properties, and to produce antisera for serological studies (*see* Section 6.6), it is necessary for the virus particles to be separated from the host and concentrated '*invitro*'. This process is referred to as *purification*, and for new viruses in particular, it is essential for characterization and identification. The purified virus should be physically and chemically undamaged by the purification procedure and free from contaminating host-material.

Since Stanley first purified tobacco mosaic virus (TMV) in 1935 (*see* Section 1.2), plant virologists have strived to purify each new virus that has been discovered. Some viruses, such as TMV and potato virus X, can be readily purified by a number of different methods since they are very stable and occur at high concentrations in their hosts. In contrast, other viruses are less stable, or occur at relatively low concentrations in their hosts, and are hence more difficult to purify.

Procedures for the purification of most known plant viruses are now available, but these methods are diverse. In this chapter, the general principles of virus purification will be outlined, together with detailed purification methods for two specific viruses (*see* Figure 5.1*a* and *b*). It must be emphasized, however, that different individual viruses, and sometimes different strains of the same virus, may require specifically different treatments at any stage in their purification.

Sometimes, if a newly isolated virus possesses characteristics which are shared with those of a well-characterized virus, it may be possible to purify it successfully by following established procedures. Often, however, each step of the purification procedure must be worked out for a new virus by trial and error.

It is therefore important, when developing a purification method, to have an adequate means of assaying the virus concentration in different preparations, at each stage of the procedure. A good local lesion host is of considerable advantage for quantitative biological assay, but if one is not available, then assays may have to be made in a host susceptible to systemic infection (*see* Section 4.4). Although biological assay is essential for comparing the amounts of infectious virus in different preparations, the time necessary (a minimum of 4 or 5 days and usually longer) for the virus symptoms to appear on the test plant can be disadvantageous. Other methods such as electron microscopy and analytical centrifugation (*see* Chapter 6), are also used frequently since they allow a comparison of virus particle concentrations, in different preparations to be made within minutes. When these physical methods alone are used, however, no information is obtained on the amounts of infectious virus present in the preparations.

Detailed information on methods of plant virus purification may be found in publications by Steere (1959), Brakke (1967), Kado and Agrawal (1972) and Noordam (1973).

5.2 Propagation of Virus for Purification

The choice of host plant in which a virus is multiplied for purification is often critical, and it may be a different species from the host used to maintain or assay the virus (*see* Chapter 4).

For purification purposes, the virus should multiply to high concentrations in the selected host. The host should be free of inhibitors, tannins, gums, latex or phenolic compounds, which might inactivate or interfere with the virus during purification, and the virus should be easily separable from the host constituents. Species of tobacco, for instance, are very suitable hosts if they are sufficiently susceptible to the virus concerned (Francki, 1964). It is also advisable to select a host species that can be easily and quickly grown from seed.

Virus concentration in the propagation host will be influenced by both the age of the host at the time of inoculation and environmental conditions, before and after virus inoculation (*see* Chapter 4). The time after inoculation that the infected plant material is harvested for purification is vitally important. Preliminary experiments need to be carried out to determine the optimal period, between inoculation and harvest, for maximum virus concentration.

Systemically infected leaf material is usually the source of virus used for purification, for the virus concentration is usually higher in these than in inoculated leaves. Normally leaves with symptoms contain more virus than those without, and if the leaves are large, it may be advantageous to remove the fibrous midrib before the virus is extracted.

5.3 Partial Purification Procedure

The first stage of the purification process is to extract the virus particles from the host cells, and to separate the particles from the heavier host constituents. This and all subsequent stages of the purification process must be carried out at temperatures of 3° to 5°C. The separation process is referred to as *clarification* and when this is completed, the virus can be concentrated into a small volume that is referred to as a '*partially purified preparation*' (*see* Figure 5.1). This preparation will not be completely free of host constituents, but may be clean enough for some diagnostic studies. For many purposes, however, the partially purified preparation must be further separated from the remaining host constituents by the methods described in Section 5.4.

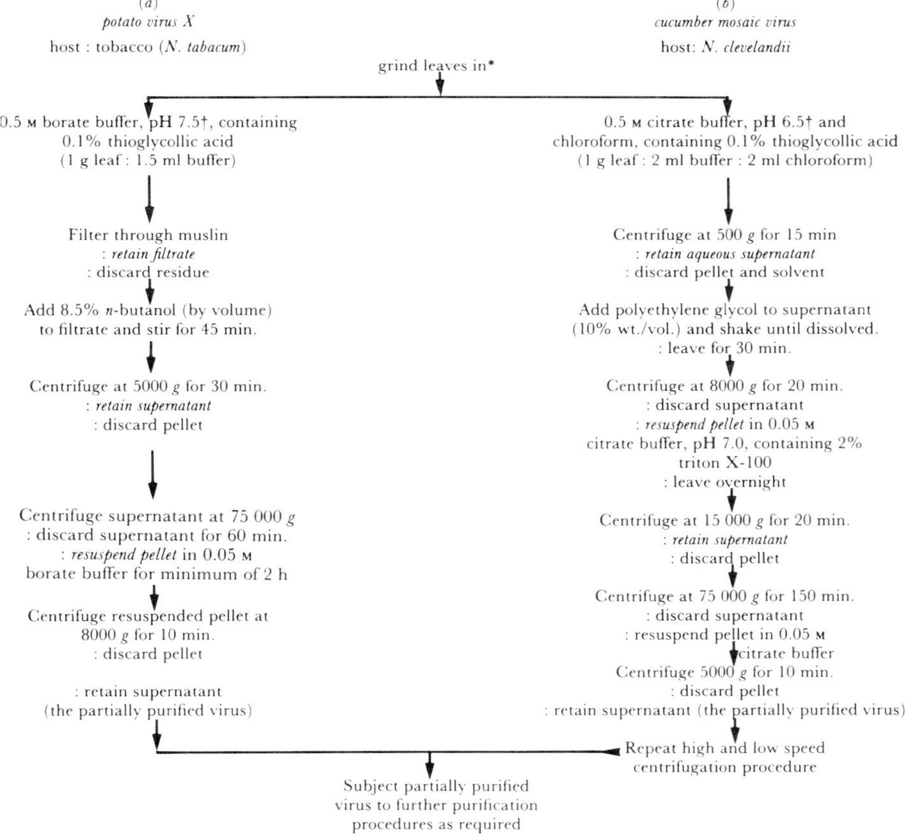

Fig. 5.1 Procedures for the partial purification of potato virus X (*a*) and cucumber mosaic virus (*b*).

5.3.1 Virus extraction and clarification

Extraction

The virus is extracted from the infected leaf by grinding in a suitable liquid at low temperature (3° to 5°C). The leaf may be ground in a pestle and mortar, but because large quantities of leaf material (often hundreds or thousands of grams) are normally used, a kitchen blender, mincer or specially constructed liquidizer is frequently used to produce the sap homogenate. The latter will contain the virus particles and host constituents ranging from large pieces of fragmented tissues to the smaller fractions of chloroplast, ribosomes, soluble cell proteins and low molecular weight solutes. Some of these cell constituents will be capable of inhibiting or inactivating the released virus (*see* Section 4.3). Consequently, precautions must be taken to ensure that the extraction liquid used, is a suitable buffer containing additives that will minimise virus loss and inactivation.

Table 5.1 Buffers and additives frequently used in virus purification extraction medium

Buffers		
Potassium phosphate	(0.01–0.5M : pH 7.0–8.0)	
Sodium borate	(0.05–0.5M : pH 7.6–8.5)	
Sodium acetate	(0.1–0.5M : pH 4.5–6.2)	
Sodium citrate	(0.1–0.5M : pH 6.0–7.4)	
Tris-HCl	(0.1M : pH 7.2–8.4)	
Solvents	*Reducing agents*	
n-butanol	thioglycollic acid	(0.1–0.5%)*
Carbon tetrachloride	2-mercaptoethanol	(0.2–1.0%)
Chloroform	sodium sulphite	(0.1–0.3%)
Diethyl ether	ascorbic acid	(0.1–0.3%)
ethanol		
Chelating agents		
EDTA (disodium ethylene-diamine-tetra-acetate, (0.01–0.1M : pH 7.5)		
DIECA (sodium diethyl-dithiocarbamate, 0.01–0.02M)		
Additives for other purposes†		
Bentonite clay (1–15%)		
PVP (polyvinyl-pyrrolidone, 1%)		
Activated charcoal		
Urea (1M)		
Triton X-100 (1–5%)		
Tween-80 (1–2%)		

*Percentage added to total volume of extraction medium.
†*See* Section 5.3.1.

Extraction buffers

The choice of buffer used for extraction (*see* Table 5.1) will depend on the particular virus being purified and can usually be determined only be preliminary experiments. A borate buffer may be preferable to a phosphate buffer for one virus (Tomlinson, 1963), or *vice versa*. The strength of the buffer (i.e. its molarity) may also be critical, as some viruses are unstable in low molarity buffers while the opposite is true for others. A molarity of between 0.1 M and 0.5 M is frequently used but a buffer of lower molarity may be used during later stages of purification.

The pH of the buffer is also important. For every virus there is a pH at which the particles have no net charge (known as the *isoelectric point*), and at this pH the particles may precipitate. To avoid precipitation during the initial extraction process this pH value must be avoided. Most viruses have an isoelectric point on the acid side of neutrality, so neutral or slightly alkaline buffers are normally used for extraction (Brakke, 1967). The pH of the buffer must not be too alkaline, however, or the bonding between the virus protein and its nucleic acid will be broken and the virus particle will dissociate. Optimal pH values for individual viruses must be determined by preliminary experimentation, but normally they range between 7.0 and 8.5.

Extraction medium additives

Reducing agents, such as thioglycollic acid, sodium sulphite or mercaptoethanol, are frequently added to the extraction buffer at low concentrations (around 0.1%) to prevent virus inactivation by oxidation (*see* Table 5.1). Such compounds may also prevent the absorption of host constituents to virus particles. Sodium sulphite may also serve to reduce the action of phenol oxidase and the addition of cysteine has been shown to have a similar action (Pierpoint, 1966). Additives such as hide powder (Brunt and Kenton, 1963) PVP (polyvinylpyrrolidone) and PEG (polyethylene glycol) (Kosuge, 1965) have also been found to reduce virus inactivation by binding with the phenols.

Chelating agents such as EDTA (ethylene diamine tetra-acetate) used at a concentration of about 1% of 0.05 M (pH 7.5) are often added to assist in the removal of host ribosomes and polyribosomes. EDTA may also prevent some virus aggregation by chelating bivalent cations and prevents the oxidation of polyphenols. Ribosomes may also be absorbed by the use of bentonite clay, a compound that will also absorb fraction 1 host protein, fragmented chloroplasts and ribonucleases (Dunn and Hitchborn, 1965). Pigments and other host material may also be removed by the addition of activated charcoal.

The inclusion of *detergents* such as Igepon T-73 (Brakke, 1959) Triton X-100 (Van Oosten, 1972) or Tween-80 (Leiser and Richter, 1978) in the extraction buffer, has helped in the release of virus particles from the host constituents and reduced particle aggregation. A similar effect may also be obtained by the addition of 0.5 to 1.0 M urea (Damirdagh and Shepherd, 1970).

Clarification

Having produced a sap homogenate containing virus and host cell debris, the next stage is to separate the virus from the debris. The homogenate may be initially filtered through muslin to remove the larger cell debris, but the first major step is to subject the sap to low speed centrifugation (1,000 to 10,000 × *g* for 5 to 15 mins). This is sufficient to sediment the larger plant debris, but not the virus particles (*see* Figure 5.1*a*). The pellet of host debris is discarded and the aqueous phase containing the virus and smaller host-cell contaminants is usually further clarified by the addition of an organic solvent. In any extraction process, only a proportion of the total virus will be separated from the host debris during this first cycle of procedures, and sometimes if virus concentration is relatively low in the host plant, it may be necessary to repeat the initial extraction procedure. The pellet from the first centrifugation is resuspended in fresh buffer solution, stirred and then subjected to a second low speed centrifugation. The aqueous phases from the first and second cycles may then be combined before proceeding further.

Solvents such as ethanol, butanol, chloroform, ether and carbon tetrachloride are most frequently used in the next stage of the clarification process (*see* Table 5.1). More than one solvent may be combined in the same purification procedure and the concentration at which the solvent is used is variable (often between 20 to 50% of the total extract volume). The solvent causes the larger host constituents to coagulate, but leaves the virus in solution. Vigorous stirring of the sap with the solvent (for 10 to 30 mins) will sometimes help to improve the clarification process. Some solvents may cause virus inactivation, in this context chloroform is less harsh than some of the other solvents mentioned.

The homogenate of extracted sap and solvent is subjected to further low speed centrifugation (5,000 to 10,000 × *g* for 10 to 20 mins), following which the homogenate separates into three layers. The densest layer contains the organic solvent and plant materials such as chlorophyll and waxes, the next layer contains the bulk of the plant debris and the lightest, the aqueous phase, contains the virus. The latter is carefully removed and retained and the remainder discarded.

In many purification procedures (*see* Figure 5.1*b*), organic solvents may be added to the buffer together with other additives before the infected leaf is homogenized (Steere, 1956). Additional clarification is obtained with some viruses, if infected leaf material is deep frozen before the virus is extracted; other viruses however, are denatured by deep freezing.

Acidification of the sap homogenate may also be used to clarify the extract of some viruses (Matthews, 1960). By careful control of the pH of the homogenate (usually between pH 3.5 and 5.5) some host proteins may be precipitated leaving the virus particles in solution. The sediment is then removed by low speed centrifugation and discarded. The virus may also be precipitated by acidification at its isoelectric point. Although effective with some viruses when applied to others, this method can cause irreversible particle aggregation (Francki, 1966).

5.3.2 Concentration of the virus

Following the separation of the virus from the bulk of the host-plant debris, the next stage of the purification process is to concentrate the virus and to separate it from soluble, low molecular weight host contaminants. This may be achieved in several ways.

High-speed centrifugation

Ultracentrifugation is the most commonly used method for concentrating the virus. Using this technique, the aqueous phase containing the virus (*see* Figure 5.1*a*) is centrifuged in an angle rotor (*see* Plate 5.1). During this high speed centrifugation (about 75,000 × *g* for 1½ to 2 h)

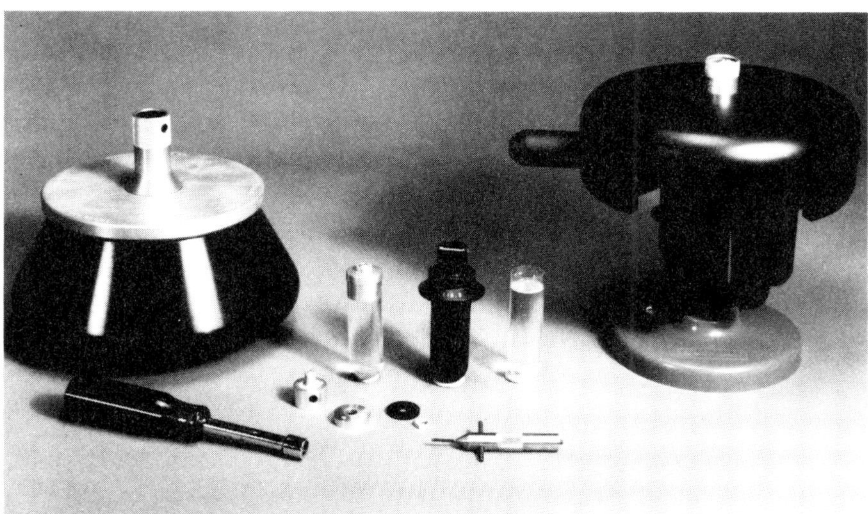

Plate 5.1 Centrifuge rotors and tubes used in virus purification. An angle rotor (*left*) and a swing-out bucket rotor (*right*).

the virus particles sediment against the sloping outer walls of the tubes in the angled rotor, and slide down to form a pellet in the bottom of the tube. The low molecular weight plant material is left in solution (Long *et al.*, 1976). Immediately the centrifugation run is finished, the aqueous liquid is discarded and the virus pellet is resuspended in a small volume of dilute buffer and allowed to stand for some hours.

Finally, the partially purified, concentrated preparation may be further clarified by low speed centrifugation (5,000 × *g* for 10 mins) after which, the aqueous phase containing the virus is carefully removed and the sediment discarded (*see* Figure 5.1). The preparation may be subjected to further cycles of high and low speed sedimentation and clarification (a process commonly referred to as *differential centrifugation*), to increase the purity of the virus preparation (Schumaker and Rees, 1972).

High speed centrifugation is not suitable for concentrating all viruses; some are broken down by the stresses of the gravitational forces involved and others become highly aggregated (Tremaine *et al.*, 1976).

Another method frequently used to concentrate the virus following the initial clarification process, is to precipitate it with polyethylene glycol (PEG) (Hebert, 1963) (*see* Figure 5.1*b*). PEG is a hydrophilic compound that is fully soluble in water and is usually used in the form which has a molecular weight of 6,000. A mixture of PEG and clarified aqueous solution containing the virus (at 6 g PEG/1,000 ml of solution) is stirred for 2 hours at 3 to 4°C and then centrifuged at 15,500 × *g* for 20 mins (Walkey *et al.*, 1972). The PEG pellet (containing the precipitated virus) is then resuspended in a small volume of suitable buffer. A further low speed centrifugation (9,000 × *g* for 10 min) will remove denatured protein and coagulated debris resulting from the PEG treatment leaving the concentrated virus in solution. The virus is finally pelleted from the aqueous phase by high speed (75,000 × *g* for 2 h) centrifugation and the pellet resuspended and further clarified using the differential centrifugation procedure described above.

Viruses may also be concentrated from clarified solutions by salt precipitation, usually using a concentrated solution (⅓ saturation) of ammonium sulphate. The mixture should be thoroughly shaken and allowed to stand for some hours, after which the precipitated virus may be sedimented by low speed centrifugation and resuspended in a suitable buffer. This method is too harsh for many viruses and is, therefore, used infrequently.

5.4 Methods for Final Purification

Very pure preparations of virus are required for many biochemical studies and are advantageous for use in antiserum production (*see*

Chapter 6). It is usual, therefore, to subject the partially purified virus preparation to further treatment to remove as much as possible of the remaining host-plant contaminants. One or more of the following procedures may be used for this purpose.

5.4.1 Density gradient centrifugation

Density gradient centrifugation, is the technique most commonly used for the final purification procedure. This technique involves high speed centrifugation (50,000 to 70,000 × *g*) through a density gradient along a horizontal axis. This is achieved by centrifuging the tube holding the gradient in a rotor with a swing-out bucket (*see* Plate 5.1) (Brakke, 1960, 1964). The gradient is usually composed of sucrose and is often linear ranging from 10 to 40% (*see* Figure 12.1). In this procedure using velocity centrifugation, the virus is separated from other contaminating components according to their differing sedimentation coefficients. Alternatively, the different components may be separated by isopycnic centrifugation, in caesium chloride or caesium sulphate gradients, which separates the components according to their differing buoyant densities. For some purposes, exponential gradients may be preferable to linear gradients, and caesium chloride may be used instead of sucrose for certain viruses. Gradients may be prepared manually by layering solutions of decreasing density on top of one another and allowing the different solutions to diffuse into each other overnight (or at least for several hours). Alternatively, they may be mixed mechanically and used immediately (Stace-Smith, 1965).

A small sample of virus (0.5 to 1.0 ml) is carefully layered on to the surface of the gradient and the tube centrifuged. During centrifugation, the virus and other host contaminants move along the gradient at different rates, because of differences in their sedimentation coefficients. After centrifugation, the virus layer can be observed in a diffuse beam of light as a dense, opalescent band within the centrifuge tube. The virus can be collected manually with a hypodermic syringe or by using a photometric scanner and fraction collector. The sucrose or other salt is removed from the virus preparation by dialysis or the virus preparation is diluted in buffer and re-pelleted by high speed centrifugation.

If large volumes of partially purified virus requires further purification, this may be carried out on a density gradient in a special type of centrifuge rotor called a *zonal rotor* (Schumaker and Rees, 1972). The techniques of zonal centrifugation have recently been described by Griffith (1979).

5.4.2 Gel-chromatography

Final purification may also be carried out by procedures based on molecular sieving in chromatography columns (Van Regenmortel, 1962, Francki, 1972). The columns are filled with gel-beads such as agarose (Sepharose) and dextran (Sephadex). The beads consist of macromolecules which are cross-linked to form a network of polysaccharide chains with pores of controlled size. A sample to be fractionated is placed on top of the column and caused to move through it by a flow of buffer under pressure. Molecules which are too large to enter the pores of the beads move quickly down the column in the flow of buffer whereas smaller molecules diffuse into the beads to a greater or lesser extent and move more slowly down the column. The virus may be detected and collected after elution using a UV absorbance monitor and fraction collector.

Similar columns, filled with glass beads of controlled pore size instead of gels, have also been successfully used to produce highly purified preparations of certain viruses (Barton, 1977).

5.4.3 Other methods

Since all virus particles carry a positive or negative charge on their outer protein surface (except at their isoelectric point) the particles will migrate in an electric field. This movement is known as *electrophoresis*, and at a suitable pH, the amount of movement can be sufficient to separate virus particles from other contaminants (Van Regenmortel, 1964, 1982). This technique is not, however, extensively used in plant virus purification.

Antisera prepared against healthy host-plant proteins may be used to remove host contaminants during the final stages of virus purification (Gold, 1961; Van Regenmortel, 1982). After incubation of partially purified virus with antiserum for some hours, precipitated host proteins are removed by low speed centrifugation, leaving the virus in solution.

5.5 References

Barton, R. J. (1977). An examination of permeation chromography on columns of controlled pore glass for routine purification of plant viruses. *J Gen Virol* **35**, 77–87.

Brakke, M. K. (1959). Dispersal of aggregated barley stripe mosaic virus by detergents. *Virol* **9**, 506–21.

Brakke, M. K. (1960). Density gradient centrifugation and its application to plant viruses. *Adv Virus Res* **7**, 193–224.

Brakke, M. K. (1964). Non ideal sedimentation and the capacity of sucrose gradient columns for virus in density-gradient centrifugation. *Arch Biochm Biophys* **107**, 388–403.

Brakke, M. K. (1967). Density-gradient centrifugation; Miscellaneous problems in virus purification. In *Methods in Virology* (ed. Maramorosch, K. and Koprowski, H.), Vol 2. pp. 93–118 and 119–36. Academic Press: New York.

Brunt, A. A. and Kenton, R. H. (1963). The use of protein in the extraction of cocoa swollen-shoot virus from cocoa leaves. *Virol* **19**, 388–92.

Damirdagh, I. S. and Shepherd, R. J. (1970). Purification of the tobacco etch and other viruses of the potato Y group. *Phyt* **60**, 132–42.

Dunn, D. B. and Hitchborn, J. H. (1965). The use of bentonite in the purification of plant viruses. *Virol* **25**, 171–92.

Francki, R. I. B. (1964). Inhibition of cucumber mosaic virus infectivity by leaf extracts. *Virol* **24**, 193–9.

Francki, R. I. B. (1966). Some factors affecting particle length distribution in tobacco mosaic virus preparations. *Virol* **30**, 388–96.

Francki, R. I. B. (1972). Purification of viruses. In *Principles and techniques in plant virology* (ed. Kado, C. I. and Agrawal, H. O.), pp. 293–335. Van Nostrand Reinhold: Princeton.

Gold, A. H. (1961). Antihost serum improved plant virus purification. *Phyt* **51**, 561–5.

Griffith, O. M. (1979). *Techniques of preparative, zonal and continuous flow ultracentrifugation*. pp. 1–50. Beckman Instruments, Inc.

Hebert, T. T. (1963). Precipitation of plant viruses by polyethylene glycol. *Phyt* **53**, 362.

Kado, C. I. and Agrawal H. O. (1972). *Principles and techniques in plant virology*. Van Nostrand Reinhold: Princeton.

Kosuge, T. (1965). Biochemical aspects of virus inhibitor *in vivo* and *in vitro*. Proc. Int. Conf. Virus and Vector on Perennial Hosts. Univ. Calif. Davis, pp. 247–54.

Leiser, R. M. and Richter, J. (1978). Reinigung une einige Eigenschaften des Kartoffel-Y-Virus. *Arch Phytopathol Pflanzenschutz* **14**, 337–50.

Long, D. G., Borsa, J. and Sargent, M. D. (1976). A potential artifact generated by pelleting viral particles during preparative ultracentrifugation. *Biochim Biophys Acta* **451**, 639–42.

Lot, H., Marrou, J., Quiot, J. B. and Esvan, C. (1972). A contribution to the study on cucumber mosaic virus (CMV) 11. Quick method of purification. *Ann Phytopathol* **4**, 25–38.

Matthews, R. E. F. (1960). Properties of nucleoprotein fractions isolated from turnip yellow mosaic virus preparations. *Virol* **12**, 521–39.

Noordam, D. (1973). Identification of plant viruses: methods and experiments. *Cen Agric Pub Doc* Wageningen, pp. 207.

Pierpoint, W. S. (1966). The enzymic oxidation of chlorogenic acid and some reactions of the quinone produced. *Biochem. J.* **98**, 567–80.

Schumaker, V. and Rees, A. (1972). Preparative centrifugation in virus research. In *Principles and techniques in plant virology* (ed. Kado, C. I. and Agrawal, H. O.), pp. 336–68. Van Nostrand Reinhold: Princeton.

Stace-Smith, R. (1965). A simple apparatus for preparing sucrose density gradient tubes. *Phyt* **55**, 1031.

Stanley, W. M. (1935). Isolation of a crystalline protein possessing the properties of tobacco mosaic virus. *Sci* **81**, 644–5.

Steere, R. L. (1956). Purification and properties of tobacco ringspot virus. *Phyt* **46**, 60–9.

Steere, R. L. (1959). The purification of plant viruses. *Adv Virus Res* **6**, 1–73.

Tomlinson, J. A. (1963). Effect of phosphate and borate on the infectivity of some viruses during purification. *Nature* **200**, 93–4.

Tomlinson, J. A. and Walkey, D. G. A. (1967). Effects of ultrasonic treatment on turnip mosaic virus and potato virus X. *Virol* **32**, 267–78.

Tremaine, J. H., Ronald, W. P. and Valcic, A. (1976). Aggregation properties of carnation ringspot virus. *Phyt* **66**, 34–9.

Van-Oosten, H. J. (1972). Purification of plum pox (sharka) virus with the use of Triton-X-100. *Neth J Plant Pathol* **78**, 33–44.

Van Regenmortel, M. H. V. (1962). Purification of a plant virus by filtration through granulated agar. *Virol* **17**, 601–2.

Van Regenmortel, M. H. V. (1964). Purification of plant viruses by zone electrophoresis. *Virol* **23**, 495–502.

Van Regenmortel, M. H. V. (1982). *Serology and immunochemistry of plant viruses.* Academic Press: New York.

Walkey, D. G. A., Stace-Smith, R. and Tremaine, J. H. (1972). Serological, physical, and chemical properties of strains of cherry leaf roll virus. *Phyt* **63**, 566–71.

5.6 Further Selected Reading

Brakke, M. K. (1967). Density gradient centrifugation, pp. 93–118; Miscellaneous problems in virus purification, pp. 119–36. In *Methods in virology* (ed. Maramorosch, K. and Koprowski, H.), Academic Press: New York.

Francki, R. I. B. (1972). Purification of viruses. In *Principles and techniques in plant virology* (ed. Kado, C. I. and Agrawal, H. O.), pp. 295–335. Van Nostrand Reinhold: Princeton.

Matthews, R. E. F. (1981). *Plant virology*. Academic Press: New York.

6 *Virus Identification*

6.1 Introduction

Correct identification of the virus causing a disease in the field is essential, if adequate control measures are to be found. Symptoms are, on their own, usually insufficient to allow positive identification. The symptoms may result from the presence of more than one virus, or alternatively, several different viruses may individually, cause similar symptoms in the same crop plant.

In this chapter the various procedures and techniques commonly used for the diagnosis of virus diseases are described. Some of these procedures may be carried out *in situ* or on crude sap extracts prepared from the diseased plant, while others require highly purified preparations of the virus. Most of the techniques described can be carried out in a well equipped plant virus laboratory, although the applied worker may require assistance from colleagues in biochemical laboratories, for protein and nucleic-acid analysis.

At an early stage of diagnosis it is essential to determine if the disease symptoms are caused by a single virus, or a complex of two or more viruses. This can usually be done using the electron microscope to ascertain how many types of particle are present (*see* Section 6.5), by examining the symptoms induced in a range of laboratory test plants following sap inoculation, and by serology (*see* Section 6.6). If a mixture of viruses is found, then the individual viruses must be separated. This may be achieved in a number of ways:

(*a*) The viruses may have different host ranges, and may be separated by inoculation on to a number of plant species.

(*b*) Two viruses may infect the same host species, but only one of them may infect the plant systemically (*see* Section 3.2).

(*c*) Two viruses may cause different types of local lesions on inoculated leaves and if cultures are established from these lesions by single lesion transfer, pure cultures of the separate viruses may result (*see* Section 2.3).

(*d*) The individual viruses in the mixture may differ in their modes of transmission. One, for instance, may be aphid transmitted, and the other not.

Once pure cultures have been established from the original mixture, or it has been shown that the original field disease is caused by a single virus, then the procedures to identify the virus or viruses may be started. At this stage it is advisable to store the pure isolate or isolates using methods described in Section 4.5, so that if contamination of the culture should occur during diagnosis, it is possible to return to the original culture.

If a virus has a distinctive shape or size, this information combined with positive serological tests may be diagnostic, but for other viruses, many characteristics may need to be known before the identity can be confirmed. These studies must be particularly extensive if the virus is thought to be new and previously undescribed. Conclusive evidence must de obtained that the virus is not merely a variant strain of an already recognized virus. The literature is full of examples of inadequately described viruses that have been called 'new', but which were later shown to be viruses previously isolated and described from another host.

Finally, Koch's postulates should be satisfied by reinoculating the virus to the initial host plant, to induce the original disease symptoms (*see* Glossary).

Guidelines for the identification and characterisation of plant viruses have recently been provided (Hamilton *et al.*, 1981).

6.2 Mode of Transmission, Host Range and Symptoms

For the early workers with plant viruses these characteristics were amongst the more readily available features by which a virus could be identified. The method by which a virus is transmitted is still important to the virologist, for further studies may be difficult or seriously impaired if a virus is not readily sap-transmitted by mechanical inoculation. Failure of sap-transmission may in itself, be a useful diagnostic feature (*see* Chapter 4). In addition to mechanical sap transmission, however, many viruses have alternative methods of natural transmission that may be helpful for diagnoses (*see* Chapter 7). Transmission of a virus by a specific vector or group of vectors (for example the nematode transmitted *Nepovirus group*, *see* Chapter 2) may immediately indicate the type of the virus under investigation and suggest its probable identity.

The host range of an unknown virus and the symptoms it produces are often important clues to its identity, but these characteristics should be treated with caution. Frequently the type of symptom produced may be dependent on the particular strain of the virus concerned (*see* Section 2.3) or upon the cultivar of the host plant used. Increasing knowledge on the variation that exists in host plants for response to virus infection, and the comparable variation that exists in different virus strains (*see* Section 10.3.4), emphasizes the need to exercise caution during diagnostic studies. It must also be remembered that environmental conditions may effect symptom expression (*see* Section 3.5.2). The symptoms produced by a particular virus in a specific host may nevertheless, be a useful guide to its identity. Some related viruses may produce a similar range of symptoms in one particular host. This may indicate that the unknown virus belongs to this group of viruses. For instance, viruses of the nepo-group frequently produce distinct local lesions on the inoculated leaves of *Chenopodium* spp, followed by systematic flecking and necrosis of the apical area. The same viruses may also cause characteristic ringspotting-symptoms in tobacco species (*see* Plate 3.2).

Other viruses may have a very specific host range that may greatly assist their identification. The host range of celery mosaic virus for instance, is restricted to species of the *Umbelliferae* family. Consequently, if a rod-shaped virus of around 750 nm in length is isolated from celery and can be sap transmitted only to certain other umbelliferous species, the diagnosis of the virus can be considered to be well advanced.

It may be concluded, therefore, that host range and studies of symptoms are useful in that they help to create a picture of the general characteristics of an unknown virus. However they rarely result in conclusive diagnosis. Host range studies may also provide useful information on the best hosts for propagating, assaying and maintaining a newly isolated virus (*see* Section 4.2.1).

A useful range of test plants for virus diagnosis have been listed by Hollings (1983), and detailed information on symptoms produced by plant viruses have been described in the 'C.M.I./A.A.B. Descriptions of Plant Viruses' (*see* Section 12.8) and in books by Smith (1972) and Kurstak (1981).

6.3 *In vitro* Properties in Crude Sap

For many years plant virologists used three simple tests to obtain an indication of the stability of a virus and its concentration. These tests were carried out with crude sap extracts from the infected plant and were designed to determine the temperature at which the virus is inactivated, referred to as the *thermal inactivation point* (TIP), the *dilution end-point* of the virus (DEP), and the *longevity* of the virus *in vitro* (LIV). All three tests were carried out following a standard procedure (Bos *et al.*, 1960; Noordam, 1973). These tests are now not considered to be of any great diagnostic value (Hamilton *et al.*, 1981), but they are quite valuable in giving an indication of stability and concentration of a virus in sap, which may be helpful in developing purification procedures.

6.3.1 Thermal inactivation temperature

The TIP is determined by grinding infected leaf material in distilled water or 0.01 M potassium phosphate buffer, pH 7.0, and heating 0.5 ml aliquots of the extracted crude sap in thin-walled glass tubes in a water-bath. Each sample is heated for 10 min over a range of temperatures usually separated by 5°C. After heating, the sample is immediately cooled by plunging the tube in ice-cold water. The sap is then inoculated to a suitable test plant. The temperature at which virus infectivity is lost is quoted as the TIP.

Thermal inactivation points for different viruses vary greatly. Tomato spotted wilt virus for instance, is inactivated at 40–46°C (Ie, 1970) and tobacco mosaic virus at 90°C (Zaitlin and Israel, 1975), but the TIPs for most viruses range between 55° and 70°C.

6.3.2 Dilution – end point

The lowest dilution at which sap from an infected plant can infect a mechanically inoculated test plant, is known as the *dilution end-point*. The infected sap is diluted in distilled water or 0.01 M phosphate buffer in a series of ten-fold dilutions. The DEP for different viruses may vary from 10^{-1} to 10^{-7}.

6.3.3 Longevity in crude sap

The test to determine the LIV is usually carried out by storing aliquots of the infected crude-sap at 20°C and assaying individual samples after increasing periods of time. The storage time after which the sap looses its infectivity is then quoted as its *longevity in vitro*. This may be as short as an hour with tulare apple mosaic virus (Yarwood, 1955) or as much as a year for tobacco mosaic virus (Zaitlin and Israel, 1975).

The longevity of infected sap stored between 0° and 2°C is also quoted by some workers (Hollings, 1983).

6.4 Cross-Protection Tests

Plants systematically infected with one strain of a virus, frequently will not develop additional symptoms when inoculated with a second strain of the same virus (McKinney, 1929; Salaman, 1933). This resistance phenomenon forms the basis of cross-protection tests, although the mechanism responsible is not fully understood.

In the early years of plant virology, cross-protection tests were regularly used to provide evidence for virus identity and strain relationships. The phenomenon is well illustrated by viruses such as nepo viruses, that cause ringspot symptoms in tobacco spp. Following initial inoculation, the host plant develops severe systemic symptoms, but the younger leaves later show no symptoms even though virus can be detected in them. When these symptomless leaves are inoculated with either the same, or a closely related virus, no symptoms develop. However, if an unrelated virus is inoculated they may develop local and systemic symptoms.

Experience has shown, however, that cross-protection by related virus strains does not always occur, and that there are many exceptions to this phenomenon (Gibbs and Harrison, 1976). Consequently, tests of this sort are now used infrequently for virus identification.

6.5 Electron Microscopy

6.5.1 Introduction

Visual observation of the shape and size of the virus particle is a basic requirement for identification. In many instances it may provide a rapid method of identifying the group to which an unknown virus belongs. In the case of rod-shaped virus particles, length and morphology are particularly characteristic of specific taxonomic groups (*see* Chapter 2), and the outline of isometric viruses is often of diagnostic significance. Isometric viruses may for example, be round and smooth (as is the case with cucumo and bromoviruses), round and knobbly (tombus and tymoviruses), ovoid or imperfectly spherical (ilarviruses) or angular (nepo and comoviruses). Similarly, viruses belonging to the rhabdovirus group have distinct bullet-shaped particles of characteristic size.

Virus particles are only visible in the electron microscope (*see* Figure 1.1), and this instrument is expensive. Although most plant virus laboratories have access to an electron microscope, when one is not available, material may need to be sent elsewhere for examination. Suitably packed material can survive a week or more in the post, but a more satisfactory method of sending specimens for examination, is to prepare pre-coated electron microscope grids with a mixture of virus infected sap and negative stain (*see* Section 12.6).

Electron microscopy of viruses may be carried out on purified preparations, which is often necessary if the fine details of virus structure are to be studied, or on crude extracts of infected sap. Examination of crude sap preparations enables information on virus morphology to be obtained within minutes.

The use of the electron microscope for diagnostic serology and cytopathology is discussed in Sections 6.6.3 and 6.7, respectively.

6.5.2 Procedures for particle examination

Early studies on virus particles were carried out using *metal shadowing* techniques (Williams and Wycoff, 1944). These procedures required relatively pure virus preparations, which were sprayed on to the specimen holding grid. This grid was made of copper mesh and was about 3 mm in diameter and pre-coated with a suitable film to support the preparation. After the preparation had dried, the grid was held at an angle in a vacuum, and exposed to vapour of an electron dense heavy metal, such as gold, platinum or uranium. The heavy metal accumu-

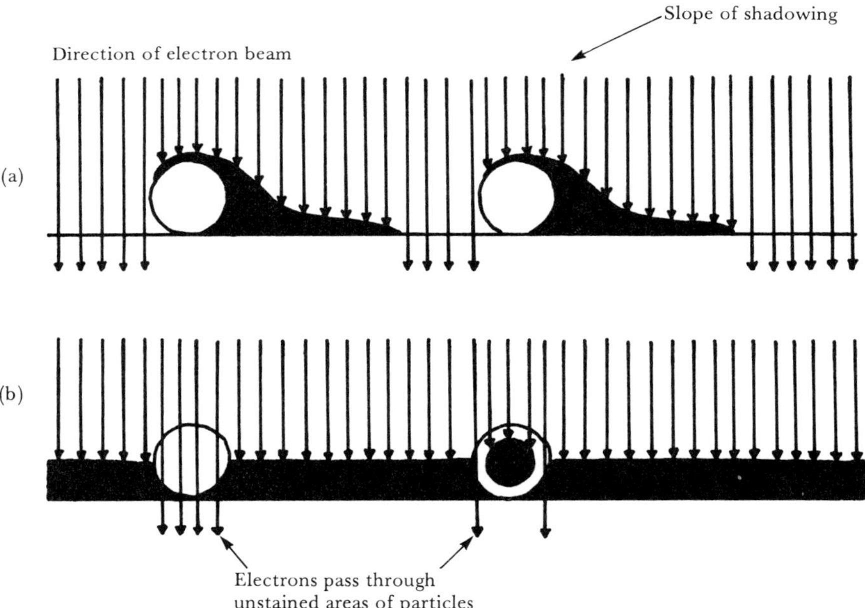

Fig. 6.1 Diagram showing the difference between a shadowed (*a*) and negatively stained (*b*) virus preparation during electron microscopy (based on Bos, 1983). (*a*) The electron beam is unable to penetrate the metal particles that have accumulated around the virus particles; (*b*) the electron beam passes through the shell of an empty virus particle, but not through the particle whose contents have absorbed the electron dense negative stain.

lated against the protruding virus particles, so that an electron-translucent shadow area formed on the side of the particles away from the shadowing slope (*see* Figure 6.1*a* and Plate 6.1). The shadowing technique allowed the shape and size of the virus to be determined, but was time consuming and did not allow observation of the fine structure of the particle.

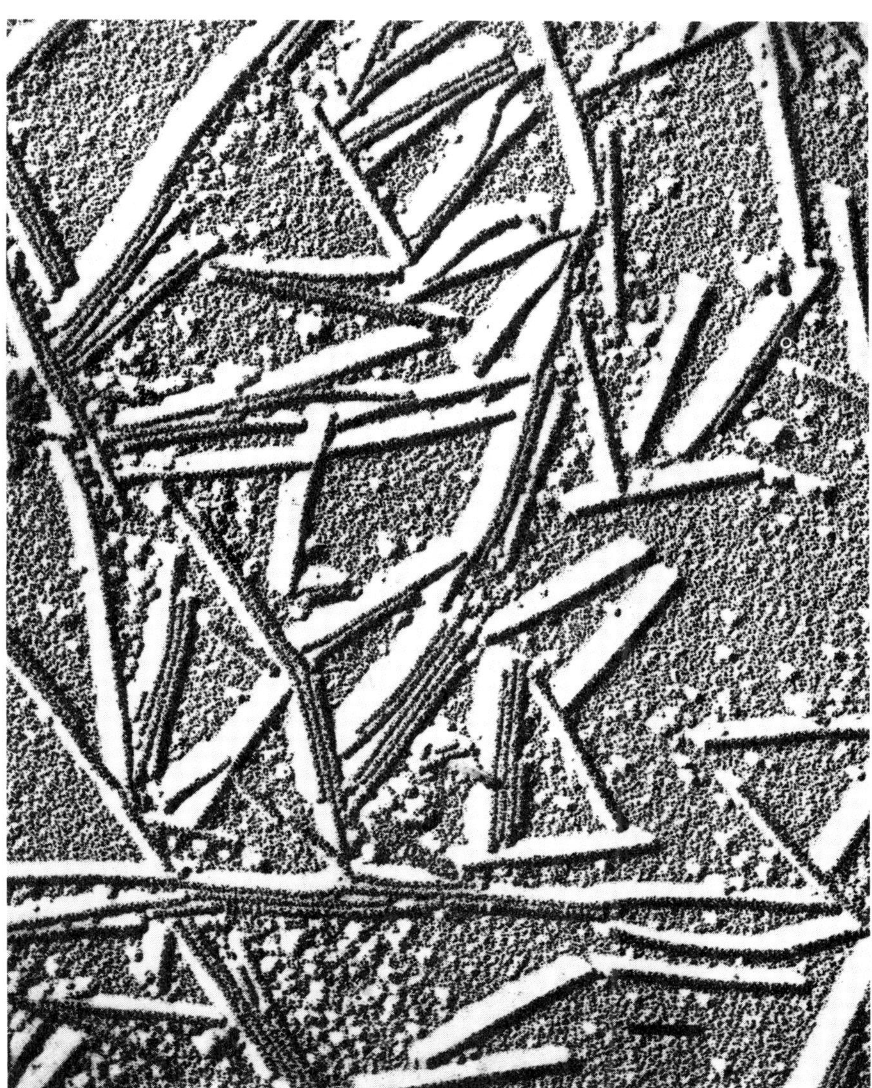

Plate 6.1 An electron micrograph of tobacco mosaic virus particles shadowed with vapour of a heavy metal, magnification bar = 100 nm (courtesy of G. J. Hills).

A major advance in the electron microscopy of plant viruses occurred with the development of *negative contrast staining* by Brenner and Horne (1959). This technique is now universally used, and procedures have been developed that allow simple and rapid examination of virus infected leaf material, without the necessity of using purified virus preparations. The latter may still be required for examining the fine structure of some negatively stained viruses, or for observing certain viruses that occur in very low concentrations in the infected plant (*see* Plate 2.1).

The technique of negative staining is described in detail in Section 12.6 and involves the mixing of the virus preparation with a solution of electron-dense stain such as sodium phosphotungstate (PTA), at a concentration of around 2% (wt/volume). The mixture is placed on an electron microscope grid that has been pre-coated with carbon or other support film. Excess liquid is removed with filter paper and the mixture allowed to dry. The grid may then be examined in the electron microscope, where the particles are seen in negative contrast as the beam of electrons pass through the virus particles, but not through the electron-dense background stain (*see* Figure 6.1*b*). The stain may penetrate some particles, allowing some of the internal structure to be seen.

Some workers prefer to coat their E.M. grids with Colloidion (cellulose nitrate), Formvar or another plastic, but such supports are frequently strengthened with carbon. Although PTA is a suitable stain for many viruses, it causes some viruses to break down and for these, alternative stains such as uranyl acetate, or ammonium molybdate may be used.

With modern electron microscopes, for most practical purposes, resolution down to 1 nm can be readily obtained, although greater resolution may be obtained if special procedures are used. Most viruses can be seen at a magnification of ×10,000 to 30,000, although magnification up to ×200,000 may be required to study fine structure.

The techniques for examining a virus in crude-sap extracts by negative staining are generally referred to as '*quick-dip*' procedures. This term is derived from the *quick leaf-dip* procedure first described by Brandes (1957), in which a drop of infected leaf sap, obtained by squeezing sap from the freshly cut surface of a leaf, is examined. The '*epidermal-strip*' method of Hitchborn and Hills (1965) modified this procedure by placing a strip of epidermal tissue, peeled from the undersurface of an infected leaf, on to a drop of negative stain. Later Walkey and Webb (1968) found that sap from '*squash homogenates*' of apical meristems was a useful source of particles of nepoviruses. Today, many modifications of these procedures are used to obtain infected sap for staining, and individual workers often have their own preferred procedure. Experience has also shown that the optimum material may vary, depending upon the virus and the host species concerned. Usually it is better to take a piece of leaf from an area showing symptoms and in the case of nepoviruses, the shoot tips often contain high concentration of particles. In some cases the older leaves may contain a higher concentration of virus than the younger leaves (e.g. turnip mosaic virus in mustard).

6.5.3 Measurement of particles

Magnifications of not less than ×200,000 are required for accurate measurement of particles, and the method and chemicals used to prepare the virus for E.M. examination should be given when a new virus is described. Particle size and morphology can be markedly affected by the procedure used for virus purification and the length of rod-shaped viruses particularly, can be affected by the presence of Mg^+ ions (Govier and Woods, 1971).

The magnification of the microscope must be carefully calibrated and this can be done by mixing a suitable internal standard with the virus preparation. Tobacco mosaic virus particles may be used for this

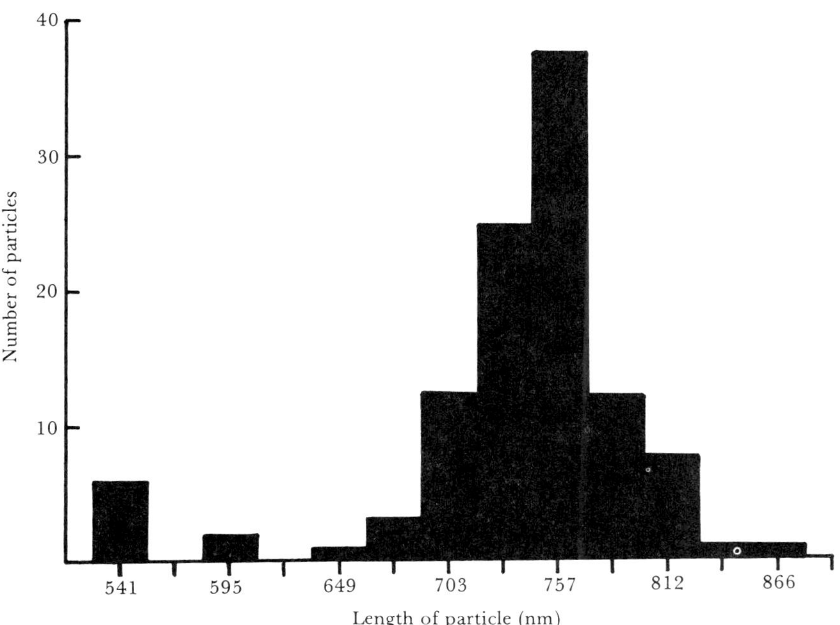

Fig. 6.2 A histogram of the length distribution of lettuce mosaic virus particles (based on Tomlinson, 1964).

purpose, as they are generally accepted to have a modal length of 300 nm (Bos, 1975) and a helix pitch of 2.3 nm (Zaitlin and Israel, 1975). Alternatively crystals of catalase with a lattice spacing of 8.6 nm (Wrigley, 1968) or a diffraction gradient may be used.

A minimum of at least 100 particles should be measured and this may be done directly from the E.M. screen using a binocular microscope and micrometer eyepiece, or from a photomicrograph. When measurements of rod-shaped particles are given it is usual to present the data in the form of a histogram (*see* Figure 6.2) and the overall particle length is quoted as either the *modal* length (i.e. the value that occurs most frequently) or the arithmetical mean.

6.6 Serology

6.6.1 Introduction

Serological tests may be decisive in the final identification of an unknown virus and important for studying the relationships between related virus isolates and strains. Such tests are based upon the binding capacity that individual *antibodies* have for their own specific (homologous) *antigens*.

An *antigen* is a substance, particularly a protein, that is capable of inducing an *immune response* when introduced into an appropriate animal. The ability of an antigen to induce an immune response is usually referred to as *immunogenicity* and substances that are capable of inducing an immune response are called *immunogens*. The antigen may enter the animal either by infection with a pathogenic agent or artificially by injection. The alien antigen provokes the production of the antibodies in certain lymphatic cells of the animal (Van Regenmortel, 1982). The capacity of an antigen to react specifically with an antibody is referred to as *antigenic reactivity*, and is of course the main mechanism of acquired immunity in animals and man against infectious disease, the study of which is referred to as *immunology*. The antibodies circulate in the blood stream and are capable of binding with and immobilizing any of the same antigen that re-enters the blood stream.

Most plant viruses are effective antigens when injected into a suitable animal (usually a rabbit) and stimulate the production of antibodies that can be used in various serological tests. The serum containing the antibodies is separated from the remaining blood components and is referred to as *antiserum*. Detailed information on the serology and immunochemistry of plant viruses has recently been provided by Van Regenmortel (1982).

6.6.2 Preparation and storage of antisera

Highly purified virus is essential for antiserum production. This is so that the resulting antiserum does not contain a large amount of antibody against the host plant protein. The antiserum is prepared by injecting the purified virus suspension either intravenously or intramuscularly (or both) into the experimental animal. Rabbits are the normal choice, although mice, chickens, goats and even horses have been used. There is little reliable information on the relative merits of different immunization procedures, as few workers have compared the effectiveness of their own procedures with those of other laboratories, and individual animals vary greatly in their immunogenic response (Van Regenmortel, 1982).

The number of injections given will, however, affect the specificity of the antiserum produced. If only one or two injections are given, only antibodies to major antigenic determinants (epitopes) are produced resulting in a highly specific antiserum. If, however, six or more injections are given, antibodies to both major and minor antigenic determinants will result and a broader ranged antiserum will be produced.

If intramuscular injections are used the virus is mixed before injection with *Freund's incomplete adjuvant*, a substance which contains an emulsifier and mineral oil, which allows slow release of the virus into the blood stream within the animal. Details of a reliable procedure for the production of antiserum are given in Section 12.5.1. In the author's laboratory, antiserum is routinely produced by four to six, weekly intramuscular injections, followed by bleeds ten to fourteen days after the last injection. After removal from the rabbit, the blood is allowed to clot overnight at room temperature and the serum is carefully separated from the clot. The serum is then centrifuged at low speed (2,000–5,000 × *g* for 5 min) to remove any remaining corpuscles, and the resulting supernatant stored.

The antiserum may be stored in glycerol (1 vol glycerol/1 vol antiserum), at −20°C, freeze-dried or preserved by adding 0.02 to 0.1% sodium azide. In the author's experience however, sodium azide is only suitable for short storage periods.

For some serological tests, such as ELISA (*see* Section 6.6.3), it is advantageous to use the purified protein components (the γ-globulins) of the complete antiserum. These may be purified by ammonium sulphate precipitation or other methods (*see* Van Regenmortel, 1982).

6.6.3 Serological tests used for virus identification

(a) Introduction

A visible *precipitation* or *precipitin* reaction occurs when adequate quantities of antibody and antigen are combined (*see* Plate 6.2). In such reactions the antigen and antibody bind together to form an insoluble lattice (*see* Figure 6.3). In some tests the term *agglutination* is used instead of precipitation. This term is usually restricted to reactions involving large clumps of reactants, such as in the *latex particle* test when the antibodies are attached to latex particles prior to mixing with the virus or in the *slide agglutination* test, when the virus is attached to cells and cell debris which clump together when mixed with antiserum on a microscope slide.

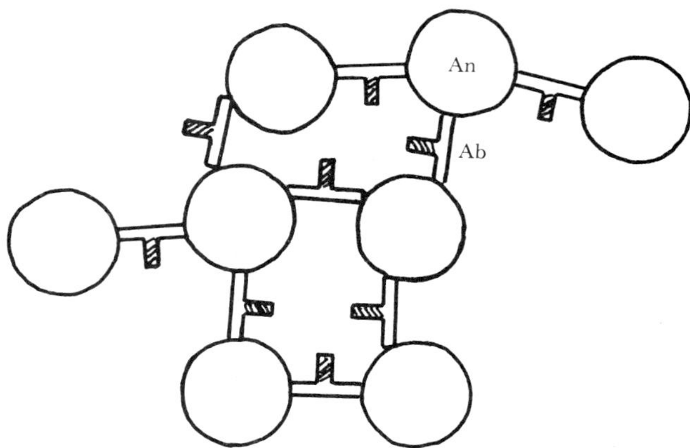

Fig. 6.3 Diagram of an antibody-antigen reaction. When sufficient numbers of antigen (An) and antibody (Ab) molecules combine to form a lattice, a visual, insoluble precipitate results.

The *titre* of an antiserum is the highest dilution of the antiserum that will react with its own homologous virus. For example, an antiserum with a titre of 1024 will react when diluted 1/1024 (dilutions are usually made in 0.9% NaCl solution), and such an antiserum contains eight times more antibodies than one with a titre of 128.

Although purified virus antigen is essential for injection in the initial preparation of a specific antiserum, infected crude sap may be used as the antigen source in some of the following serological tests.

(b) Precipitin tests

Tests involving the visual precipitation of the antigen and antibody components may be carried out in liquid or gel systems. The latter are referred to as immunodiffusion tests.

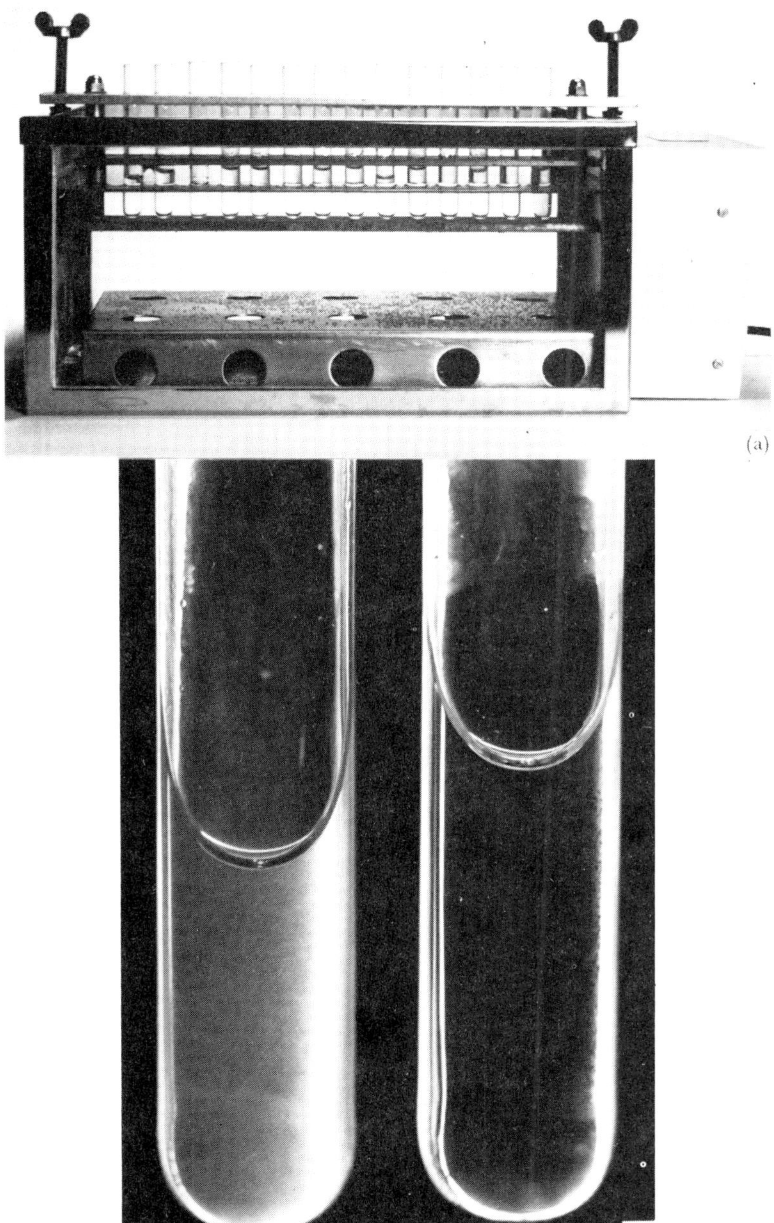

Plate 6.2 Precipitation tube test.
(*a*) Heated water-bath used to incubate the tubes containing the antibody/antigen reactants; (*b*) a flocculent precipitate formed by the antibody/antigen reactants (*left*), and control tube (*right*).

Precipitin-tube tests

These tests are normally used for determining the titre of an antiserum, and for comparing the relationships of different viruses and virus strains by titre values. Tests are carried out in small thin-walled glass tubes (about 7 mm diameter). Usually a two-fold dilution series of one reactant (normally the antiserum) is added to a constant dilution of the other reactant. A 0.9% solution of NaCl is normally used to dilute the antiserum or antigen. An aliquot of 0.5 ml of purified virus and a similar volume of antiserum are mixed in the tube, then incubated at 36°C in a waterbath (*see* Plate 6.2). The tubes are placed in the waterbath so that half the contents are below the surface of the water and the other half above. This allows convection currents to speed up the rate of precipitation. Precipitation is observed at regular intervals (say 10 min) until no further changes are seen. The highest dilution at which precipitation has occurred gives the titre.

Isometric viruses produce a fine granular precipitate and elongated viruses, a more flocculent precipitate (*see* Plate 6.2). The precipitate is best observed by holding the tube over a light source against a black background.

Relationships between viruses can be studied by comparing antiserum titres in precipitin tube tests. The titre of an antiserum reacted against the virus antigen used to prepare it (referred to as the *homologous reaction*) is compared with the titre of the same antiserum when it is reacted against a related, but not necessarily identical virus (referred to as the *heterologous reaction*). The more closely two viruses are related, the closer the titre values for these homologous and heterologous reactions will be (*see* Table 6.1).

Precipitin-ring tests

Virus relationships may also be studied in tube tests by carefully layering antigen on to a volume of antibody in a 3 to 6 mm diameter

Table 6.1 An example of antiserum titres in homologous and heterologous virus reactions

	*Antiserum titre**		
	Antiserum virus A	*Antiserum virus B*	*Antiserum virus C*
Virus A	1 024†	1 024	256
Virus B	1 024‡	1 024	256
Virus C	256	256	1 024

*The titres show that viruses A and B are closely related and possibly serologically identical, but that virus C is more distantly related to viruses A and B.
†Signifies the homologous reaction and ‡ the heterologous reaction.

glass tube, a ring of precipitate is formed at the interface of the two layers if the reaction is positive. The antiserum is normally diluted between 10 and 30% with glycerine and NaCl (for further details *see* Whitcomb and Black, 1961).

Microprecipitin-tests

These tests are very economical in the use of both antiserum and antigen, and are quite sensitive. Single drops of the reactants are mixed in the bottom of a petri-dish and the mixture covered with a cover slip (Noordam, 1973), or a layer of mineral oil (Van Slogteren, 1955) to prevent drying out. The formation of a precipitate is observed using a microscope.

(c) Immunodiffusion-tests

In these tests the antibody-antigen reaction is carried out in a gel instead of liquid. The reactants are allowed to diffuse through the gel and combine (Ackers and Steere, 1967). Agar gels at a concentration of 0.7 to 1.0% are usually used.

The smaller, isometric viruses will readily diffuse through the gels without pretreatment, but the larger rod-shaped viruses must be degraded into smaller units before they will diffuse to give a successful reaction. They may be broken down either by chemical treatment (Purcifull and Shepherd, 1964; Purcifull and Gooding, 1970) or physically by ultrasonic treatment (Tomlinson and Walkey, 1967).

The immunodiffusion techniques most frequently used have been extensively reviewed by Ouchterlony, 1968; Crowle, 1973; and Ouchterlony and Nilsson, 1978.

One type of immunodiffusion test that has been used for virus identification is the single, *radial diffusion* (Oudin, 1952; Mancini *et al.*, 1965). In this test the antibody (or antigen) is added to the liquid gel before it sets. A well is then cut in the gel, and a solution containing the antigen (or antibody) is added to the well. The antigens (or antibodies) diffuse out into the gel, and a halo of precipitation is formed around the well if the reaction is positive. The disadvantage of this technique is the relatively large amount of reactants that must be added to the gel.

A second type of immunodiffusion technique, the *gel double-diffusion* test (often referred to as the *Ouchterlony test*), is by far the most widely used by plant virologists.

In this test the gel initially contains neither reactant. The antibody and antigen are added to wells cut in the gel and are allowed to diffuse towards each other (*see* Plate 6.3*a*). When the reactants meet, a precipitation line is formed where serologically optimal proportions of the reactants occur. The design of the wells holding the reactants in the gel, may be varied to suit the experimental requirement (Ouchterlony,

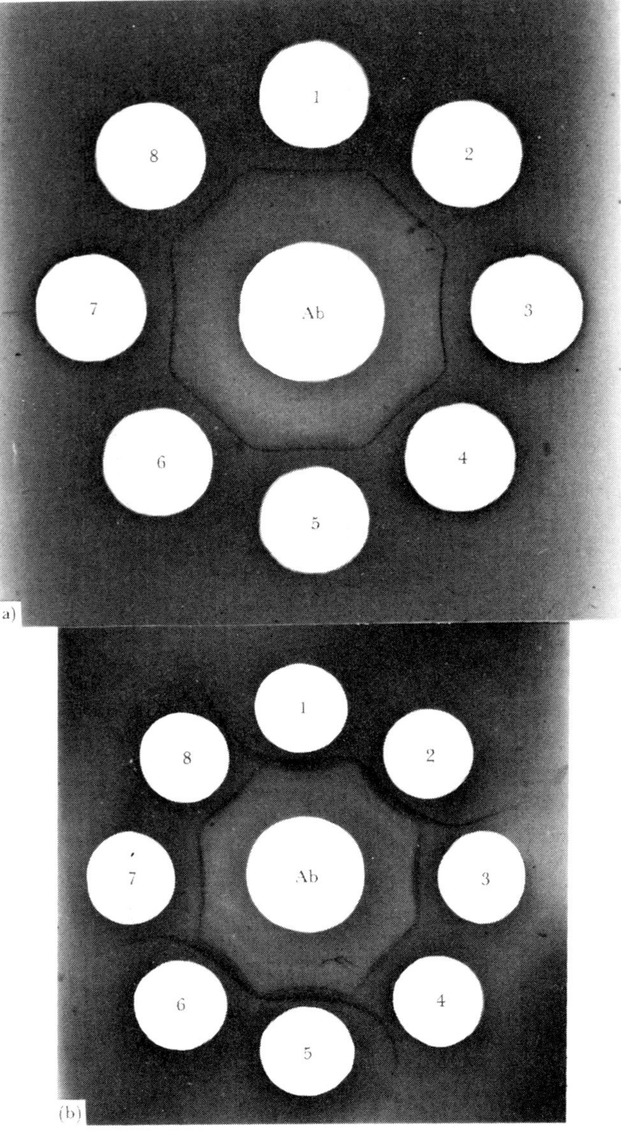

Plate 6.3 Gel double-diffusion test. The antiserum has been placed in the central well (AB) and the antigens in the outer wells (1 to 8).
(*a*) The outer wells contain virus samples with identical antigens, resulting in a continuous (confluent) precipitation line; (*b*) the outer pairs of wells *1*, *2* and *5*, *6* contain one virus strain and the wells *3*, *4* and *7*, *8* another. The two virus strains share some common antigens, but also have other distinct antigens which result in the spur-precipitation reaction (courtesy of Agriculture Canada Research Station, Vancouver).

1968; Crowle, 1973) and depending upon the concentration of the virus involved, either infected crude sap extracts or partially purified preparations may be used as the antigen source. A practical method for carrying out a gel double-diffusion test is described in Section 12.5.3.

The pattern of precipitation lines obtained in these tests will depend upon the antigenic proteins of the virus, the presence of antibodies to these antigens, and the relative size and concentration of these components (Van Regenmortel, 1982). If the antigen consists of large, complete virus particles (such as tobacco mosaic virus rods) the precipitation lines will form very close to the antigen well. The smaller, isometric particles will diffuse further towards the antiserum well before precipitation. Frequently, the antigen preparations also contain even smaller, healthy host antigens, such as fraction 1 protein, and these will diffuse still further and form precipitation lines close to the antiserum well, if the antiserum is also contaminated with antibodies to the healthy plant proteins. The position of the precipitation lines will also alter if concentration of the reactants is unbalanced. The zone of precipitation will broaden and move away from the well containing the excess reactant, but if the concentration of reactants is optimal, a thin, distinct precipitation line is formed (Van Regenmortel, 1982) (*see* Plate 6.3*a*). Gel diffusion tests can also be used to examine the relationships between virus isolates. If two virus isolates are placed in adjacent wells (*see* Figure 6.4*a*) a continuous, *confluent* precipitation line between them and the antiserum well, indicates total absorption of the diffusing antibodies (A) by both sets of antigens (a). This reaction shows that the two virus isolates are serologically identical. If, however, a *spur* (or *partial fusion*) precipitation line is formed between the two virus isolates (*see* Figure 6.4*b*), this indicates that one of the antigens (a) is failing to precipitate some of the antibodies (B) diffusing from the antiserum well, which therefore pass through the precipitation lattice. These unprecipitated antibodies (B) are, however, precipitated by other antigenic proteins (b) diffusing radially from one antigen well (ab) (*see* Plate 6.3*b*). A spur reaction indicates that the two viruses share at least one common antigenic protein but that not all their antigenic-proteins are identical. The spur may arise if viruses differ in their antigenic proteins, or if a single protein has antigenic differences due to different antigenic determinants (epitopes). Since many plant viruses have a single capsid polypeptide species, the latter situation is quite common. Thus, if a spur reaction is formed, the two virus isolates concerned can be considered to be related, but not serologically identical.

When tests such as these are conducted it must be remembered that the antiserum may be composed of antibodies formed against more than one antigenic protein, and that they will diffuse through the gel as separate entities (e.g. A and B antibodies as in Figure 6.4*b*). In

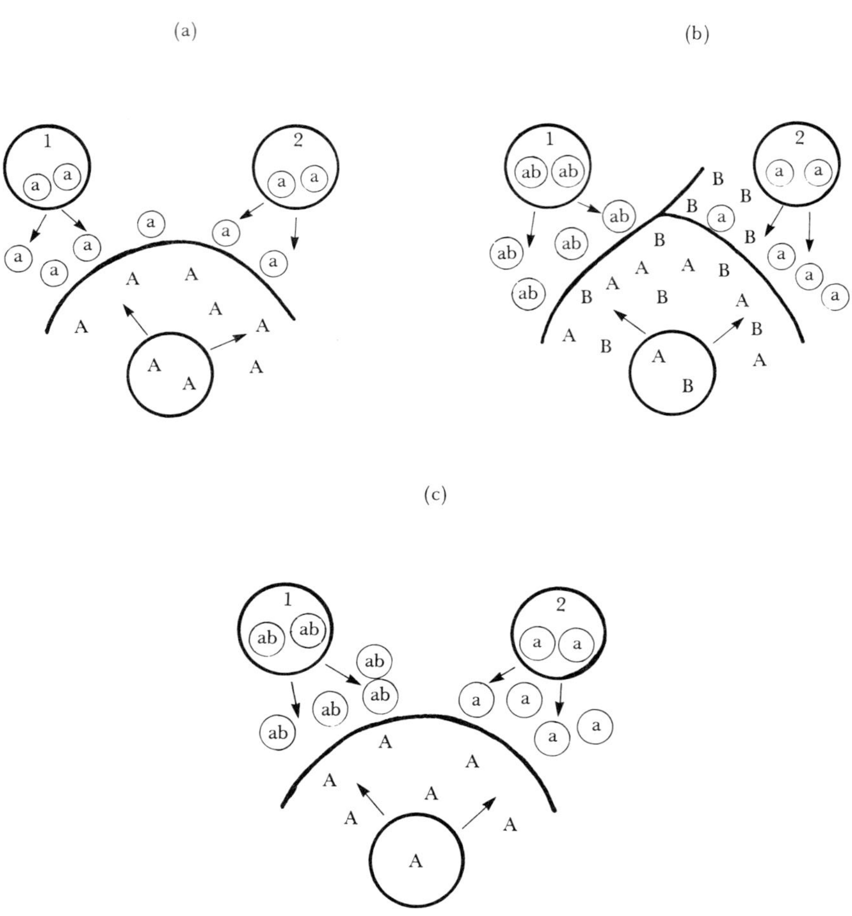

Fig. 6.4 The explanation for spur formation on precipitation lines in agar gel double-diffusion tests. (*a*) The antigen wells (1 and 2) contain virus with identical antigenic determinants (*a*), which react with the antibodies (*A*) diffusing from the antiserum well; (*b*) antigen wells (1 and 2) contain different virus strains with different antigenic determinants (*ab* or *a*). The *ab* antigens react with the *A* or *B* antibodies diffusing from the antiserum well, but the *a* antigens only react with the *A* antibodies, allowing the *B* antibodies to diffuse onwards to meet more *ab* antigens diffusing out radially from well 1 and to react to form the spur; (*c*) no spur precipitation line is formed because the *ab* antigenic determinants (well 1) diffuse as a complete particle and are precipitated by the *A* antibodies, as are the *a* antigens (well 2).

contrast, the virus antigen, which may carry more than one antigenic protein (e.g. ab, Figures 6.4*b* and *c*), diffuses as a complete entity. Consequently, if only antiserum against one virus isolate (a) is used, no spur precipitation line would be formed (*see* Figure 6.4*c*). It is important, therefore, that antiserum against both virus strains is tested (i.e. the homologous and heterologous antisera) when the relationships of two strains is being studied.

(d) Agglutination tests

In serological *agglutination* tests the antibody or virus antigen is absorbed on to larger particles. A positive reaction causes these larger particles to clump and so the antibody–antigen reaction is visibly amplified.

In one such test, called the *slide-agglutination* or *chloroplast-agglutination* test, virus-infected crude sap is mixed with antiserum on a microscope slide. If the reaction is positive the chloroplasts and other sap debris clump together. This test has been particularly useful in the past for the rapid detection of virus-infected potatoes in the field (Van Slogteren, 1955).

Another useful agglutination test is the *latex particle* test, in which either the antigen or antibody is absorbed on to polystyrene latex particles. When mixed with the corresponding reactant, a positive reaction is indicated by visual clumping of the latex particles. The test is highly sensitive and reported to be 100 to 1,000 times more sensitive than microprecipitin or immunodiffusion tests (Koenig *et al.*, 1979), and can be carried out with lower concentrations of reactants than those required for precipitin tests.

Erythrocytes (i.e. blood cells) have been used in the same way in tests called *haemagglutination* tests, and bentonite and barium sulphate have also been employed in agglutination tests (Van Regenmortel, 1982).

(e) Electron microscope serology

Techniques involving the visualization of serological reactions in the electron microscope are highly sensitive (*see* Table 6.2) and have

Table 6.2 Approximate sensitivity of various methods of virus detection

Method	*Minimum detectable virus concentration* (ng)
Gel diffusion	1 000
Precipitin tube	500
Electron microscopy	100
Host infectivity	100
ELISA	1
EM serology	1

become important methods for virus identification in recent years (Derrick, 1973; Milne and Luisoni, 1975; Lesemann, 1983). They have the great advantage of requiring only very small amounts of antiserum and antigen, and the virus may be used directly in crude sap homogenates.

Various terms have been used to describe these techniques, but *electron microscope (EM) serology* or *immunosorbent electron microscopy* (ISEM) are recommended (Roberts *et al.*, 1982) and these are synonymous in their use with immunoelectron microscopy (Hamilton *et al.*, 1981). Derrick (1973) was the first to use EM serology when he 'trapped' viruses on to antibody-coated grids (ACG) (*see* Section 6.6.3). In this procedure the ACG is covered with a sap-homogenate (or other preparation of the virus), so that the antibodies trap virus particles from the solution, with the result that a higher concentration

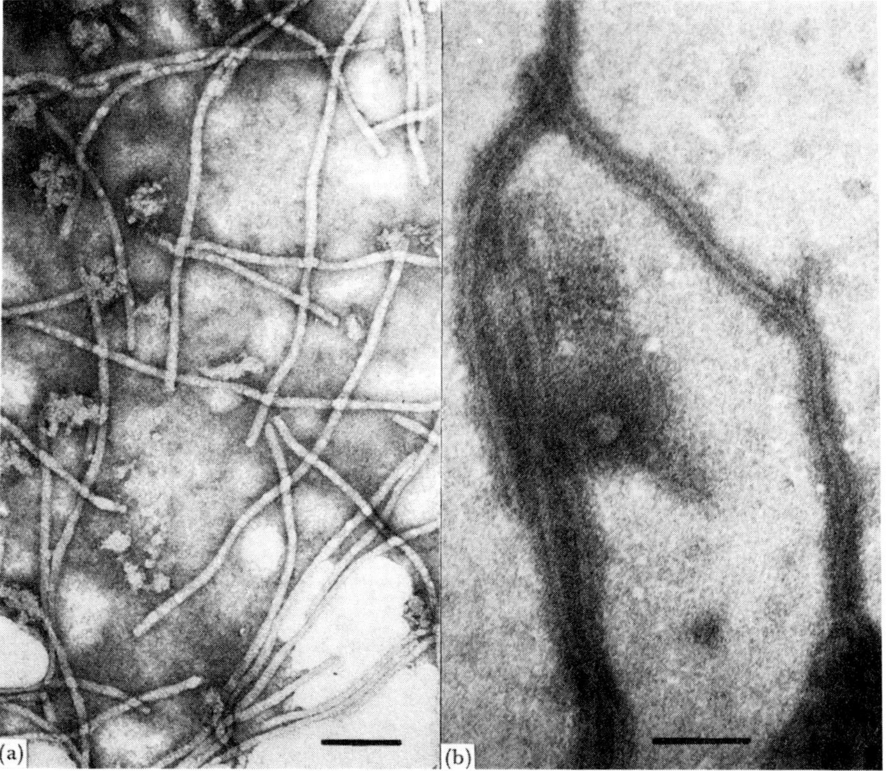

Plate 6.4 An electron microscope serology 'decoration' test.
(*a*) Undecorated bean yellow mosaic virus particles, magnification bar = 100 nm; (*b*) BYMV particles decorated by their specific antibodies, magnfication bar = 100 nm (courtesy of M. J. W. Webb).

of particles become attached to the grid than is normally possible by the usual 'quick-dip' methods. The ACG procedure is usually referred to as the 'trapping' technique and is particularly useful when a virus occurs at low concentrations in the host plant.

Two other EM serology techniques are also used to identify plant viruses (Milne and Luisoni, 1975). The first involves the mixing of virus particles and antisera prior to their being placed on a grid for EM examination. This method results in the virus particles being linked into groups by the antibodies and is generally referred to as '*clumping*'. In the second technique, the antibodies are added to a grid that already has virus particles attached, so that the particles become coated or 'decorated' with the antibodies (*see* Plate 6.4). Besides its use for virus

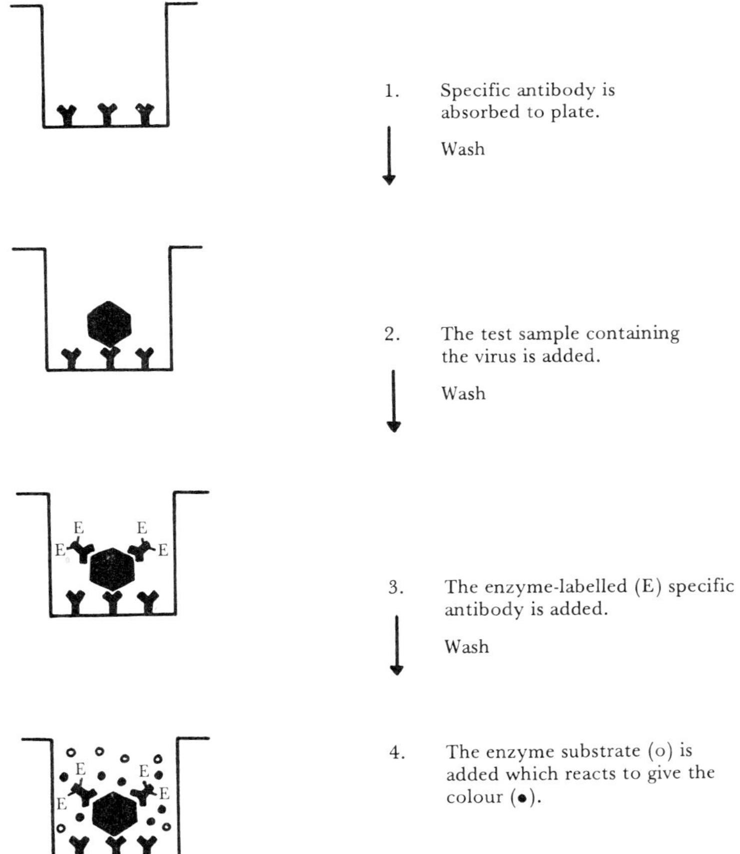

Fig. 6.5 Diagram of the procedures used for enzyme immuno-assay (ELISA) (based on Clark and Adams, 1977).

identification, the 'decoration' procedure has also been used to demonstrate degrees of relationship between viruses (Walkey and Webb, 1984). A comparison of the 'trapping', 'clumping' and 'decorating' procedures suggests that the latter technique may be the most specific for some viruses (Milne and Lesemann, 1978).

Increased sensitivity may sometimes be obtained by 'trapping' particles on the grid and then 'decorating' them with antibodies (Noel *et al.*, 1978).

(f) Enzyme-linked immunosorbent assay (ELISA)

Although EM serology procedures are highly sensitive and useful for virus identification, they are not really practical if large numbers of plant samples have to be tested. In such cases, identification is more readily accomplished by using a highly sensitive test called *enzyme-linked immunosorbent assay* (ELISA).

In this procedure the sensitivity of detection of the antibody–antigen reaction is increased by attaching either of the two reactants to a minute quantity of enzyme. An enzyme substrate is then added, and the resulting colour reaction may be quantitatively measured enabling virus to be detected in very low concentrations (*see* Table 6.2). The *ELISA* procedure was first developed by Voller and his co-workers (Voller *et al.*, 1976; Voller and Bidwell, 1977) and numerous variations of ELISA may now be used for plant virus identification (Koenig and Paul, 1983). Of these the '*double antibody sandwich*' method (Clark and Adams, 1977) is the most commonly used. Wells of a polystyrene plate (*see* Plate 6.5) are first coated with γ-globulin purified from the antiserum; the test sample of virus is then added to the absorbed antibody and the enzyme-labelled antibody then added to the 'trapped' virus (*see* Figure 6.5). The attached enzyme subsequently digests an added enzyme-substrate which results in a colour change. This colour change may be recorded by visual examination or can be measured quantitatively with a colorimeter. Alkaline phosphatase is the enzyme most frequently used for the antibody-conjugate and this is normally detected using the substrate *p*-nitrophenylphosphate. The hydrolysis of the substrate is usually stopped by adding sodium hydroxide (NaOH) to the wells of the plate before the colour reaction is measured.

6.7 Cytopathology

Various *inclusion bodies* (*see* Section 3.3.2) resulting from virus infection are characteristic of individual viruses and groups of viruses. These inclusion bodies can be seen with the light microscope in epidermal strips or sections (Christie and Edwardson, 1977; Fraser and Matthews, 1979). Using the electron microscope they can be observed in

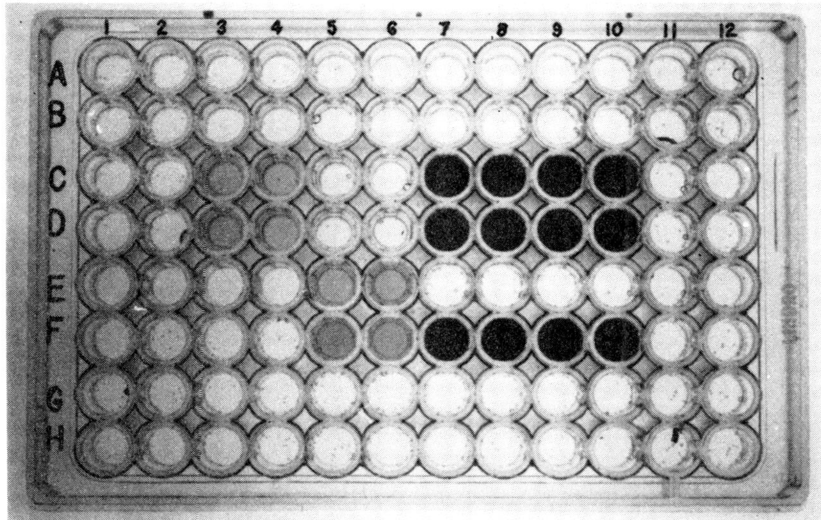

Plate 6.5 An enzyme-linked immunosorbent assay (ELISA) test plate. The darker colours (e.g. in wells C 7, 8, 9 and 10) indicate a positive reaction.

ultra-thin sections (Edwardson and Christie, 1978) and 'squash homogenates' (Walkey and Webb, 1970).

The diagnostic value of the cylindrical inclusions formed by many of the potyviruses has been illustrated by Edwardson (1974), while Hamilton *et al.* (1981) list nine different virus groups that produce diagnostic inclusion bodies. The use of virus-induced inclusions for identification purposes has been reviewed by Edwardson and Christie (1978).

An area of diagnostic cytopathology that has yet to be fully developed is the use of *immunocytological* procedures to identify viruses in ultra-thin sections in the electron microscope. These techniques are technically difficult, involving the use of labelled-antibodies to detect virus in the sections. Electron dense ferritin particles or enzymes have been used for labelling (Kurstak *et al.*, 1977) and more recently colloidal gold labelling has given promising results (Beesley *et al.*, 1982).

6.8 Sedimentation Properties

Information on the sedimentation properties of viruses can help in virus characterization and identification, by indicating the number of distinct, sedimenting components, and their *sedimentation coefficients* and *buoyant densities*.

Table 6.3 Examples of sedimentation coefficient ($S_{20}W$) values of various isometric viruses

Viruses with a single sedimenting component		*Viruses with two sedimenting components*		*Viruses with three sedimenting components*	
Brome mosaic (bromovirus gp.)	87†	Turnip yellow mosaic (tymovirus gp.)	54,115	Cowpea mosaic (comovirus gp.)	58, 95, 115
Cucumber mosaic (cucumovirus gp.)	99	pea enation mosaic (PEMV gp.)	100, 120	Black raspberry mosaic (ilarvirus gp.)	81, 89, 98
Southern bean mosaic (sobemovirus gp.)	115			Raspberry ringspot (nepovirus gp.)	50, 91, 125
Tomato bushy stunt (tombusvirus gp.)	140				
Cauliflower mosaic (caulimovirus gp.)	202				
Carnation ringspot (dianthovirus gp.)	135				
Tobacco necrosis (TNV gp.)	112–133				
Maize streak (geminivirus gp.)	76*				

*value for pairs of particles.
†sedimentation coefficients are measured in Svedberg units (S).

These properties may be studied by using analytical ultracentrifugation (Markham, 1967), or by gradient centrifugation in a preparative centrifuge (Brakke, 1967, *see* Section 5.4.1). Usually purified virus preparations are required for these studies, although clarified crude sap may be used in the analytical ultracentrifuge to obtain preliminary information on the number, and sedimentation rates of different virus components.

The *sedimentation coefficient* of a virus is the rate of sedimentation per unit centrifugal field measured in *Svedberg units* (S) and corrected for factors such as medium viscosity and temperature, to what the sedimentation would be in water at 20°C (referred to as $S_{20}W$) (Matthews, 1981). Most workers, however, determine the sedimentation coefficient in sucrose at about 4°C and simply refer to it as the *S* value without correction. The *S* values for most viruses are between 50 and 200S (*see* Table 6.3), but are as high as 1,000*S* for certain rhabdoviruses. Most viruses with multipartite genomes (*see* Chapters 1 and 2) have nucleoprotein particles of two or more types. The particles often have the same diameter, but different RNA contents. This results in different sedimentation properties. Preparations of some viruses

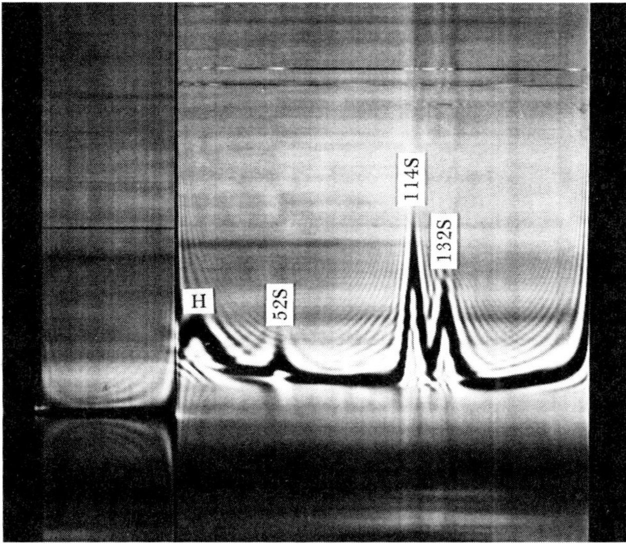

Plate 6.6 Photograph taken with the Schlieren optics of an analytical ultra-centrifuge showing the relative proportion of the three components (52 S, 114 S, and 132 S) of cherry leaf roll virus. The peak on the left (H) represents host constituents (courtesy of Agriculture Canada Research Station, Vancouver).

(such as nepoviruses) are characterized by the presence of empty protein shells, which consequently have a low *S* value. Raspberry ringspot virus, for example, has three kinds of isometric particles with the same diameter. Each particle has a different sedimentation coefficient (i.e. 52, 92 and 130 *S*), and the three particle types contain respectively: 0, 30 and 44% RNA (Murant *et al.*, 1972). Using the analytical ultracentrifuge, the three distinct components of the virus may be seen (*see* Figure 6.6), and the height of the peaks indicates the relative amount of each component present in the preparation (*see* Plate 6.6). Methods for calculating sedimentation coefficients have been described by Schumaker and Rees (1972) and Trautman and Hamilton (1972).

The *buoyant density* of a virus is also useful information to obtain if a new virus is being characterized. It is measured in a caesium chloride (CsCl) or caesium sulphate (Cs_2SO4) gradient by *equilibrium-zonal centrifugation* (i.e. centrifugation of a virus in a gradient, until the virus stops at a level where the density of the medium equals the density of the virus). The buoyant density of the virus components is calculated from the refractive indices of collected fractions, using tables relating refractive index and density (Anderson and Anderson, 1973). Most viruses have a buoyant density of between 1.2 and 1.6 g/cm^3 and this is again correlated with the particle's nucleic acid content.

6.9 Electrophoretic Mobility

The movement of a virus in an electric field is known as its *electrophoretic*

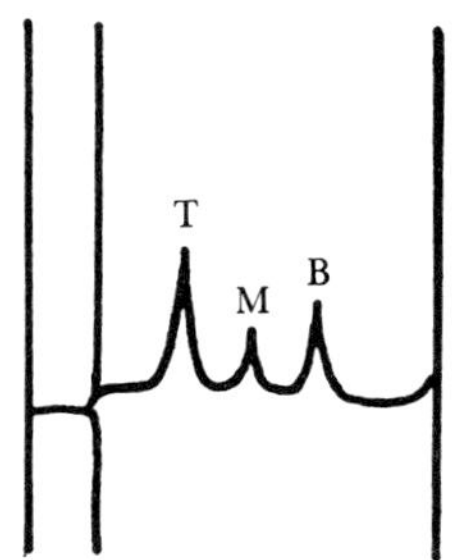

Fig. 6.6 Diagram of the Schlieren pattern for the separation of the components of the tripartite raspberry ringspot virus. The three separately encapsidated genomic components of raspberry ringspot virus, may be separated by analytical centrifugation into top (*T*), middle (*M*) and bottom (*B*) components. The sedimentation coefficients of the three components are 52, 92 and 130S, respectively. The meniscus is on the left of the diagram (based on Murant *et al.*, 1972).

Table 6.4 The movement of different strains of cherry leafroll virus following electrophoresis in agar gel at pH 6.5

Virus strain	*Movement* (mm)*
Cherry	−5
Rhubarb	−6
Golden elderberry	−8
Dogwood	+2
Elm	+4

*− Signifies movement towards cathode and + towards anode. (Information based on Walkey 1973).

mobility. This movement depends on the charge/mass/(c/m) ratio and shape of the virus, and the c/m varies with the pH of the suspension medium.

Since different strains of a virus may vary in their overall net charge, electrophoretic mobility is often useful in distinguishing related strains (Ginoza and Atkinson, 1955; Walkey *et al.*, 1973) (*see* Table 6.4). Experiments can be carried out in agar, buffered between pH 7.5 and 8.6, but the virus's *isoelectric point* (i.e. the pH of zero net charge) must be avoided, since no movement will occur at this pH. The virus is placed in a well, cut in agar on a glass slide (*see* Plate 6.7). A current is passed through the gel for several hours, which causes the virus to move out from the well into the agar. After the current is switched off, the position and movement of the virus may be located by the use of antiserum placed in a trough cut in the agar, parallel to the movement of the virus.

6.10 Chemical Composition

For complete characterization and identification of any new virus, it is necessary to analyse its nucleic acid and coat protein. The procedures required to carry out this analysis require expertise in biochemical techniques, that may be beyond the scope of individuals working in a laboratory concentrating on practical field problems. In this situation, collaboration with colleagues in a suitably equipped laboratory will be necessary. These procedures are generally carried out using highly purified virus preparations.

6.10.1 Nucleic acid analysis

First, the type of nucleic acid must be determined. The presence of viral *RNA* or *DNA* can be determined by the buoyant density of the nucleic acid in caesium salt gradients (Birnie and Rickwood, 1978), its sensitivity to pancreatic RNase or DNase, or by its base composition (Hamilton *et al.*, 1981). Secondly, information must be obtained as to

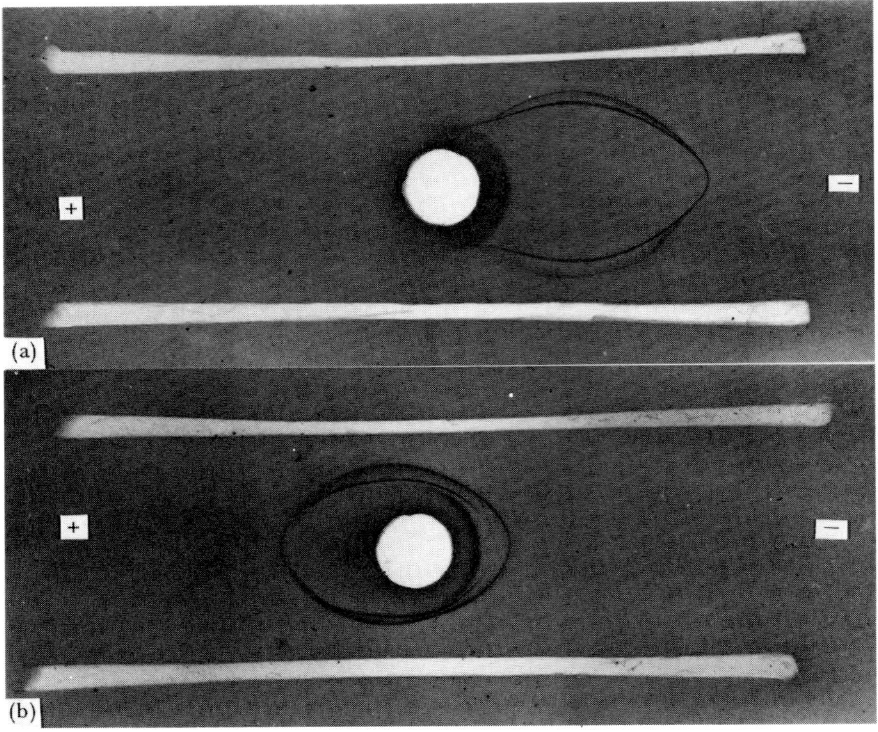

Plate 6.7 Electrophoretic mobility test using different strains of cherry leaf roll virus. (*a*) Precipitation lines developed in agar gel buffered at pH 6.5 showing the movement of the golden elderberry strain towards the cathode after electrophoresis; (*b*) movement of the dogwood strain towards the anode after electrophoresis at pH 6.5 (courtesy of Agriculture Canada Research Station, Vancouver).

whether the nucleic acid is *single-stranded (ss) or double-stranded (ds)*. This can be determined by melting techniques (Shepherd *et al.*, 1970) or methods involving nucleases and gel-electrophoresis (Morris and Dodds, 1979; Luisoni *et al.*, 1979). Gel electrophoresis may also be used to estimate the number and molecular weight of the polynucleotides (Loening, 1969; Peacock and Dingman, 1967) (*see* Plate 6.8).

Reviews on the isolation and properties of plant virus nucleic acids have been written by Hull (1979), Zaitlin (1979) and Lane (1979).

6.10.2 Coat protein analysis

Information on the molecular weights of and the number of polypeptides that the virus particle contains, is useful in identifying the group to which a new virus may belong (*see* Table 6.5). For this

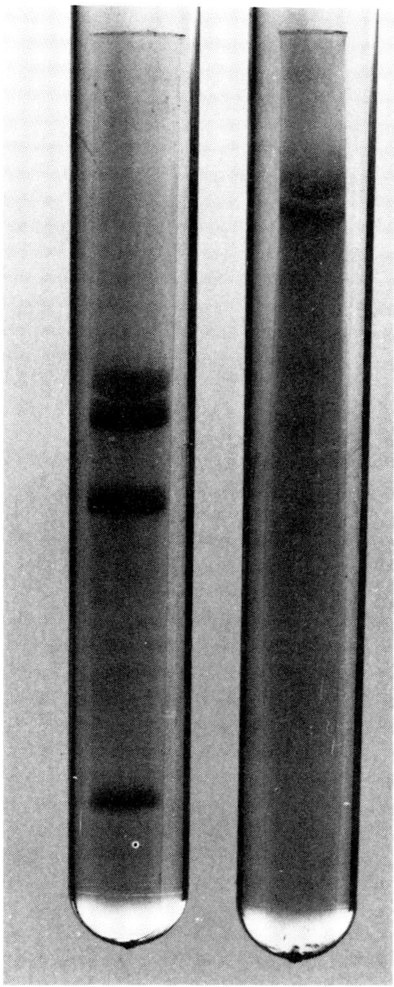

Plate 6.8 Electrophoresis of cherry leaf roll virus RNA in 5% polyacrylamide gel (*right*) and of the standard, brome mosaic virus (*left*) (courtesy of Agriculture Canada Research Station, Vacouver).

analysis, the virus is usually dissociated by SDS (sodium dodecyl sulphate) treatment, and the number of polypeptides and their molecular weight determined by polyacrylamide gel-electrophoresis of the dissociated particle (Laemmli, 1970; Maizel, 1971). The molecular weight of the unknown virus protein being estimated by comparing its mobility in the gel, with that of other proteins of known molecular weight (Shapiro *et al.*, 1967). The values obtained should be treated with caution, however, as errors may sometimes occur using this SDS procedure (Hamilton *et al.*, 1981).

Finally the particle composition in respect of the relative percentage of protein and nucleic acid should be calculated. This may be determined by various methods, but an accurate and reliable method, is hydrolysis of the RNA and determination of the total nitrogen content of the virus preparation (Tomlinson *et al.*, 1983).

Table 6.5 Examples of the approximate molecular weights of the capsid protein of various groups of isometric viruses

Virus group	*Molecular weight* ($\times 10^{-3}$)
Alfalfa mosaic	24.5
Bromovirus	20
Comovirus	22 and 42
Cucumovirus	24.5
Ilarvirus	27
Nepovirus	55
Pea enation mosaic	22
Tobacco necrosis	30
Tombusvirus	41
Tymovirus	20
Sobemovirus	30

6.11 Molecular Hybridization Analysis

The technique of nucleic acid hybridization is a very recent development for the detection and identification of complete or partial viral genomes. *Molecular hybridization analysis*, which is also referred to as the '*spot hybridization*' or '*dot blot*' technique, has been shown to be a highly sensitive and specific procedure for identifying RNA or DNA viruses (Abu-Samah and Randles, 1983; Gould and Symons, 1983; Maule *et al.*, 1983) and plant viroids (Palukaitis and Symons, 1978; Owens and Diener, 1981).

The technique involves the production of complementary DNA (cDNA) using highly purified preparations of the viral nucleic acid or viroid concerned. The cDNA is labelled with radioactive ^{3}H or ^{32}P and then hybridized with crude sap samples from the virus infected plant to be tested. Prior to hybridization the sap samples are 'dotted' and then baked on to a nitrocellulose membrane. Autoradiography is then used to detect the samples which show positive hybridization.

Once a cDNA probe has been prepared to a specific virus, the technique may be used to rapidly screen large numbers of crude sap samples. The limit of its sensitivity has been reported to be as low as 5 to 20 pg virus per spot sample (Maule *et al.*, 1983).

6.12 References

Abu-Samah, N. and Randles, J. W. (1983). A comparison of Australian bean yellow mosaic virus isolates using molecular hybridization analysis. *Ann Appl Biol* **103**, 97–107.

Ackers, G. K. and Steere, R. L. (1967). Molecular sieve methods. In *Methods in virology* (ed. Maramorosch, K. and Koprowski, H.), Academic Press: New York. pp. 325–67.

Anderson, N. G. and Anderson, N. L. (1973). In *Handbook of biochemistry. Selected data molecular biology* (ed. Harte, R. A.), C.R.C. Press: Cleveland.

Beesley, J. E., Orpin, A. and Adlam, C. (1982). A comparison of immunoferritin, immunoenzyme and gold-labelled protein A methods, for the localization of capsular antigen on frozen thin sections of the bacterium, *Pasteurella haemolytica. Histochem. J.* **14**, 803–10.

Birnie, G. D., and Rickwood, D. (1978). *Centrifugal separations in molecular and cell biology*. Butterworths: London.

Bos, L. (1975). The application of TMV particles as an internal magnification standard for determining virus particle sizes with the electron microscope. *Neth J Plant Pathol* **81**, 168–75.

Bos, L. (1983). *Introduction to plant virology*. Longman: Harlow.

Bos, L., Hagedorn, D. J. and Quantz, L. (1960). Suggested procedures for international identification of legume viruses. *Tijdschr Plantenziekter* **66**, 328–43.

Brakke, M. K. (1967). Density gradient centrifugation. In *Methods in Virology* (ed. Maramorosch, K. and Koprowski, H.), Academic Press: New York. pp. 93–118.

Brandes, J. (1957). Einen elektronmikroskopische Schnellmethode zum Nachweis faden-und stäbschenförmiger Viren insbesondere in Kartoffeldunkelkeimen. *Nachrbe deut Pflanzenschutzdienst* **9**, 151–2.

Brenner, S. and Horne, R. W. (1959). A negative staining method for high resolution electron microscopy of viruses. *Biochim Biophys Acta* **34**, 103–10.

Christie, R. G. and Edwardson, J. R. (1977). Light and electron microscopy of plant virus inclusions. Florida Agricultural Experiment Station Monograph **9**, p. 150. Gainesville, Florida.

Clark, M. F. and Adams, A. N. (1977). Characteristics of the microplate method of enzyme-linked immunosorbent assay for the detection of plant viruses. *J Gen Virol* **34**, 475–83.

Crowle, A. J. (1973). *Immunodiffusion*. Academic Press: New York.

Derrick, K. S. (1973). Quantitative assay for plant viruses using serologically specific electron microscopy. *Virol* **56**, 652–3.

Edwardson, J. R. (1974). Some properties of the potato virus Y group. *Fla Agric Exp Stn Monogr Ser* **4**, 1–398.

Edwardson, J. R. and Christie, R. G. (1978). Use of virus-induced inclusions in classification and diagnosis. *Ann Rev Phyto* **16**, 31–55.

Fraser, L. and Matthews, R. E. F. (1979). Strain-specific pathways of cytological change in individual Chinese cabbage protoplasts infected with turnip yellow mosaic virus. *J Gen Virol* **45**, 623–30.

Gibbs, A. J. and Harrison, B. D. (1976). *Plant virology*. Edward Arnold:

London.

Ginoza, W. and Atkinson, D. E. (1955). Comparison of some physical and chemical properties of eight strains of tobacco mosaic virus. *Virol* **1**, 253–60.

Gould, A. R. and Symons, R. H. (1983). A molecular biological approach to relationships among viruses. *Ann Rev Phyto* **21**, 179–99.

Govier, D. A. and Woods, R. D. (1971). Changes induced by magnesium ions in the morphology of some plant viruses with filamentous particles. *J Gen Virol* **13**, 127–32.

Hamilton, R. I., Edwardson, J. R., Francki, R. I. B., Hsu. H. T., Hull, R., Koenig, R. and Milne, R. G. (1981). Guidelines for the identification and characterisation of plant viruses. *J Gen Virol* **54**, 223–41.

Hitchborn, J. H. and Hills, G. J. (1965). The use of negative staining in the electron microscopic examination of plant viruses in crude extracts. *Virol* **27**, 528–40.

Hollings, M. (1983). Virus diseases. In *Plant Pathologist's pocketbook* (ed. Johnston, A. and Booth, C.), Commonwealth Agricultural Bureaux: Slough. pp. 46–77.

Hull, R. (1979). The DNA of plant DNA viruses. In *Nucleic acids in plants* (ed. Hall, T. C. and Davies, J. W.), CRC Press: Florida.

Ie, T. S. (1970). Tomato spotted wilt virus CMI/AAB Descriptions of plant viruses No. 39.

Koenig, R. and Paul, H. L. (1983). Detection and differentiation of plant viruses by various ELISA procedures. *Acta Hort* **127**, 147–58.

Koenig, R., Fribourg, C. E. and Jones, R. A. C. (1979). Symptomatological, serological and electrophoretic diversity of isolates of Andean potato latent virus from different regions of the Andes. *Phyt* **69**, 748–52.

Kurstak, E. (1981). *Handbook of plant virus infections*. Elsevier/North Holland: London.

Kurstak, E. Tyssen, P., and Kurstak, C. (1977). Immunoperoxidase technique in diagnostic virology and research: Principles and applications. In *Comparative diagnosis of viral diseases* (ed. Kurstak, E. and Kurstak, C.). Academic Press: New York. pp. 404–48.

Laemmli, U. K. (1970). Cleavage of structural proteins during the assembly of the head of bacteriophage T4. *Nature* **227**, 680–5.

Lane, L. C. (1979). The nucleic acids of multipartite, defective and satellite plant viruses. In *Nucleic acids in plants*, Vol 2 (ed. Hall, T. C. and Davies, J. W.), pp. 65–110. CRC Press: Florida.

Lesemann, D. E. (1983). Advances in virus identification using immunosorbent electron microscopy. *Acta Hort* **127**, 159–74.

Loening, U. E. (1969). The determination of the molecular weight of ribonucleic acid by polyacrylamide gel electrophoresis. The effects of changes in conformation. *Biochem J* **113**, 131–8.

Luisoni, E., Boccardo, G., Milne, R. G. and Conti, M. (1979). Purification, serology and nucleic acid of oat sterile dwarf virus subvirus particles. *J Gen Virol* **45**, 651–8.

McKinney, H. H. (1929). Mosaic diseases in the Canary Islands, West Africa and Gibraltar. *J Ag Res* **39**, 557–78.

Maizel, J. V. (1971). Polyacrylamide gel electrophoresis of viral proteins.

Methods in Virology **5**, 180–246.

Mancini, G., Carbonara, A. O and Heremans, J. F. (1965). Immunochemical quantitation of antigens by single radio immunodiffusion methods for the immunochemical quantitation of antigens. *Immunochem* **2**, 235–54.

Markham , R. (1967). The ultracentrifuge. *Methods in Virology* **2**, 3–39.

Matthews, R. E. F. (1981). *Plant virology*. Academic Press: New York.

Maule, A. J., Hull, R. and Donson, J. (1983). The application of spot hybridization to the detection of DNA and RNA viruses in plant tissues. *J Virol Methods* **6**, 215–24.

Milne, R. G., and Lesemann, D. E. (1978). An immunoelectron microscope investigation of oat sterile dwarf and related viruses. *Virol* **90**, 299–304.

Milne, R. G. and Luisoni, E. (1975). Rapid high resolution immune electron microscopy of plant viruses. *Virol* **68**, 270–4.

Murant, A. F., Mayo, M. A., Harrison, B. D., and Goold, R. A. (1972). Properties of virus and RNA components of raspberry ringspot virus. *J Gen Virol* **16**, 327–38.

Morris, T. J. and Dodds, J. A. (1979). Isolation and analysis of double-stranded RNA from virus-infected plant and fungal tissue. *Phyt* **69**, 854–8.

Noel, M. C., Kerlan, C., Garnier, M. and Dunez, J. (1978). Possible use of immune electron microscopy (IEM) for the detection of plum pox virus in fruit trees. *Ann Phytopathol* **10**, 381–6.

Noordam, D. (1973). Identification of plant viruses: Methods and experiments. *Cen Agric Pub Doc Wagenigen.*

Ouchterlony, O. (1968). *Handbook of Immunodiffusion and immunoelectrophoresis.* Ann Arbor Scientific Publications: Michigan.

Ouchterlony, O. and Nilsson, L. A. (1978). Immunodiffusion and immunoelectrophoresis. In *Handbook of experimental immunology* (ed. Weir, D. M.), Blackwell: Oxford.

Oudin, J. (1952). Specific precipitation in gels and its application to immunochemical analysis. *Methods Med Res* **5**, 335–78.

Owens, R. A. and Diener, T. O. (1981). Sensitive and rapid diagnosis of potato spindle tuber viroid disease by nucleic acid hybridization. *Sci* **213**, 670–71.

Palukaitis, P. and Symons, R. H. (1978). Synthesis and characterization of a complementary DNA probe for chrysanthemum stunt viroid. *F.E.B.S. Lett.* **92**, 268–72.

Peacock, A. C. and Dingman, C. W. (1967). Resolution of multiple ribonucleic Acid species by polyacrylamide gel electrophoresis. *Biochemistry* **6**, 1818–27.

Purcifull, D. E. and Gooding, G. V. (1970). Immunodiffusion tests for potato Y and tobacco etch viruses. *Phyt* **60**, 1036–9.

Purcifull, D. E. and Shepherd, R. J. (1964). Preparation of the protein fragments of several rod-shaped plant viruses and their use in agar-gel diffusion tests. *Phyt* **54**, 1102–8.

Roberts, I. M., Milne, R. G. and Van Regenmortel, M. H. V. (1982). Suggested terminology for virus/antibody interactions observed by electron microscopy. *Intervirol* **18**, 147–49.

Salaman, R. N. (1933). Protective inoculation against a plant virus. *Nature* **131**, 468.

Schumaker, V., and Rees, A. (1972). Preparative centrifugation in virus research. In *Principles and Techniques in Plant Virology* (ed. Kado, C. I. and Agrawal, H. O.), Van Nostrand Reinhold: New York.

Shapiro, A. L., Vinuela, E. and Maizel, J. V. (1967). Molecular weight estimation of polypeptide chains by electrophoresis in SDS-polyacrylamide gels. *Biochem Biophys Res Commun* **28**, 815–20.

Shepherd, R. J., Bruening, G. E. and Wakeman, R. J. (1970). Double stranded DNA from cauliflower mosaic virus. *Virol* **41**, 339–49.

Smith, K. M. (1972). *A textbook of plant virus diseases.* Longman: London.

Tomlinson, J. A. (1964). Purification and properties of lettuce mosaic virus. *Ann Appl Biol* **53**, 95–102.

Tomlinson, J. A. and Walkey, D. G. A. (1967). Effects of ultrasonic treatment on turnip mosaic virus and potato virus X. *Virol* **32**, 267–78.

Tomlinson, J. A., Faithfull, E. M., Webb, M. J. W., Fraser R. S. S. and Seeley, N. D. (1983). Chenopodium necrosis; a distinct strain of tobacco necrosis virus isolated from river water. *Ann Appl Biol* **102**, 135–47.

Trautman, R. and Hamilton M. G. (1972). Analytical ultracentrifugation. In *Principles and techniques in plant virology* (ed. Kado, C. I. and Agrawal, H. O.), pp. 491–530. Van Nostrand Reinhold: New York.

Van Regenmortel, M. H. V. (1982). *Serology and immunochemistry of plant viruses*, Academic Press: New York.

Van Slogteren, D. H. M. (1955). Serological micro-reactions with plant viruses under paraffin oil. *Proc. conf. potato virus diseases*, 1954, pp. 51–4.

Voller, A., and Bidwell, D. E. (1977). Enzyme immunoassays and their potential in diagnostic virology. In *Comparative diagnosis of viral diseases* (ed. Kurstak, E. and Kurstak, C.), Academic Press: New York. pp. 449–57.

Voller, A., Bidwell D. E., Clark, M. F. and Adams, A. N. (1976). The detection of viruses by enzyme-linked immunosorbent assay (ELISA). *J Gen Virol* **33**, 165–7.

Walkey, D. G. A. and Webb, M. J. W. (1968). Virus in plant apical meristems. *J Gen Virol* **3**, 311–13.

Walkey, D. G. A. and Webb, M. J. W. (1970). Tubular inclusion bodies in plants infected with viruses of the NEPO type. *J Gen Virol* **7**, 159–66.

Walkey, D. G. A., and Webb, M. J. W. (1984). The use of a simple electron microscope serology procedure to observe relationships of seven potyviruses. *Phytopathol Z* **110**, 319–27.

Walkey, D. G. A., Stace-Smith, R. and Tremaine, J. H. (1973). Serological, physical and chemical properties of strains of cherry leaf roll virus. *Phyt* **63**, 566–71.

Whitcomb, R. F. and Black, L. M. (1961). A precipitin ring tube test for estimation of relative soluble-antigen concentrations. *Virol* **15**, 508–9.

Williams, R. C. and Wycoff, R. G. W. (1944). The thickness of electron microscopic objects. *J Appl Phys* **15**, 712–16.

Wrigley, N. C. (1968). The lattice spacing of crystalline catalase as an internal standard length in electron microscopy. *J Ultrastruct Res* **24**, 454–64.

Yarwood, C. E. (1955). Mechanical transmission of apple mosaic virus. *Hilgardia*, **23**, 613–28.

Zaitlin, M. (1979). The RNA's of monopartite viruses. In *Nucleic acids in plants*

(ed. Hall, T. C. and Davies, J. W.), CRC Press: Florida. pp. 31–64.
Zaitlin, M. and Israel, H. W. (1975). Tobacco mosaic virus CMI/AAB. Descriptions of plant viruses No. 151.

6.13 Further selected reading

Hamilton, R. I., Edwardson, J. R., Francki, R. I. B., Hsu, H. T., Hull, R., Koenig, R. and Milne, R. G. (1981). Guidelines for the identification and characterisation of plant viruses. *J Gen Virol* **54**, 223–41.
Hollings, M. (1983). Virus diseases. In *Plant pathologist's pocketbook* (ed. Johnson, A. and Booth, C.), Commonwealth Agricultural Bureaux: Slough. pp. 46–77.
Keonig, R. and Paul, H. L. (1983). Detection and differentiation of plant viruses by various ELISA procedures. *Acta Hort* **127**, 147–58.
Kurstak, E. (1981). *Handbook of plant virus infections*, Elsevier/North Holland: London.
Lesemann, D. E. (1983). Advances in virus identification using immuno-sorbent electron microscopy. *Acta Hort* **127**, 159–74.
Matthews, R. E. F. (1981). *Plant virology*. Academic Press: New York.
Smith, K. M. (1972). *A textbook of plant virus diseases*. Longman: London.
Van Regenmortel, M. H. V. (1982). *Serology and immunochemistry of plant viruses*. Academic Press: New York.

7 *Virus Transmission by Biological Means*

7.1 Introduction

In order to survive under natural conditions plant viruses, being obligate parasites, must be spread from time to time from one susceptible host to another. If the virus is infecting an annual or short-lived plant the transmission must be frequent, but if the virus infects a tree or other long-lived plant, then less frequent transmissions are necessary.

Plant viruses are unable to penetrate the cuticle of their host and establish infection by their own processes, and infection can only be initiated by the virus entering the tissues through a wound (*see* Section 4.2). For some viruses this process is achieved through another organism, which carries the virus from an infected to a healthy plant. The organism carrying the virus is referred to as a *vector*. For other viruses the entry process is avoided altogether when a virus is seed transmitted, or if infected vegetative propagules are taken from an infected parent plant.

The experimental mechanical transmission of viruses using sap from an infected plant, has already been discussed in Chapter 4, and in this chapter only the various biological or natural means of mechanical transmission are considered. It should be mentioned, however, that many of the other biological methods of transmission covered in this chapter, are frequently used in laboratory studies.

Some plant viruses have only one normal method of natural transmission, but many have more than one. Either one or more methods may be important in the epidemiology of any individual virus. A complete understanding of the mode of transmission of a virus in its various hosts, is essential for the experimental study of the virus and the development of methods for its eventual control.

7.2 Mechanical Transmission in the Field

Transmission of viruses in the field by natural mechanical damage to the plant tissues is relatively rare, and probably of very minor economic importance. It mainly occurs with very stable viruses that multiply to high concentrations in the host plants. Potato virus X may be transmitted from infected to healthy potato plants when their leaves rub together in wind (Loughnane and Murphy, 1938) and through root contact (Roberts, 1946). Similarly, glasshouse soil, contaminated with debris from tobacco mosaic virus infected tomato plants, may cause infection in young tomato seedlings (Broadbent, 1976). In this case, infection probably occurs as a result of virus entering the root cells through abrasion of the tissues, as the roots grow through the soil.

A more common means of mechanical transmission in the field is through normal horticultural practices. Tobacco mosaic virus may be transmitted in tomato and tobacco crops by contaminated hands, clothing and tools, and many other viruses may be transmitted by unsterilized tools during pruning procedures and when cuttings are taken. The importance of using sterilized tools and clean hands for taking cuttings was clearly demonstrated in experiments with pink (*Dianthus allwoodii*) cuttings (Abdul Magid, 1981). Up to 36% infection with carnation ringspot virus occurred, if a sterilized knife was passed once through an infected shoot prior to it being used to remove a cutting from a healthy plant, but all cuttings were healthy, if the sterilized blade was used directly. The results of these experiments with carnation ringspot and other carnation viruses are shown in Table 7.1.

Table 7.1 Effect of unsterilized knives and hands on the virus status of pink (*Dianthus allwoodii*) cuttings

	Incidence of transmission in cuttings taken by		
Virus	*Unsterilized knife*	*Flame sterilized knife*	*Unsterilized hands*
Carnation etched ring	2/36* (6)†	0/36 (0)	0/38 (0)
Carnation latent	7/36 (19)	0/36 (0)	3/38 (8)
Carnation ringspot	13/36 (36)	0/36 (0)	5/38 (13)
Carnation vein mottle	4/36 (11)	0/36 (0)	2/28 (5)

*Number of plants infected/number of cuttings taken.
†Percentage infection (data from Abdul Magid, 1981).

7.3 Transmission by Grafting

Grafting is an ancient horticultural practice in which a union is established between the cut tissues of two different plants. There are many different ways in which the graft may be established (Garner, 1958; Bos, 1967) and one of the most common is the union between the shoot portion of one plant, referred to as the *scion* and the root-bearing portion of another, called the *stock* (Figure 7.1*a*). If either the scion or the stock is infected, the virus will probably pass into the healthy portion and establish infection.

Grafting has been widely used in plant virology, especially in the early years of the science to carry out experimental laboratory transmissions. This method was particularly useful for viruses that could not be mechanically sap transmitted and for which no other natural method of transmission was known. Grafting is not used as

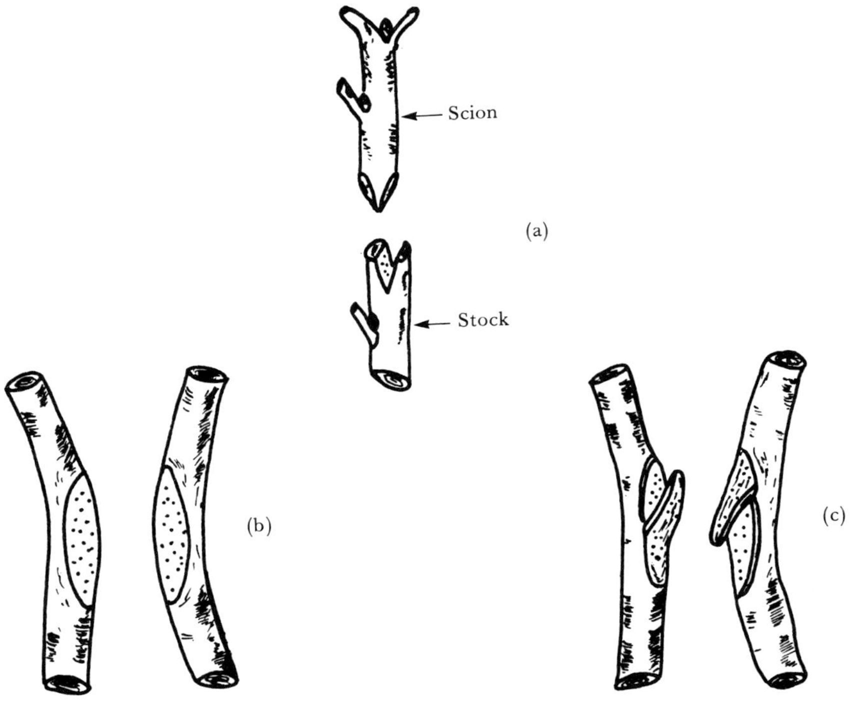

Fig. 7.1 Techniques for graft transmission. (*a*) *Wedge graft*; (*b*) *sliced approach graft* in which two components have been sliced to expose the cambium; (*c*) *tongued approach graft* in which surface contact is increased by an additional cut downwards on the one component and upwards on the other.

frequently today by plant virologists, but is still useful when studying some viruses that infect woody species.

As early as the seventeenth century Dutch tulip growers recognized that desirable flower break symptoms (*see* Plate 1.1), later shown to be caused by virus infection, could be passed from one tulip to another by grafting bulbs together. Similarly, many other horticulturalists have inadvertently transmitted viruses by grafting infected scions or root-stocks. Consequently, widespread infection has occurred in many economically important fruit crops such as apple, pear, cherry, plum, grapevine and citrus, as well as in many ornamental shrubs such as roses.

Virus transmission by grafting occurs most readily when a good union is established between the cambial cells of the scion and stock, and this is best achieved if the scion and stock are of closely related species. A good graft union is not essential however, for virus transmission can also occur if dissimilar species, such as *Chenopodium amaranticolor* and grapevine, are united by approach grafts (Figure 7.1*b* and *c*) in which only callus is produced at the grafted surfaces (Cadman *et al.*, 1960). The time required for virus to establish itself in healthy tissues following a successful graft, may vary from several days to months. Symptoms appear rapidly in herbaceous plants, and in many woody plants a dormancy period may be necessary before leaf symptoms appear.

Graft transmission of viruses in nature is probably uncommon, but may occur through chance grafting of roots as they grow together. Apple mosaic virus has been reported to be transmitted in this way in apple trees (Hunter *et al.*, 1958).

7.4 Insect Transmission

7.4.1 Introduction

Insects are by far the most important group of plant virus vectors, both in terms of the number of viruses transmitted and in the economic importance of the diseases concerned. Of 381 species of animals reported to transmit plant viruses, approximately 94% belong to the phylum *Arthropoda* and 6% to the phylum *Nematoda*, and of the arthropod vectors approximately 99% are insects (Harris, 1981).

Over 70% of all insect vectors of plant viruses belong to the order *Homoptera* and the aphids (family *Aphididae*) are the most important vectors of this group. The leafhoppers (*Cicadellidae*), plant hoppers (*Delphacidae*) and tree hoppers (*Membracidae*) are also important vectors and other vectors in the order, include whiteflies (*Aleyrodidae*) and mealy bugs (*Pseudococcidae*). Although there are examples of plant virus vectors in various other insect groups, only the beetles (*Coleoptera*) and thrips (*Thysanoptera*) are of significant importance.

In other arthropoda groups, only the mites (order *Acarina* – family *Eriophyidae*) belonging to the class *Arachnida* are important plant virus vectors. A comprehensive review of arthropod vectors of plant viruses has been carried out by Harris (1981).

Usually viruses which are transmitted by vectors in one of these major taxonomic groups are not transmitted by vectors from another group, but exceptions to this rule do occur. *Tobacco ringspot* virus for instance, has been reported to be transmitted by thrips (Messieha, 1969) and spider mites (Thomas, 1969), as well as by its usual nematode vector. The question of vector specificity is discussed in greater detail in the following sections.

7.4.2 *Aphid transmission*

(a) Basic characteristics of aphid transmission

Aphid transmission of plant viruses may be conveniently divided into three basic types, *non-persistent*, *semi-persistent* and *persistent*. Some virologists prefer to refer to viruses that are transmitted in a non-persistent manner as being *stylet-borne* and those transmitted in a persistent way as *circulative*. Both sets of terms are frequently encountered in the literature.

Non-persistent viruses

Viruses which are transmitted in a non-persistent manner are of considerable economic importance (*see* Table 7.2) and are far more numerous than those transmitted by aphids in a semi-persistent, or persistent way. Non-persistent transmission is characterized by the following features:

(*a*) The virus is acquired by the insect after feeding on the infected plant for a very short time (referred to as the *acquisition feeding time*), often only a few minutes or seconds.

(*b*) The virus is transmitted immediately the insect transfers from the infected to a healthy plant and inserts its stylets (referred to as the *inoculation* or *test feeding period*).

(*c*) The insect rapidly (usually within four hours) loses the ability to transmit the virus after leaving the infected plant.

(*d*) Non-persistent viruses are carried on or near the mouthparts of the insect and do not multiply within the insect.

Although it is generally accepted that non-persistent transmission is essentially a passive process, in which the virus is carried as a contaminant in or on the insect's mouthparts, the mechanism of transmission is still not fully explained despite much experimentation. Other experiments have shown that the efficiency of transmission of non-persistent viruses by aphids, is increased, if the aphids are starved for a time before the acquisition feeding period on the virus infected plant (Watson, 1972). Fasting causes the aphid to make a number of brief probes into the leaf, rather than one longer feeding probe, which is more typical of an aphid that has recently been fed. Experiments have shown that non-persistent viruses are more readily acquired by the aphid, during brief probes than during longer feeds. Because non-persistent viruses can be acquired and transmitted by the aphid vector during feeding probes that can be as short as ten seconds, it is thought that the virus is taken from and inoculated into, the epidermal cells of the leaf. The presence of these viruses in such relatively superficial tissues, probably correlates with the fact that non-persistent viruses are usually readily sap transmissible by mechanical inoculation.

Persistent viruses

Viruses which are transmitted in persistent or circulative manner (*see* Table 7.2) have the following characteristics:

(*a*) A long acquisition feeding time. Although some aphids may be able to transmit a persistent virus after feeding for as little as twenty minutes on an infected plant (Watson, 1972), transmission is much more efficient if the acquisition feeding time is between six and twenty-four hours.

(*b*) A latent period, which may be twelve hours or more, is usually required following the time that the insect starts feeding on the infected leaf, before it is able to transmit the virus to a healthy plant.

(*c*) Having acquired the virus the insect retains the ability to transmit it for at least a week, but frequently much longer, and sometimes it is able to transmit the virus for the remainder of its life.

(*d*) The virus is retained through the moult of the insect a feature which is called *transstadial* transmission.

Persistent or circulative viruses can be divided into two categories, those such as barley yellow dwarf virus which do not multiply within their vector (Paliwal and Sinha, 1970), and those which do, such as lettuce necrotic yellows (O'Loughlin and Chambers, 1967) and sowthistle yellow vein (Sylvester and Richardson, 1970) viruses. The circulation of this type of virus within its vector, is thought to be through the gut wall to the haemolymph, and then to the salivary glands, from which it is transmitted in the saliva to the healthy plant as the vector feeds.

Persistent viruses that multiply within their vector are called

Table 7.2 Examples of some aphid transmitted viruses

Virus	*Vector*	*Type of transmission*	*CMI/AAB No.*
Alfalfa mosaic	Various spp.	non-persistent	229
Bean common mosaic	*Acyrthosiphon pisum**	non-persistent	73
Bean yellow mosaic	*A. pisum**	non-persistent	40
Beet mosaic	*Myzus persicae**	non-persistent	53
Citrus tristeza	*Toxoptera citricida**	non-persistent†	33
Cucumber mosaic	Various sp.	non-persistent	213
Lettuce mosaic	*M. persicae**	non-persistent	9
Potato virus Y	*M. persicae**	non-persistent	242
Soybean mosaic	*A. pisum**	non-persistent	93
Sugarcane mosaic	*Dactynotus ambrosiae**	non-persistent	88
Turnip mosaic	*M. persicae**	non-persistent	8
Barley yellow dwarf	*A. dirhodum**	persistent	32
Beet western yellows	*M. persicae**	persistent	89
Lettuce necrotic yellows	*Hyperomyzus lactucae*	persistent	26
Pea enation mosaic	*A. pisum**	persistent	257
Sowthistle yellow vein	*H. lactucae*	persistent‡	62
Beet yellows	*M. persicae**	semi-persistent	13
Parsnip yellow fleck	*Cavariella aegopodii*	semi-persistent	129

*One of several important aphid vectors of this virus.
†May also be circulative.
‡Virus multiplies in vector.

propagative viruses and are sometimes transmitted through the eggs of the infected vector to its progeny. This type of virus movement is called *transovarial* transmission and has been shown to occur with sowthistle yellow vein virus in its vector *Hyperomyzus lactucae* (Sylvester, 1969).

In contrast to non-persistent viruses in which one virus may frequently be transmitted by more than one aphid species, persistent viruses usually show a high level of specificity in their vector relationship. Also the efficiency of transmission of persistent viruses, is not increased by fasting the aphid prior to the acquisition feeding period. Persistent viruses are usually located in the host plant in, or close to the phloem cells. Consequently, some persistent viruses, such as sowthistle yellow vein virus are not sap transmissible (Peters, 1971), because the virus is not readily available when the leaf is homogenized. In contrast, others such as lettuce necrotic yellows virus, are sap transmitted (Francki and Randles, 1970).

Semi-persistent viruses

Some viruses, such as beet yellows and parsnip yellow fleck viruses, have transmission properties which are intermediary between non-persistent and persistent viruses. Basically these viruses are non-persistent in the sense that they do not circulate within their vector (Harris, 1981), but their vector retains the ability to transmit them for as long as three to four days. The virus may be acquired by the vector in as little as thirty minutes, but transmission is usually more efficient if the acquisition feeding time is several hours. In common with persistent viruses, semi-persistent viruses are usually associated with phloem cells, so that aphids have to probe into deeper tissues to acquire and inoculate the virus. Starving the vector before an acquisition feeding period does not increase the transmission efficiency of this type of virus, and these viruses again show greater vector specificity than non-persistent viruses.

The terms non-persistent, semi-persistent and persistent have been evolved as a result of studies relating to aphid transmission, and are therefore, not always directly applicable to transmission by other insect groups. Nevertheless, these terms may be encountered in literature relating to virus transmission by vectors other than aphids.

(b) Helper viruses

A number of examples now exist of non-persistent, semi-persistent and persistent viruses that can only be transmitted by aphids, if the source plant on which the vector feeds is infected with a second virus. In this type of *dependent* transmission, the second virus is referred to as the *helper* virus. Kassanis (1961) for example, showed that the aphid *Myzus persicae* could transmit potato aucuba mosaic virus only if the source

Table 7.3 Examples of aphid transmitted viruses that are dependent on a second (helper) virus for transmission

Virus	*Helper virus*	*Vector*	*Type of transmission*
Potato aucuba mosaic	Potato virus A or Y	*Myzus persicae*	Non-persistent
Potato virus C	Potato virus Y	*M. persicae*	Non-persistent
Tobacco etch (NAT strain)	Potato virus Y	*M. persicae*	Non-persistent
Parsnip yellow fleck	Anthriscus yellows	*Cavariella aegopodii*	Semi-persistent
Barley yellow dwarf (MAV strain)	Barley yellow dwarf (RPV strain)	*Rhopalosiphum padi*	Persistent
Carrot mottle	Carrot red-leaf	*C. aegopodii*	Persistent
Tobacco mottle	Tobacco vein-distorting	*M. persicae*	Persistent

plant was also infected with potato virus A. Other examples of helper virus transmission are shown in Table 7.3.

In the case of non-persistent viruses that are dependent upon a helper virus for transmission, it has been suggested that a helper component is produced in the infected source plant as a result of the joint infection, and that this component is essential for the transmission of the first virus (Govier and Kassanis, 1974).

The mechanism of dependent transmission appears to be different with other viruses, however, for studies with strain mixtures of the persistently transmitted barley yellow dwarf virus (Rochow, 1977) have shown that two strains must be inoculated together for infection to occur.

Joint inoculation of a plant with the RPV and MAV strains of the virus, results in the encapsidation of the nucleic acid of the MAV strain by the protein of the RPV strain, during simultaneous replication of the two virus strains. This process is referred to as *transcapsidation*, and in the case of barley yellow dwarf virus, results in the transmission of the MAV strain by an aphid, which will normally only transmit the RPV strain of the virus.

(c) Aphid ecology

The importance of the vector's life-cycle in determining the epidemiology of a virus cannot be over emphasized, particularly when considering control strategies. Many insect vectors, especially aphids, have very complex life-cycles and although it is beyond the scope of this book to cover the subject in detail, the importance of this aspect of virus transmission may be illustrated by the life-cycle of the peach-potato aphid, *Myzus persicae*.

Aphids frequently show a well-defined alternating generation of asexual and sexual forms, each adapted for a particular part of the life-cycle as is the case with *M.persicae* (Figure 7.2). The plant on which the sexual forms mate and lay the eggs to overwinter is called the *primary host* and is usually a tree or shrub (*Prunus* sp. in the case of *M.persicae*) and that on which the asexual generations reproduce, is called the *secondary host*. The secondary hosts are frequently herbaceous agricultural crops (a varied range of host species in the case of *M.persicae*) and asexual reproduction on these hosts is rapid and very efficient, the young being born *viviparously* (alive and active).

In the cooler, temperate regions, the eggs remain dormant on the primary host during the winter months and start to hatch when the young leaves develop in the spring. The eggs hatch to produce wingless (*apterous*) females called *fundatrix*, which in turn produce further wingless offspring (*nymphs*). These develop through five growth phases called *instars* and cast their integument at the end of each instar. This

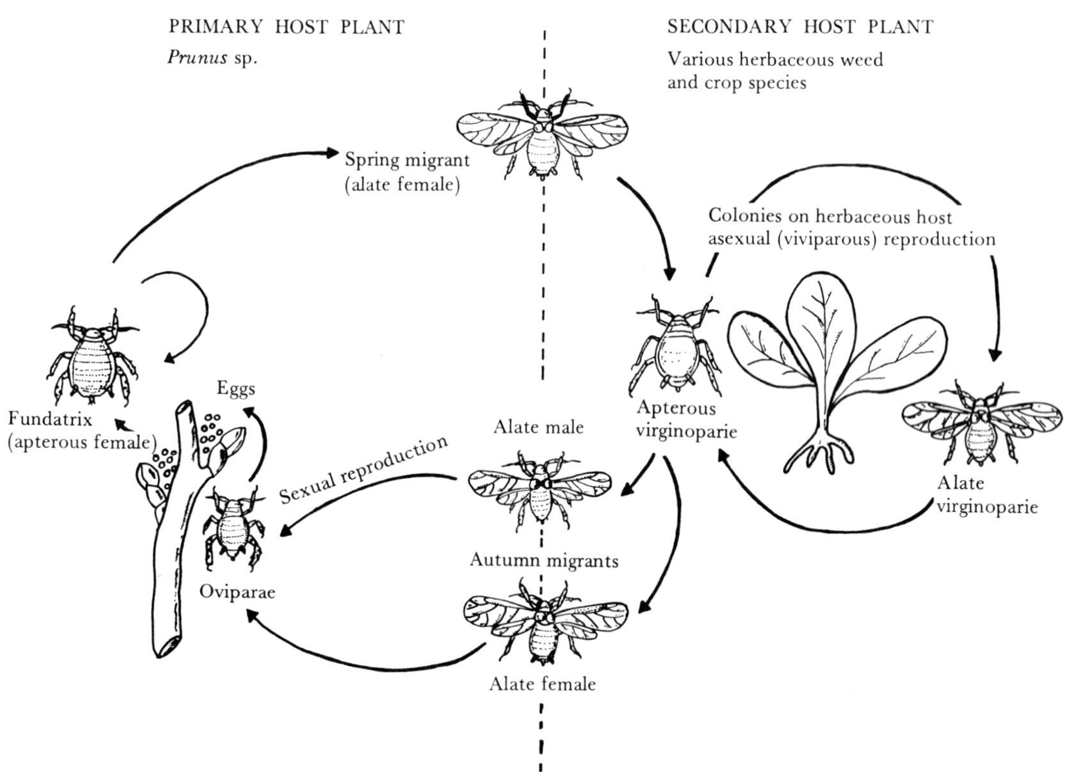

Fig. 7.2 Life-cycle of *Myzus persicae* (the peach-potato aphid).

moulting is called *ecdysis*. After one or two generations these apterous females produce offspring which develop into winged (*alate*) females called spring migrants. These migrate in the late spring or early summer from the primary to the secondary host. In tropical or arid regions eggs may be laid on the primary hosts in wooded hilly areas or near oasis, before migration to the secondary host.

The alatae colonize the secondary host and produce viviparous apterous generations called *virginoparie* and each female may produce forty to eighty nymphs, depending upon the aphid species and the environment (Watson, 1972). The nymphs rapidly develop and reproduce to form a colony. When the population becomes too large for the plant, or adverse weather conditions occur, the aphids will migrate as wingless apterae by crawling to adjacent plants or will start producing winged forms (*alate virginoparie*) which fly to other plants. It is during these movements that most viruses are spread within the crop and to new crops. When adverse weather conditions occur, in the autumn, alate male and female autumn migrants are produced which migrate back to the primary host. The female autumn migrant of *M.persicae* produces a wingless female (*the oviparae*) which lays the eggs to complete the life-cycle after mating with the alate male.

In tropical regions where day length and temperature may not stimulate the production of sexual forms, and where the alternative primary host may be absent, some aphid species maintain their colonies on secondary hosts. Similarly, during mild winters in temperate regions, viviparous females may overwinter on the secondary hosts. If this occurs, the aphid population may multiply very rapidly in the spring and cause early virus epidemics.

Aphids such as *M.persicae* which alternate between primary and secondary hosts are said to be *dioecious* and are frequently referred to as *polyphagus* because they feed on various secondary host species. In contrast, the grey cabbage aphid, *Brevicoryne brassicae*, lays its eggs and completes its life-cycle on biennial *Brassica* species, and is referred to as *monoecious* and is said to be *mono-* or *oligophagus* as it feeds on specific types of plant.

7.4.3 Leafhopper, planthopper and treehopper transmission

The leafhoppers (*Cicadellidae*) and their allies the planthoppers (-*Delphacidae*) and treehoppers (*Membracidae*) are the next most important group of insect vectors. More than thirty species of leafhopper have been reported to transmit at least thirty different viruses (Harris, 1981), and about twenty-two species of planthopper and one species of treehopper are reported to be vectors. Examples of viruses transmitted by hoppers are given in Table 7.4.

Table 7.4 Some leaf, plant and tree-hopper transmitted viruses

Virus	*Vector*	*Type of transmission*	*CMI/AAB No.*
	Leafhopper		
Beet curly top	*Circulifer tennellus*	Persistent	210
Maize chlorotic dwarf	*Graminella nigrifrons*	Semi-persistent	194
Oat blue dwarf	*Macrosteles fascifrons*	Persistent	123
Potato yellow dwarf	*Agallia constricta, Aceratagallia sanguinolenta*	Persistent	35
Rice dwarf	*Nephotettix cincticeps*	Persistent	102
Rice tungro	*N. impicticeps*	Semi-persistent	67
Wound tumour	*A. constricta*	Persistent	34
	Planthopper		
Maize mosaic	*Peregrinus maidis*	Persistent	94
Maize rough dwarf	*Laodelphax striatellus*	Persistent	72
	Treehopper		
Tomato pseudo-curly top*	*Micrutalis malleifera*	Persistent	–†

*Agent thought to be a virus.
†Not described in CMI/AAB descriptions of plant viruses.

Viruses transmitted by leafhoppers, mainly cause yellowing and leaf-rolling symptoms in the infected host plant and only a few are mechanically sap transmissible. The viruses are most concentrated in the phloem cells and their vectors feed mainly in the phloem tissues. There are no non-persistent leafhopper transmitted viruses and most, such as wound tumour virus vectored by *Agallia constricta* (Whitcomb, 1972), are transmitted in a persistent manner with the virus being circulative and frequently propagative within the vector. The virus is sucked from the infected plant into the gut of the vector, passes through the gut wall into the haemocoel, circulates to the salivary glands and is finally released and transmitted in the salivary secretion (Black, 1959).

Propagative leafhopper transmitted viruses generally have a latent period of a week or more, are retained through the moult, and the vectors frequently remain viruliferous for life. Transovarian transmission of the virus to the eggs of the vector occurs, and the virus can multiply within a viruliferous hopper even if the insect is feeding on an immune host plant. Eggs carrying viruses may overwinter, and provide a source of virus to infect spring crops, even in the absence of diseased plants. The persistence of these viruses in the hopper and their transovarian transmission, are factors that can be of considerable epidemiological importance.

In contrast to the persistent leafhopper transmitted viruses, rice tungro (transmitted by *Nephotettix cincticeps*, (Gálvez, 1971)) and maize cholorotic dwarf (transmitted by *Graminella nigrifrons*, (Gingery *et al.*, 1978) viruses are exceptional in that they behave like semi-persistent viruses. They persist for only a few days in their vectors, have no latent period and are not carried through the moult.

Many species of leafhopper may be raised in an insectary for experimental purposes, but care must be taken to ensure that the colony is initially free of viruses that may have been transmitted through the eggs. The techniques and problems associated with rearing leafhoppers have been summarized by Whitcomb (1972).

Far less research has been carried out on viruses transmitted by planthoppers, but these vectors have been shown to transmit certain reoviruses and rhabdoviruses (Harris, 1979) (*see* Table 7.4 and Chapter 2). Transmission by plant hoppers is circulative and usually propagative.

Pseudo-curly top disease of tomato is thought to be caused by a virus and is the only disease known to be transmitted by a treehopper (Simons, 1962; 1980). Studies indicate that it is circulative in its vector *Micrutalis malleifera*.

7.4.4 Whitefly transmission

Virus diseases transmitted by whitefly (Aleyrodidae) are of considerable economic importance in tropical areas, and to a lesser extent in sub-tropical and temperate regions, such as Israel and the southern states of the U.S.A. In tropical areas they transmit viruses which attack crops, that serve as major sources of protein and carbohydrate, as well as other crops of economic importance. These include cassava mosaic in Africa and India, several bean viruses throughout the tropics, tobacco leaf curl in Indonesia, Africa, India, Central and South America, and cotton leaf curl in Africa (Bird and Maramorosch, 1978).

The virus diseases transmitted by whitefly are often referred to as *rugaceous* and cause mosaic and leaf distortion symptoms in infected plants. *Bemisia tabaci* is the most important and widespread vector, and can transmit most of the known whitefly transmitted viruses that are of economic importance in the tropics. The vectors feed mainly on phloem tissues and the viruses are not usually sap transmitted by mechanical means.

Minimal acquisition feeding periods of ten to sixty minutes have been reported, but transmission efficiency increases with acquisition feeds of up to several hours (Harris, 1981). A latent period of between four and eight hours occurs with most transmissions and in general, the vector requires a longer acquisition feeding period to acquire the virus, than the feeding period required to transmit it to a healthy plant. The vectors may retain the virus for periods ranging from two to twenty-five days.

In many ways whitefly transmission resembles the persistent and circulative transmission of aphid transmitted viruses such as barley yellow dwarf virus. They do not normally multiply within their vector (i.e. they are not propagative) and transovarian transmission through the eggs has not been shown. The virus may be transmitted by both nymphs and adult whitefly. Examples of whitefly transmitted viruses are given in Table 7.5.

7.4.5 Beetle transmission

Beetles (*Coleoptera*) have been shown to transmit about forty-five different plant viruses and usually viruses that are beetle transmitted have no other vectors. At least seventy-four species of beetle have been reported to be vectors and most of these belong to the families *Chrysomelidae* and *Curculionidae* (Harris, 1981). In contrast to the sucking mouthparts of aphid and leafhopper vectors, beetles have biting mouthparts. There are four major groups of viruses, the bromoviruses, comoviruses, tymoviruses and members of the southern bean mosaic virus group (*see* Chapter 2), that are beetle transmitted (Table 7.5). Many of these viruses, especially those such as cowpea

Table 7.5 Examples of plant viruses transmitted by other insect groups

Insect group	*Virus*	*Vector*	*Nature of transmission*	*CMI/AAB No*
Whitefly	Abutilon mosaic	*Bemisia tabaci*	–†	*
	Cassava mosaic	*Bemisia tabaci*	–	90
	Cotton leaf curl	*Bemisia tabaci*	–	*
	Soybean yellow mosaic	*Bemisia tabaci*	usually persistent	*
	Sweet potato mild mottle	*Bemisia tabaci*	–	162
	Tobacco leaf curl	*Bemisia tabaci*	–	*
Beetle	Brome mosaic	*Diabrotica longicornis*	–	180
	Cowpea mosaic	*Ceratoma trifurcata*	persistent	197
	Turnip yellow mosaic	*Phyllotreta sp.*	persistent	230
	Southern bean mosaic	*C. trifurcata*	persistent	274
	Squash mosaic	*D. undecimpunctata*	persistent	43
Mealy bug	Cacao swollen shoot	*Planococcoides njalensis*	usually semi-persistent	10
	Cacao mottle leaf	*P. citri*	–	*
Thrips	Tomato spotted wilt	*Frankliniella fusca*	persistent	39

*Not described in CMI/AAB Descriptions of plant viruses.
†Information not available.

mosaic virus in the comovirus group, are widely distributed and infect economically important crops, such as bean, cowpea and soybean, in many tropical countries.

Beetle transmitted viruses are usually acquired by the vector following acquisition feeding periods of twenty-four hours or less, although some beetles have been reported to acquire virus (*acquisition threshold period*) in as little as five minutes, and in some instances a single bite on an infected leaf has made a beetle viruliferous (Fulton *et al.*, 1980). In general, however, increased acquisition feeding times result in increased transmission. No latent period for beetle transmission has been reported, and there is no experimental evidence of transovarian or transstadial transmission.

The retention of virus by beetles falls into two basic categories, some vectors remaining viruliferous for one to two days, and others for seven to twenty-one days (Walters, 1969). In the case of viruses which are retained in their vector for several days or longer, the greater the duration of the acquisition feeding period, the longer the vector remains viruliferous (Walters and Henry, 1970). The mechanism of virus transmission by beetles is not fully understood, although virus concentration is frequently high in regurgitated food (Harris, 1981). There is no evidence at present to show that virus multiplies within the vector, but southern bean mosaic virus has been shown to pass through the gut wall into the haemocoel of its vector *Ceratoma trifurcata* (Slack and Scott, 1971) and it has been reported that virus can be detected in the blood of some beetles (Fulton *et al.*, 1980). The blood could therefore, act as a reservoir for the virus, but how the virus gets to the mouthparts has yet to be determined.

7.4.6 Mealy bug transmission

Only the mealy bugs (*Pseudococcidae*) of the various *Coccoidea* families have been reported to be virus vectors, and of these, nineteen species have been reported to transmit six viruses (Harris, 1981). Of the viruses transmitted by mealy bugs, those affecting the cacao tree *Theobroma cacao* are the most important (*see* Table 7.5). *Planococcoides njalensis* and *P.citri* transmit most viruses in the group, including cacao swollen shoot virus.

The vectors feed on the phloem cells of the host plant, but are not very efficient vectors as they are not particularly mobile and rely on crawling to move from plant to plant. Virus transmission by mealy bugs has the characteristics of semi-persistent aphid transmission, except that a starvation period prior to the acquisition feeding period, increases the transmission efficiency of the vector. Acquisition feeding times of forty-eight to seventy-two hours give the best transmission, although transmission has occurred following a minimal acquisition feeding period of five to seven hours (Harris, 1981). Infection of the host plant can occur following transmission feeding periods as short as fifteen minutes, but three to four hour feeding periods increase transmission efficiency. The insect can retain the virus for a maximum of three to four days and nymphs are more effective vectors than the adults.

7.4.7 Transmission by thrips and other insects

Only tomato spotted wilt virus is reported to be transmitted by species of thrips (*Thysanoptera*). Four species of thrips belonging to the family *Thripidae*, including *Frankliniella fusca* (*see* Table 7.5), transmit the virus in a persistent and circulative manner. Transmission efficiency increases with acquisition feeding times from fifteen minutes to four days, there is a latent period of four to sixteen days and the vector can remain viruliferous for life. Only the nymphs can acquire and transmit the virus and transovarian passage of the virus has not been reported.

Plant virus transmission has also been reported by vectors in other orders of Insecta, although these vectors are insignificant compared with those already described. In the order *Diptera*, two species of leaf-miner fly belonging to the genus *Liriomyza* have been shown to transmit sowbane mosaic and tobacco mosaic viruses (Zitter and Tsai, 1977 and 1980). The mechanism of transmission by leaf-miners is not fully understood, but is thought to be non-circulative and associated with the egg laying and feeding of the adult fly. When the fly cuts the leaf epidermis with its ovipositor, plant sap exudes upon which the fly may feed, suggesting that both the ovipositor and the fly's mouthparts may become contaminated with virus, which can be transmitted to a healthy plant.

Finally two species of lace bugs belonging to the family *Piesmidae* are known to be vectors. *Piesma cinereum* is reported to transmit sugar beet savoy virus in a circulative manner (Schneider, 1964), and beet leaf curl virus is both circulative and propagative in its vector *P.quadratum* (Proeseler, 1980).

7.4.8 Mite transmission

In the class *Arachnida*, Eriophyid mites have been shown to be vectors of plant viruses. Of the three proven examples of mite transmitted viruses (*see* Table 7.6), wheat streak mosaic virus (WSMV) transmitted by *Aceria tulipae* has been the most extensively studied (Slykhuis, 1955; Orlob, 1966), WSMV has been found to occur in high concentrations in the midgut of the vector, and virus particles have been observed in the body cavity around the intestine (Takahashi and Orlob, 1969) and in the salivary glands (Paliwal, 1980). This suggests that transmission is probably circulative, but there is no evidence that the virus is propagative.

Table 7.6 Examples of mite transmitted viruses

Virus	*Vector*	*Persistent in vector (days)*	*CMI/ABB No.*
Ryegrass mosaic	*Abacarus hystrix*	1	86
Wheat spot mosaic	*Aceria tulipae*	13	–*
Wheat streak mosaic	*A. tulipae*	9	48

*Not described in CMI/AAB Descriptions of plant viruses (*see* Slykhuis, 1972).

WSMV can be acquired following an acquisition feeding time of fifteen minutes and can be transmitted following an inoculation feeding period of similar duration. It persists in the vector for up to nine days and is transstadial, but there is no evidence of transovarian transmission (Slykhuis, 1972). In contrast, ryegrass mosaic virus is retained by its vector *Abacarus hystrix* for only one day (Mulligan, 1960).

Mites are difficult to work with experimentally as they are delicate and easily dessicated. They are only 0.25 mm in length and since a ×10 hand lens is required to observe them, they are easily overlooked on an infected plant. Wind is the main means of their dispersal in nature, and because of their small size, ever a light breeze is sufficient to dislodge them and carry them away like dust particles.

7.5 Nematode Transmission

7.5.1 Introduction

Free living (ectoparasitic), soil-inhabiting eelworms belonging to the phylum *Nematoda*, represent an interesting and important group of plant virus vectors. All the nematode vectors belong to the order *Dorylaimida* and include species from the genera *Trichodorus*, *Paratrichdorus*, *Longidorus* and *Xiphinema* (*see* Table 7.7).

The viruses transmitted by eelworms are divided into two distinct groups, the *tobraviruses* and the *nepoviruses* (*see* Chapter 2). The tobraviruses include members of the tobacco rattle group and were formerly known as the *netuviruses*, a term which stood for nematode transmitted tubular-shaped viruses. Tobraviruses are transmitted by eelworms of the genera *Trichodorus* and *Paratrichodorus*. In contrast, the nepovirus group represent nematode transmitted viruses with

Table 7.7 Examples of nematode transmitted viruses

Virus group	*Virus*	*Vector*	*Particle shape*	*CMI/AAB No.*
Tobravirus	Gladiolus notch-leaf	*Paratrichodorus* and *Trichodorus* spp.	Tubular	–*
	Pea early browning	*Paratrichodorus* and *Trichodorus* spp.	Tubular	120
	Tobacco rattle	*Paratrichodorus* and *Trichodorus* spp.	Tubular	12
Nepovirus	Arabis mosaic	*Xiphinema diversicaudatum*	Isometric	16
	Artichoke Italian latent	*Longidorus apulus*	Isometric	176
	Cacao necrosis	*L. sp.*	Isometric	–
	Cherry leaf roll	*X. coxi, X. diversicaudatum*	Isometric	80
	Cherry rasp leaf	*X. americanum*	Isometric	159
	Grapevine chrome mosaic	*X. index*	Isometric	–
	Grapevine fan leaf	*X. index, X. italiae*	Isometric	28
	Mulberry ring-spot	*L. martini*	Isometric	142
	Myrobalan latent ringspot	*L. sp.*	Isometric	160
	Peach rosette mosaic	*X. americanum*	Isometric	150
	Raspberry ringspot	*L. elongatus, L. macrosoma*	Isometric	198
	Strawberry latent ringspot	*X. diversicaudatum*	Isometric	126
	Tobacco ringspot	*X. americanum*	Isometric	17
	Tomato black ring	*L. attenuatus, L. elongatus*	Isometric	38
	Tomato ring-spot	*X. americanum*	Isometric	18

*Not described in CMI/AAB descriptions of plant viruses

polyhedral shaped particles and are vectored by *Longidorus* and *Xiphinema* species.

7.5.2 Mode of transmission

Virus transmission by nematodes is believed to be of a non-circulative nature and there is no evidence that the virus multiplies within the vector. The vectors have probing mouthparts consisting of a single central stylet or spear. In *Xiphinema* and *Longidorus* species the stylet (often called the *odontostyle*) is hollow and its basal region is called the *odontophore* (*see* Figure 7.3). Muscles which connect with the oesophagus, allow the odontostyle to be thrust forward to penetrate the plant cells during feeding (*see* Plate 7.1). In *Trichodorus* species the odontostyle is a modified, curved tooth, through which the plant cell contents are sucked into the pharynx (Hooper, 1978).

A transmission mechanism similar to the ingestion–egestion mechanism suggested for the non-persistent transmission by aphids, is also proposed for virus transmission by nematodes (Harris, 1981). The virus-laden plant material is thought to be ingested by the eelworm,

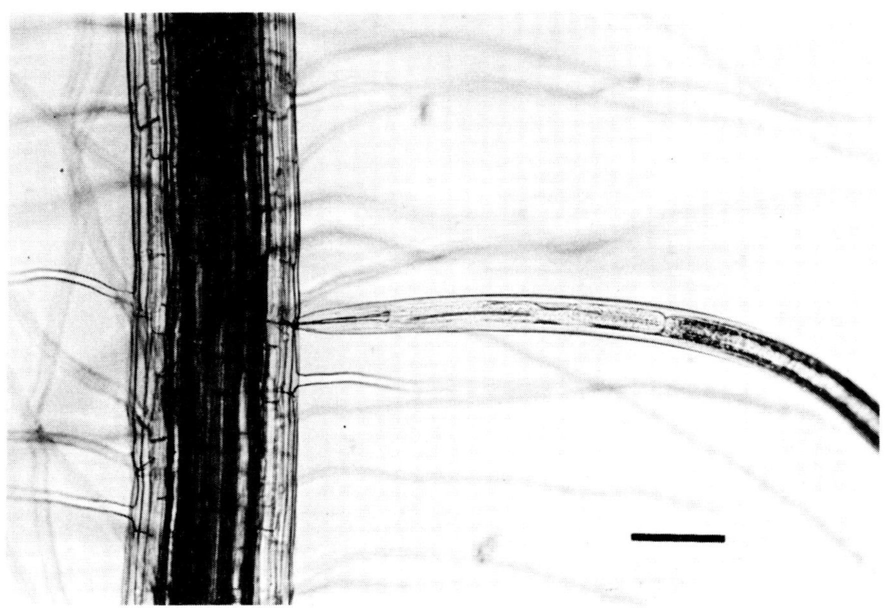

Plate 7.1 A *Xiphinema diversicaudatum* eelworm feeding on a root of ryegrass (*Lolium perenne*). Note the penetration of the root's epidermal layers by the stylet, magnification bar = 100 μm (courtesy of W. M. Robertson).

and the virus absorbed on to the internal surfaces of the anterior regions of the eelworm's digestive tract (Taylor and Robertson, 1970*b*) (*see* Figure 7.3). Virus transmission then occurs by the back flow of material, including virus released from the absorption sites, when the

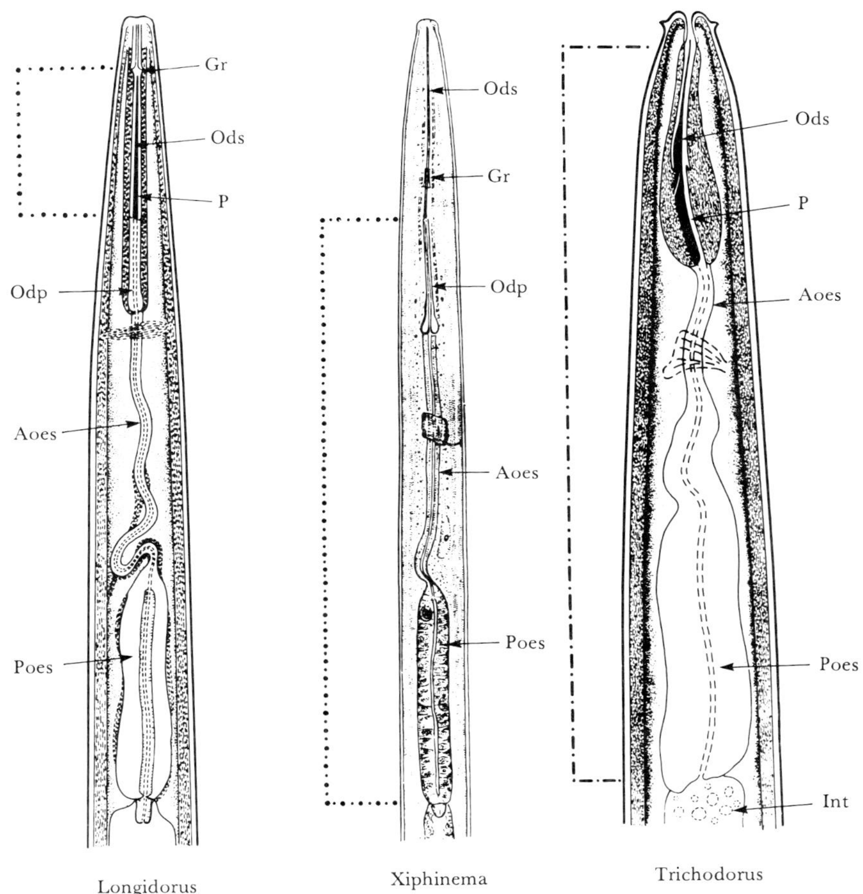

The dotted lines indicate the areas of virus retention in the gut.
Int. = intestine; Poes = posterior oesophagus; Aoes = anterior oesophagus;
P = pharynx; Odp = odontophore; Ods = odontostyle; Gr = guide ring.

Fig. 7.3 Mouthparts and anterior gut region of *Longidorus*, *Xiphinema* and *Trichodorus* species of nematode (based on Taylor and Robertson, 1969, 1970*a*, *b*). The dotted lines indicate the areas of virus retention in the gut.

Int. = intestine; Poes. = posterior oesophagus; Aoes. = anterior oesophagus; P. = pharynx; Odp. = odontophore; Ods. = odontostyle; Gr. = guide ring.

eelworm feeds on the roots of a nearby healthy plant. This back flow or egestion occurs during the initial phases of feeding, when saliva is secreted from the oesophageal bulb and passes forward through the oesophagus carrying the virus with it.

7.5.3 Tobraviruses

The most important tobravirus, tobacco rattle virus, has a wide host range and causes corky ringspot disease of potato tubers and diseases of various ornamental bulbs (Harrison, 1970). Other viruses in the group (*see* Table 7.7) include pea early browning which is locally important in England and the Netherlands (Harrison, 1973). The adult *Trichodorus* and *Paratrichodorus* eelworms are about 2 mm long and feed on young roots in the root hair region. They feed mainly on epidermal cells and the damage caused by feeding results itself in browning and stunting of the roots. Tobacco rattle virus is usually transmitted following acquisition feeds of fifteen minutes to one hour or more, but transmission efficiency increases with feeds up to forty-eight hours (Das and Raski, 1968; Ayala and Allen, 1968). There is no evidence of a latent period. The virus can persist in its vector for months or even years (Van Hoof, 1970), and is retained on the cuticular lining of the pharynx and oesophagus (Taylor and Robertson, 1970*a*).

Tobacco rattle virus can also survive at an infected field site because of its wide host range which includes many weeds such as *Stellaria media* (Noordam, 1956), which frequently overwinter. In addition, the virus can survive in infected seed of the weed *Viola tricolor* (Cooper and Harrison, 1973). Spread of virus from an infected site can occur in several ways. First, it may spread through movement of the viruliferous eelworm itself, but the distances involved are likely to be small, perhaps only a metre or so a year. Dissemination over greater distances is likely to occur, if the viruliferous eelworms are moved during soil cultivation, in wind-blown soil or in soil around the roots of transplanted plants. Secondly, establishment of infection at new sites may result from the transplanting of infected plants, or through the dispersal of infected seed, to soils where the vector is present.

7.5.4 Nepoviruses

Viruses belonging to the nepovirus group are more numerous than the tobraviruses (*see* Table 7.7) and cause important diseases of various soft and tree fruits, including fanleaf disease of grapevine.

Although the vectors of nepo- and tobra-viruses share many common features, some differences do occur. the *Xiphinema* and *Longidorus* species are between 2–12 mm in length and have longer odontostyles than the *Trichodorus* vectors. The longer feeding parts allow them to penetrate further into the plant tissues and to reach the vascular tissues. *Xiphinema* species, such as *X.diversicaudatum* and *X.index*, feed mainly on root tips and cause small galls to form (Flegg, 1968). Other species, such as *X.americanum*, feed along the sides of young roots and cause cortical necrosis, whilst *Longidorous* species generally feed at the root tips and cause root stunting.

X.index can acquire grapevine fanleaf virus within fifteen minutes feeding on an infected plant and inoculate the virus after feeding fifteen minutes on a healthy plant (Das and Raski, 1968). *X.americanum* has been shown to transmit tomato ringspot virus following acquisition and inoculation feeding periods of one hour, although transmission efficiency increases with acquisition feeding times of twenty-four hours or more (Teliz *et al.*, 1966). In general, viruses appear to persist longer in *Xiphinema* than in *Longidorus* species (Harrison, 1977). Grapevine fanleaf virus can persist for up to eight months in *X.index* (Taylor and Raski, 1964), arabis mosaic virus for eight months in *X.diversicaudatum* and tobacco ringspot virus eleven months in *X.americanum* (Bergeson and Athow, 1963). Virus is retained on the cuticular lining of the guiding sheath in *Longidorus elongatus* (Taylor and Robertson, 1969), and on the lining of the oesophagus in *Xiphinema* spp. (Taylor and Robertson, 1970*b*).

Besides being retained in the vector for long periods, the survival of nepoviruses is also influenced by the large number of wild plants, including many hedgerow trees, that they can infect. *Sambucus nigra* (Elder) and *Prunus spinosa* (Sloe) for example, are important in the epidemiology of arabis mosaic virus (AMV) and its vector *X.diversicaudatum*. Areas of AMV-infected strawberries have been shown to be associated with root zones from adjacent hedgerows (Harrison and Winslow, 1961; Pitcher and Jha, 1961). Most nepoviruses are also transmitted in the seed of infected plants (Lister and Murant, 1967; Murant and Lister, 1967) and several nepoviruses have been shown to be pollen transmitted (*see* Table 7.9). This too may influence the epidemiology of these viruses.

The factors governing the spread of nepoviruses from one site to another are identical to those controlling the dissemination of tobraviruses, except that seed transmission is more important with nepoviruses. Some nepoviruses may also be spread long distances by certain horticultural practices. Grapevine fanleaf virus for instance, has been spread to and throughout many countries by the transportation of rooted cuttings in soil containing viruliferous *X.index*.

7.6 Fungal Transmission

7.6.1 Introduction

There are at least eleven examples of viruses that are transmitted by soil inhabiting fungi and two other fungal transmitted diseases whose agents are not yet known (*see* Table 7.8). The latter includes lettuce big vein, an important disease of lettuce in Europe and elsewhere, which in the past has been called a virus, but for which no virus particles have yet been observed.

These diseases are transmitted by vectors belonging to two groups of obligate parasites. *Olpidium* spp. (*Chytridales*), transmit tobacco necrosis (TNV), satellite and cucumber necrosis viruses, which have isometric particles; and *Polymyxa* and *Spongospora* spp. (*Plasmodiphorales*) transmit a number of viruses including wheat mosaic and potato mop top viruses, which have rod-shaped particles. The two diseases with unknown agents, lettuce big vein and tobacco stunt, are transmitted by *Olpidium brassicae*.

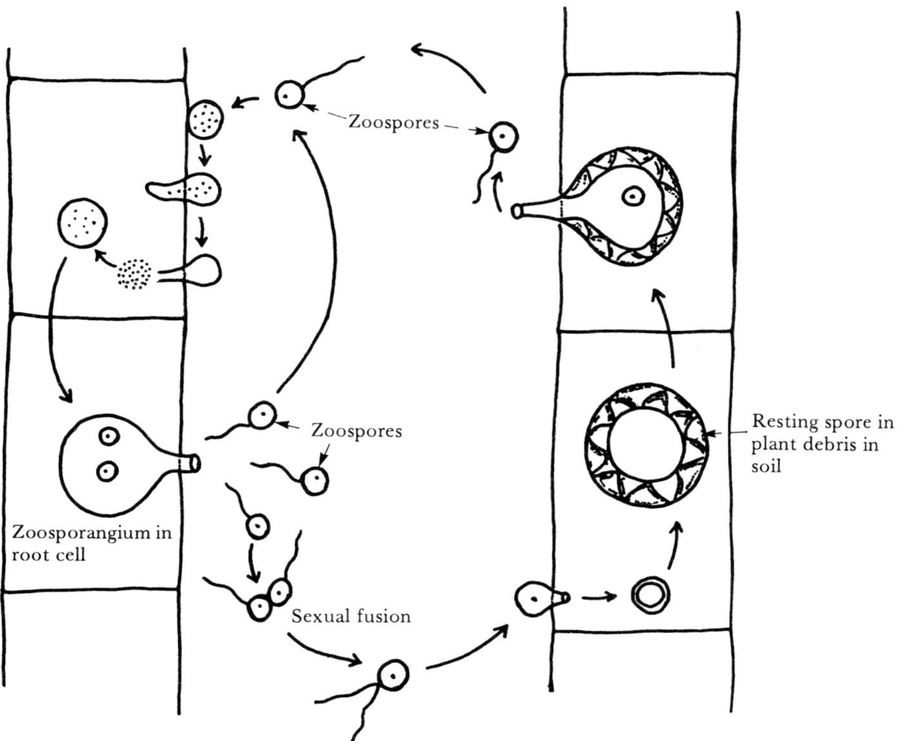

Fig. 7.4 Diagram of a generalized life-cycle of *Olpidium*.

7.6.2 Mode of transmission

To understand the mechanism of virus transmission by fungi, it is essential that the life-cycle of the vector is fully determined. To date,

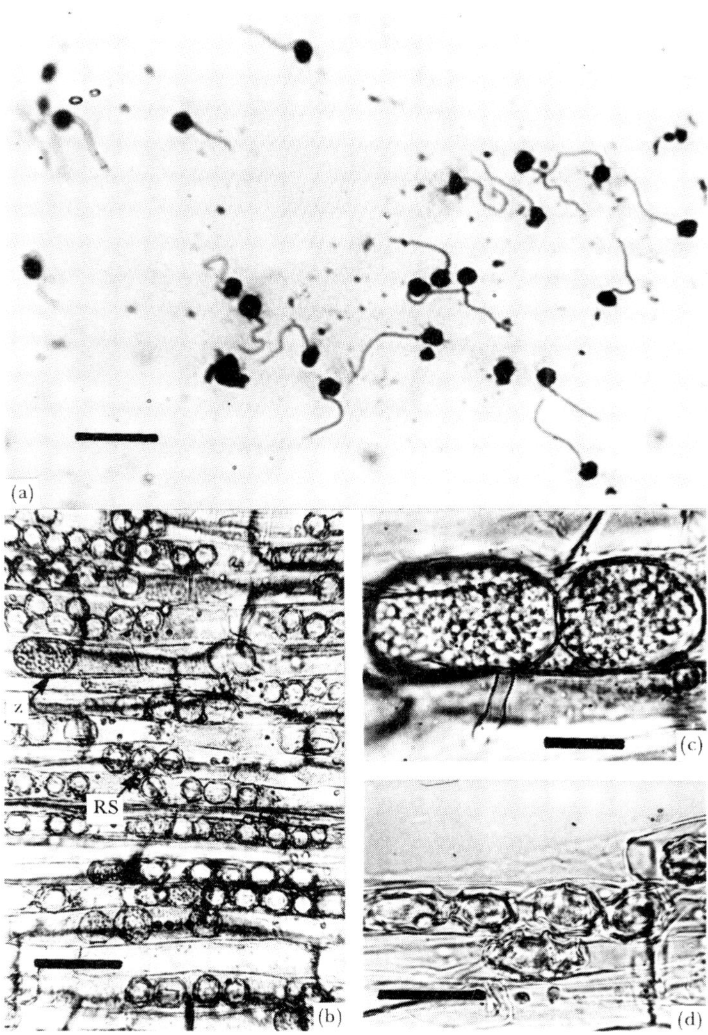

Plate 7.2 Various stages in the life-cycle of *Olpidium brassicae* isolated from lettuce. (*a*) Uniciliate zoospores, magnification bar = 5 μm; (*b*) root epidermis showing zoosporangia (Z) and resting spores (RS), magnification bar = 20 μm; (*c*) two zoosporangia, one with an exit tube, magnification bar = 40 μm; (*d*) thick-walled resting spores in root cells, magnification bar = 20 μm (courtesy of J. A. Tomlinson).

most work has been carried out on *Olpidium brassicae* and an outline of its life history is shown in Figure 7.4.

The fungus infects cells close to the root epidermis and produces spore-forming bodies called zoosporangia (*see* Plate 7.2). These produce exit tubes, through which uniciliate mobile zoospores (*see* Plate 7.2*a*) are liberated into the soil water surrounding the root. These zoospores may do one of two things. First, they may swim through the soil water and attach themselves to the surface of another root, withdraw their cilia and produce thin-walled zoospore cysts. These cysts produce an infection canal after about two hours, which penetrates the wall of the root cell, allowing protoplasm from the zoospore to enter the cell. After a further two to three days, the protoplasm from the zoospore produces a thallus which develops within the root cell to form a zoosporangium, which in turn produces more zoospores that are liberated through one or more exit tubes.

Alternatively, the liberated zoospores may fuse in pairs to form a zygote that penetrates the root cells and produces thick-walled resting sporangia (*see* Plate 7.2). The resting spores are resistant to drying, and may remain in decaying root debris for long periods before eventually germinating to produce new zoosporangia and zoospores. The other fungal vectors have similar, although not identical life-cycles.

Virus transmission by these fungi occurs in two basic ways. In the first, the virus particles are carried on the surface of the zoospore and are not transmitted in the resting spore. This group includes the viruses transmitted by *Olpidium* species: cucumber necrosis (Dias, 1970*a* and *b*), tobacco necrosis (Teakle, 1972) and TNV-satellite (Kassanis and Macfarlane, 1968). Virus particles present in the soil water become attached to the outer walls of the zoospores and cilia (Temmink *et al.*, 1970) and appear to pass with the protoplasm of the zoospore into the root cell through the zoospore infection canal. These viruses probably have only a transient association with their *Olpidium* vector and do not pass into the resting spore during its formation. It is also possible for tobacco necrosis virus to infect the root even if the *Olpidium* vector fails to multiply in the root cell (Kassanis and Macfarlane, 1964).

The viruses transmitted by Olpidium show considerable specificity, in that *O.brassicae* will not transmit cucumber necrosis virus, and *O.cucurbitacearum* will not transmit tobacco necrosis virus.

In contrast the viruses transmitted by *Polymyxa* and *Spongospora* species are transmitted through the resting spore of their vectors. Wheat mosaic, barley yellow and other allied viruses (*see* Table 7.8) are acquired by the fungus during its colonization of virus-infected roots, and not by uptake of virus particles by zoospores in suspension (Rao and Brakke, 1969; Tamada, 1975). The virus may remain viable in the resting spores for long periods and when the resting spore germinates

Table 7.8 Examples of fungal transmitted viruses

Virus	*Vector*	*Particle shape*	*CMI/AAB No.*
Cucumber necrosis	*Olpidium cucurbitacearum*	Isometric	82
Tobacco necrosis	*O. brassicae*	Isometric	14
Satellite	*O. brassicae*	Isometric	15
Barley yellow mosaic	*Polymyxa graminis*	Filamentous rod	143
Beet necrotic yellow vein	*P. betae*	Straight rod	144
Oat mosaic	*P. graminis*	Filamentous rod	145
Potato mop top	*Spongospora subterranea*	Straight rod	138
Rice necrosis mosaic	Unknown	Filamentous rod	–*
Wheat mosaic	*P. graminis*	Straight rod	77
Wheat spindle streak mosaic	*P. graminis*	Filamentous rod	167
Wheat yellow mosaic	Unknown	Filamentous rod	–
Other agents			
Lettuce big vein	*O. brassicae*	–†	–
Tobacco stunt	*O. brassicae*	–	–

*Not described in CMI/AAB descriptions of plant viruses.
†Not known.

the virus is transmitted by the zoospores to new roots. Experiments with wheat mosaic virus and *P.graminis* have shown that zoospores attach themselves to new roots within thirty minutes of being released from the fungus and that the virus may subsequently enter some root cells within four hours (Rao and Brakke, 1969).

The agents causing big-vein disease of lettuce (Campbell and Grogan, 1964) and tobacco stunt (Hidaka and Tagawa, 1962) are also carried internally in the zoospore and are transmitted in the resting spores.

7.6.3 Survival and spread

The survival of *Olpidium brassicae* and TNV is helped by both fungus and virus having wide host ranges. TNV can also survive in decaying plant debris and soil water (Harrison, 1977), while the resting spores of *O.brassicae* provide the fungus with an alternative means of survival. The virus is probably spread long distances by the transplanting of infected plant material, and short distances by movement of soil, root fragments and in drainage water.

The survival of viruses transmitted by *Polymyxa* and *Spongospora* depends to a larger extent on the persistent nature of the association between virus and vector. In general, these viruses have narrow host ranges. Barley yellow mosaic virus will not infect oats or wheat, and wheat spindle streak mosaic virus will not infect oats or barley. Their survival depends, therefore, to some extent upon cropping practices, and in Canada, wheat spindle streak mosaic virus was only found in fields that had grown winter wheat frequently (Slykhuis, 1970 and 1976). Alternatively, these viruses may survive for long periods in infected resting spores (Jones and Harrison, 1969), even if the soil is dried or stored (Slykhuis, 1970; Tamada, 1975).

Over short distances, *Polymyxa*- and *Spongospora*-transmitted viruses probably spread through the movement of zoospores or resting spores in soil water, the movement of soil during cultivation and by the movement of soil particles in the wind. In addition, potato mop top virus has been shown to be transmitted in resting spores from tubers bearing scabs of *Spongospora subterranea* (Jones and Harrison, 1969). Such tubers may be transported over long distances, perhaps the means by which the disease was brought to Europe from South and Central America (Jones and Harrison, 1972).

7.7 Transmission Through Seed and Pollen

7.7.1 *Introduction*

Virus transmission through the seed of an infected mother plant occurs in some virus/host infections, but a number of seed transmitted viruses are of considerable economic importance (Table 7.9). Of the 230 plant viruses described in the CMI/AAB list (*see* Section 12.8) by early 1982, sixty-two were reported to be seed transmitted in at least one known host plant.

The percentage of infected seed produced by an individual plant varies greatly, depending on the virus and host plant involved, and a number of other factors which are discussed later in this section. Generally, the amount of infected seed in a commercial seed lot is much lower than the percentage of infected seed originating from a single infected mother plant. This is because the infection level of a commercial seed batch is usually diluted by virus-free seed, produced from healthy plants in the same parent crop.

7.7.2 *The role of seed transmission in virus epidemiology*

Seed infection plays a major role in both the transmission and survival of a number of important virus diseases. In ecological terms, seed transmission provides an ideal starting point for the establishment of a disease in a field crop. First, it enables infection to occur at the earliest possible time in the development of the young seedling, a factor that

Table 7.9 Examples of seed and pollen transmitted viruses

Virus	*Host species*	*Percentage seed infected*	*Pollen transmission*	*CMI/AAB No.*
Transmitted on testa				
Cucumber green mottle mosaic	*Cucumis sativus*	1– 8	–	154
Tobacco mosaic	*Lycopersicon esculentum*	2– 94	–	151, 156*
Transmitted in embryo				
Alfalfa mosaic	*Medicago sativa*	10– 55	+	229
Arabis mosaic	*Chenopodium album*	80–100	–	16
	Lycopersicon esculentum	1.8		
Barley stripe mosaic	*Hordeum vulgare*	58–100	+	68
Bean common mosaic	*Phaseolus vulgaris*	18– 76	+	73
Broad bean true mosaic	*Vicia faba*	15	–	20
Cherry leaf roll	*Nicotiana rustica*	<100	+	80
Cowpea mild mottle	*Vigna unguiculata*	2– 90	–	140
Cucumber mosaic	*Stellaria media*	21– 40	–	213
Lettuce mosaic	*Lactuca sativa*	1– 14	+	9
Pea early browning	*Pisum sativum*	1– 37	–	120
Pea seed-borne	*P. sativum*	<90	–	146
Peanut clump	*Arachis hypogaea*	<20	–	235
Prunus necrotic ringspot	*Prunus sp.*	<70	+	5
Soybean mosaic	*Glycine max*	50	–	93
Squash mosaic	*Cucumis melo*	6– 20	–	43

*Tomato mosaic virus is often considered a strain of tobacco mosaic virus frequently governs the severity of virus infection in an individual plant (*see* Section 3.5). Secondly, seed infection results in individual infected seedlings being scattered widely throughout a field crop, with each infected seedling providing a virus reservoir for subsequent secondary spread (*see* Section 8.2.1).

The secondary spread of virus in the crop frequently occurs through aphid transmission, as is the case with lettuce mosaic virus (LMV) in commercial lettuce crops. Seed infection rates higher than 0.1% are likely to result in a LMV epidemic in commercial lettuce crops (Tomlinson, 1970). This can be exacerbated if successive crops are grown adjacent to one another. Consequently, most seed companies now test lettuce seed for LMV infection before packaging, and only virus-free seed should be sold.

With other viruses, seed transmission provides an ideal means by which the virus can survive the winter or other unfavourable periods. Bean common mosaic virus for instance, is seed transmitted in commercial green beans, but has few hosts other than *Phaseolus* spp. Since the bean plant is unable to overwinter in many areas where it is grown commercially, infected seed provides the only means of virus survival. The importance of virus overwintering in seed is also illustrated by viruses that infect various weed hosts. Cucumber mosaic virus, for example, has been shown to be seed transmitted in the weed *Stellaria media*, and the spring seedlings of this weed provide a reservoir of the virus which can be aphid transmitted to lettuce and other commercial crops (Tomlinson and Carter, 1970). Weed seeds, such as those of *Stellaria media*, frequently remain dormant but viable in the soil for many years, and so provide for the long-term persistence of virus during the temporary absence of a susceptible crop plant.

Seed transmission in weed species also provides a means of survival and dispersal for nepoviruses in the absence of their nematode vectors (*see* Section 7.5.4).

In addition to the importance of seed infection in the local spread of certain viruses, seed transmission is extremely important in the international transmission of plant viruses. Viruses such as bean common mosaic and pea seed-borne mosaic, have been, and are still being widely distributed to many countries through infected commercial seed. Commerical seed for use in many temperate countries, is now raised out of season in tropical and sub-tropical countries, where the climate is more favourable for seed production. Consequently, the international movement of infected seed may be widespread.

7.7.3 Modes of seed transmission

Some viruses, such as tobacco mosaic virus (TMV) in tomato seed is transmitted in or on the seed coat (testa), and transmission to the seedlings occurs when they are transplanted. The virus enters the seedling tissues through cells that are damaged in the transplanting process (Broadbent, 1965). Such external contamination of the testa may be eliminated by treating the seed with hydrochloric acid or trisodium phosphate.

Numerous other viruses may also be detected on the surface of immature seeds, but not after the seed has matured and dried. Examples of viruses that have been detected on the testa of immature seed, include barley stripe mosaic virus in barley (Inouye, 1962) and bean yellow mosaic virus in soybean (Inouye, 1973). It seems likely that testa contamination occurs with most viruses that are able to pass through the cytoplasmic connections (plasmadesmata) between the mother plant and the female gametes early in their development, but unlike TMV, few viruses are able to survive the desiccation of the testa as the seed matures.

In contrast to viruses that are carried on or in the seed coat, other viruses enter, and can be readily detected in the embryo of the seed. Embryo transmission, which may be regarded as true seed transmission, is by far the most important type of seed transmission and most of the viruses transmitted through seed are carried in the embryo. The fundamental question that has puzzled plant virologists for decades, is why some viruses are able to enter the embryo of their host and be seed transmitted, and others not. Many theories have been advanced to answer this question, but the complete explanation is not currently known.

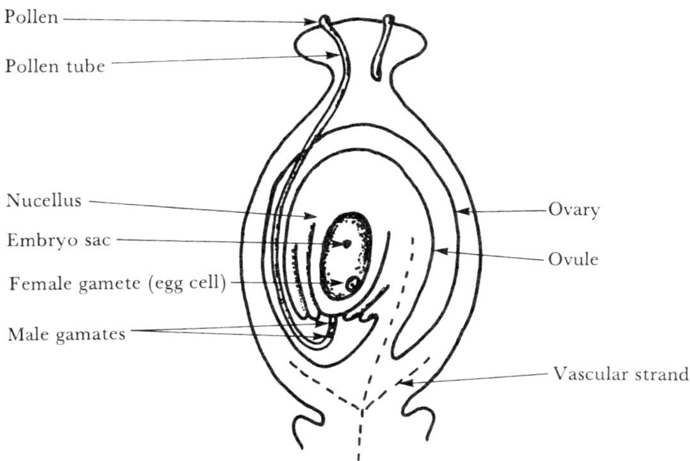

Fig. 7.5 Diagram of the anatomy of an ovule inside an ovary.

It is known that early infection of the mother plant is essential, if embryo transmission is to occur. Infection must take place before flowering and before the female gametes are formed. Therefore, it must be assumed that infection occurs through the female gametes (*see* Figure 7.5) (*see* also Section 7.7.4).

Even when early infection of the mother plant does occur, however, most viruses are still not embryo transmitted. It has been suggested (Caldwell, 1934) that passage of the virus may be prevented by marked differences in the growth rates of embryonic and endospermic tissues within the developing ovules, which results in the rupture of the plasmadesmata joining these tissues and the nucellus. Consequently, viruses that have high rates of seed transmission must have the ability either to reach and infect the embryo in the absence of plasmadesmata, or must infect the megaspore mother cell at a very early stage. Other workers have suggested that viruses may enter the meristematic tissues of the embryo (Inouye, 1966 and Crowley, 1957), but are then inactivated by an *in-vivo* mechanism similar to that which has been suggested for the restriction of virus entry into shoot meristems (*see* Chapter 11).

Some viruses, such as beet curly top, are mainly restricted to vascular tissues, and may, therefore, not be seed transmitted because of the lack of vascular connection between the mother plant and the embryo. In other cases seed transmission does not occur because infection causes the female gametes to abort. Failure to form viable seed, is common in lettuce cultivars such as Cheshunt Early Giant, infected with lettuce mosaic virus (Couch, 1955).

7.7.4 Pollen transmission

In addition to embryos becoming infected as a result of virus infection of the mother plant, female gametes may also become infected through pollination of the healthy mother plant by infected pollen (Schippers, 1963). It is probable that pollen-borne viruses enter the ovule along with the male gamete by passing through the pollen tube as it grows into the embryo sac (*see* Figure 7.5).

The rate of bean common mosaic virus transmission in bean seed is higher if a healthy female parent is fertilized with infected pollen, than if an infected mother plant is pollinated with healthy pollen. Even higher rates are obtained if both the mother plant and the pollen are infected (Medina and Grogan, 1961). Infection resulting from the fertilization of a healthy plant by infected pollen, may not be restricted to the seed. Some pollen transmission such as prunus necrotic ringspot in cherry, may result in the recipient tree becoming infected (George and Davidson, 1963).

Examples of pollen transmitted viruses are given in Table 7.9. Although reported transmissions are few compared with seed transmissions, it is possible, that as in the case of bean common mosaic virus mentioned above, infected pollen may increase the percentage of infected seed in some virus/host infections. In other virus/host relationships, infected pollen may result in poor fertilization. It has been shown that tobacco pollen infected with cherry leaf roll virus produces shorter pollen tubes which develop more slowly, than those of healthy pollen (Cooper, 1976). A high level of pollen sterility resulting in poor fertilization, has also been reported in lettuce as a result of lettuce mosaic infection (Ryder, 1964). Such pollen sterility may result from virus infection affecting the meiotic processes of the microspore, as has been observed in tomato pollen infected with tomato aspermy virus (Caldwell, 1952).

7.7.5 Factors affecting seed transmission

Infection of the mother plant before flower induction, as a pre-requisite for virus transmission through the embryo, has already been discussed. It is also known that cucumber mosaic virus is transmitted in *Stellaria media* seed at a higher rate (21 to 40%) if the mother plant is grown from infected seed, than if the mother plant is inoculated with virus at the seedling stage (3 to 21%). It has been suggested that this higher rate is also due to very early infection of the plant (Tomlinson and Carter, 1970).

Temperature also has a marked effect upon whether a virus is seed transmitted or not, with high temperatures frequently lowering the rate of preventing seed transmission. Southern bean mosaic virus, for example was transmitted in 95% of seed harvested from plants grown at 16° to 20°C, but in only 55% of seed from plants grown at 28° to 30°C (Crowley, 1959). Similarly, cherry leaf roll virus transmission was 100% and 0%, in seed collected from *Nicotiana rustica* plants grown at 20° and 30°C, respectively (Cooper, 1976).

It has already been mentioned that a virus which is seed transmitted in one host species, is often not transmitted through the seed of another species. Bean yellow mosaic virus, for example, is seed transmitted in lupin, but not in *Phaseolus* beans (Bos, 1970). In addition, seed transmission is greatly influenced by the cultivar of a particular host species, and by the strain of the virus concerned. The transmission of bean common mosaic virus varied from 1% to 75% in 51 cultivars of *Phaseolus* beans (Smith and Hewitt, 1938), and it has also been shown that individual lettuce plants of the same cultivar infected with lettuce mosaic virus, may have seed transmission rates varying from 0.2% to 14.2% (Couch, 1955).

Evidence that individual virus strains differ in their ability to be seed transmitted has been obtained for barley stripe mosaic virus in various cultivars of barley and wheat (McKinney and Greeley, 1965). In these studies seed transmission of the different strains varied from 0% to 53%.

7.7.6 *Effect upon infected seedlings*

It has been reported that seed infection does not influence the viability of lettuce seed infected with LMV (Grogan and Bardin, 1950), or pea seed infected with pea seed-borne virus (Stevenson and Hagedorn, 1970), but recent studies in the author's laboratory have shown a significant reduction in the germination of tobacco seed infected with spinach latent virus. Similarly, whether an infected seedling shows symptoms, or not, depends largely upon the virus and host concerned. In some cases, as with seedlings of *Phaseolus* beans infected with bean common mosaic virus, virus symptoms may be evident soon after germination of the seedling but in others, such as cherry leaf roll virus infected tobacco seedlings, the seedlings are symptomless and indistinguishable from their virus-free counterparts.

Latent infection of the seedling is common with many nepoviruses (Lister and Murant, 1967), but virus is readily detected in such seedlings by inoculating a healthy test seedling with sap from the infected seedling. The inoculated healthy seedling will develop the typical symptoms of the virus concerned.

7.8 Transmission by Vegetative Propagation

The widespread use of vegetative propagation for the multiplication of many horticultural crops results in the spread of viruses through propagules such as cuttings, tubers, runners and bulbs. Since infection by most viruses is completely systemic, any propagule is likely to be infected. Thus, vegetative propagation presents a very efficient method of virus spread, without the virus having the difficulty of entering and establishing infection in a new healthy plant.

Although virus spread through vegetative propagation might be expected to occur over short distances in nature by natural scattering of infected propagules such as tubers, man has been responsible for the worldwide movement of many viruses by this means.

The importance of virus infection of vegetatively propagated plants, and methods that may be used to eradicate viruses from plant clones that are totally infected, are discussed in Chapter 11.

7.9 Transmission by Dodder

Many species of dodder (*Cuscuta* spp.), a vine-like parasitic plant belonging to the family *Convolvulaceae* are able to transmit plant viruses. The various dodder species have different ranges of host plants in nature, and about twenty species have been used experimentally to transmit viruses. *Cuscuta campestris* and *C.subinclusa* have wide host ranges and have been used to transmit experimentally a number of different plant viruses (Schmelzer, 1956).

The parasite forms root-like haustoria which penetrate the host tissues to connect with the vascular system. In this respect, dodder transmission is similar to graft transmission, but whereas graft transmission may be restricted to closely related species, dodder transmission is generally less specific. Dodder transmission is used in the laboratory to transfer viruses from hosts that are difficult to work with, to more susceptible host plants. Virus transmission by dodder in nature is not of economic importance.

Transmission by dodder may be passive, for the virus does not have to multiply within the dodder plant for transmission to occur, but viruses which do multiply in dodder are more efficiently transmitted than those which do not (Bennett, 1967).

To carry out laboratory transmissions, dodder seed should be germinated on the surface of a pan of soil containing seedlings of the host plant. The germinated dodder seed is then placed in a leaf axil of the virus donor plant. Once the dodder is established on the infected plant, its shoots can be trained on to a healthy test plant. Virus symptoms usually develop in the young leaves at the apex of the test plant.

When using dodder for the experimental transmission of virus, it is first necessary to check the dodder plant for the presence of dodder latent virus. This virus is seed-borne and symptomless in infected *C.campestris* seedlings, but causes symptoms in some species of test plants parasitized by dodder (Bennett, 1967).

7.10 References

Abdul Magid, A. G. M. (1981). Investigations on viruses of pinks (*Dianthus* sp.) and their possible control. PhD thesis, University of Exeter.

Ayala, A. and Allen, M. W. (1968). Transmission of the California tobacco rattle virus (CTRV) by three species of the nematode genus *Trichodorus*. *J Ag Univ Puerto Rico* **52**, 101–25.

Bennett, C. W. (1967). Plant viruses: transmission by dodder. In *Methods in Virology*, Vol. 1 (ed. Maramorosch, K. and Koprowski, H.). Academic Press: London. pp. 393–401.

Bergeson, G. B. and Athow, K. L. (1963). Vector relationships of tobacco ringspot virus (TRSV) and *Xiphinema americanum* and the importance of this vector in TRSV infection of soybean. *Phyt* **53**, 871.

Bird, J. and Maramorosch, K. (1978). Viruses and virus diseases associated with whiteflies. *Adv Virus Res* **22**, 55–109.

Black, L. M. (1959). Biological cycles of plant viruses in insect vectors. In *The viruses* (ed. Burnet, F. M. and Stanley, W. M.) Academic Press: London. pp. 157–85.

Bos, L. (1967). Graft transmission of plant viruses. In *Methods in virology*, Vol. 1 (ed. Maramorosch, K. and Koprowski, H.). Academic Press: London. pp. 403–10.

Bos, L. (1970). Bean yellow mosaic virus. CMI/AAB Descriptions of plant viruses No. 40.

Broadbent, L. (1965). The epidemiology of tomato mosaic. XI. Seed transmission of TMV. *Ann Appl Biol* **56**, 177–205.

Broadbent, L. (1976). Epidemiology and control of tomato mosaic virus. *Ann R Phyto* **14**, 75–96.

Cadman, C. H., Dias, H. F. and Harrison, B. D. (1960). Sap-transmissible viruses associated with diseases of grapevines in Europe and North America. *Nature* **187**, 577–9.

Caldwell, J. (1934). The physiology of virus diseases in plants. V. The movement of the virus agent in tobacco and tomato. *Ann Appl Biol* **21**, 191–205.

Caldwell, J. (1952). Some effects of a plant virus on nuclear division. *Ann Appl Biol* **39**, 98–102.

Campbell, R. N., and Grogan, R. G. (1964). Acquisition and transmission of lettuce big-vein virus by *Olpidium brassicae*. *Phyt* **54**, 681–90.

Cooper, J. I. and Harrison, B. D. (1973). The role of weed hosts and the distribution and activity of vector nematodes in the ecology of tobacco rattle virus. *Ann Appl Biol* **73**, 53–66.

Cooper, V. C. (1976). Studies on the seed transmission of cherry leaf roll virus. PhD thesis, Univ. of Birmingham.

Couch, H. B. (1955). Studies on seed transmission of lettuce mosaic virus. *Phyt* **45**, 63–70.

Crowley, N. C. (1957). The effect of developing embryos on plant viruses. *Austral J Biol Sci* **10**, 443–8.

Crowley, N. C. (1959). Studies on the time of embryo infection by seed-transmission. *Virol* **8**, 116–23.

Das, S. and Raski, D. J. (1968). Vector efficiency of *Xiphinema index* in the transmission of grapevine fanleaf virus. *Nematologica* **14**, 55–62.

Dias, H. F. (1970*a*). Transmission of cucumber necrosis virus by *Olpidium cucurbitacearum*. Barr & Dias. *Virol* **40**, 828–39.

Dias, H. F. (1970*b*). The relationship between cucumber necrosis virus and its vector, *Olpidium cucurbitacearum*. *Virol* **42**, 204–11.

Flegg, J. J. M. (1968). The occurrence and depth distribution of *Xiphinema* and *Longidorus* species in south eastern England. *Nematologica* **14**, 189–96.

Francki, R. I. B. and Randles, J. W. (1970). Lettuce necrotic yellows virus. CMI/AAB, Descriptions of plant viruses No. 26.

Fulton, J. P., Scott, H. A. and Gamez, R. (1980). Beetles. In *Vectors of plant pathogens* (ed. Harris, K. F. and Maramorosch, K.). Academic Press: London. pp. 115–32.

Gálvez, C. E. (1971). Rice tungro virus. CMI/AAB, Descriptions of plant viruses No. 67.

Garner, R. J. (1958). *The grafter's handbook*. Faber and Faber: London.

George, J. A. and Davidson, T. R. (1963). Pollen transmission of necrotic ringspot and sour cherry yellows from tree-to-tree. *Can J Plant Sci* **43**, 276–88.

Gingery, R. F., Bradfute, O. E., Gordon, D. T. and Nault, L. R. (1978). Maize chlorotic dwarf virus. CMI/AAB, Descriptions of plant viruses No. 194.

Govier, D. A. and Kassanis, B. (1974). Evidence that a component other than the virus particle is needed for aphid transmission of potato virus Y. *Virol* **57**, 285–6.

Grogan, R. G. and Bardin, R. (1950). Some aspects concerning seed transmission of lettuce mosaic virus. *Phyt* **40**, 965.

Harris, K. F. (1979). Leafhoppers and aphids as biological vectors: vector-virus relationships. In *Leafhopper vectors and plant disease agents* (ed. Maramorosch, K. and Harris K. F.) Academic Press: London. pp. 217–308.

Harris, K. F. (1981). Arthropod and Nematode vectors of plant viruses. *Ann R Phyto* **19**, 391–426.

Harrison, B. D. (1970). Tobacco rattle virus. CMI/AAB, Descriptions of plant viruses No. 12.

Harrison, B. D. (1973). Pea early-browning virus. CMI/AAB. Descriptions of plant viruses No. 120.

Harrison, B. D. (1977). Ecology and control of viruses with soil-inhabiting vectors. *Ann R Phyto* **15**, 331–60.

Harrison, B. D. and Winslow, R. D. (1961). Laboratory and field studies on the relation of arabis mosaic virus to its nematode vector *Xiphinema diversicaudatum* (Micoletzky). *Ann Appl Biol* **49**, 621–33.

Hidaka, Z. and Tagawa, A. (1962). The relationship between the occurrence of tobacco stunt disease and *Olpidium brassicae*. Ann Phytopathol Soc Jpn **27**, 77–8.

Hooper, D. J. (1978). Structure and classification of nematodes. In *Plant Nematology* (ed. Southey, J. F.) HMSO: London. pp. 3–45.

Hunter, J. A., Chamberlain, E. E. and Atkinson, J. D. (1958). Note on transmission of apple mosaic by natural host root grafting. *NZ J Agric Res* **1**, 80–2.

Inouye, T. (1962). Studies on barley stripe mosaic in Japan. *Ber Ohara Inst Landw Biol.* **11**, 413–96.

Inouye, T. (1966). Some experiments on the seed transmission of barley stripe mosaic viruses in barley with electron microscopy. *Ber Ohara Inst Landw Biol* **13**, 111–22.

Inouye, T. (1973). Characteristics of cytoplasmic inclusions induced by bean yellow mosaic virus. *Nogaku Kenkyu*, **54**, 155–71.

Jones, R. A. C. and Harrison, B. D. (1969). the behaviour of potato mop-top virus in soil, and evidence for its transmission by *Spongospora subterranea* (Wallr.) Lagerh. *Ann Appl Biol* **63**, 1–17.

Jones, R. A. C. and Harrison, B. D. (1972). Ecological studies on potato mop-top virus in Scotland. *Ann Appl Biol* **71**, 47–57.

Kassanis, B. (1961). The transmission of potato aucuba mosaic virus by aphids from plants also infected with potato viruses A or Y. *Virol* **13**, 93–7.

Kassanis, B. and Macfarlane, I. (1964). Transmission of tobacco necrosis virus by zoospores of *Olpidium brassicae*. *J Gen Microbiol* **36**, 79–93.

Kassanis, B. and Macfarlane, I. (1968). The transmission of satellite viruses of tobacco necrosis virus by *Olpidium brassicae*. *J Gen Virol* **3**, 227–32.

Lister, R. M. and Murant, A. F. (1967). Seed transmission of namatode-borne viruses. *Ann Appl Biol* **59**, 49–62.

Loughnane, J. B. and Murphy, P. A. (1938). Dissemination of potato viruses X and F by leaf contact. *Sci Proc R Dublin Soc* **22**, 1–5.

McKinney, H. H. and Greeley, L. W. (1965). Biological characteristics of barley stripe mosaic virus strains and their evolution. *US Dep Agric Tech Bull* **1324**, 1–84.

Medina, A. C. and Grogan, R. G. (1961). Seed transmission of bean common mosaic virus. *Phyt* **51**, 452–6.

Messieha, M. (1969). Transmission of tobacco ringspot virus by thrips. *Phyt* **59**, 943–5.

Mulligan, T. E. (1960). The transmission by mites, host range and properties of ryegrass mosaic virus. *Ann Appl Biol* **48**, 575–9.

Murant, A. F. and Lister, R. M. (1967). Seed-transmission in the ecology of nematode-borne viruses. *Ann Appl Biol* **59**, 63–76.

Noordam, D. (1956). Waardplanten en toetsplanten van het ratelvirus van detabak. *Tijdschr Plantenziekten*, **62**, 219–25.

O'Loughlin, G. T. and Chambers, T. C. (1967). The systemic infection of an aphid by a plant virus. *Virol* **33**, 262–71.

Orlob, G. (1966). Feeding and transmission characteristics of *Aceria tulipae* as vector of wheat streak mosaic virus. *Phytopathol Z* **55**, 218–38.

Paliwal, J. C. (1980). Fate of viruses in mite vectors and non-vectors. In *Vectors of plant pathogens* (ed. Harris, K. F. and Maramorosch, K.) Academic Press: London. pp. 357–73.

Paliwal, Y. C. and Sinha, R. C. (1970). On the mechanism of persistence and distribution of barley yellow dwarf virus in an aphid vector. *Virol* **42**, 668–80.

Peters, D. (1971). Sowthistle yellow vein virus. CMI/AAB, Descriptions of plant viruses No. 62.

Pitcher, R. S. and Jha, A. (1961). On the distribution and infectivity with arabis mosaic virus of a dagger nematode. *Plant Path* **10**, 67–71.

Proeseler, G. (1980). Peismids. In *Vectors of plant pathogens* (ed. Harris, K. F. and Maramorosch, K.) Academic Press: London. pp. 97–113.

Rao, A. S. and Brakke, M. K. (1969). Relation of soil-borne wheat mosaic virus and its fungal vector, *Polymyxa graminis*. *Phyt* **59**, 581–7.

Roberts, F. M. (1946). Underground spread of potato virus X. *Nature* **158**, 663.

Rochow, W. F. (1977). Dependent virus transmissions from mixed infections. In *Aphids as virus vectors* (ed. Harris, K. F. and Maramorosch, K.) Academic Press: London. pp. 253–69.

Ryder, E. J. (1964). Transmission of common lettuce mosaic virus through the gametes of the lettuce plant. *Plant Dis R* **48**, 522–4.

Schippers, B. (1963). Transmission of bean common mosaic by seed of *Phaseolus vulgaris* L. cultivar Beka. *Acta Bot Neerl* **12**, 433–97.

Schmelzer, K. (1956). Beitrage zur Kenntnis der Ubertragbarkeit von Viren durch Cuscuta-Arten. *Phytopathol Z* **28**, 1–56.

Schneider, I. R. (1964). Studies on the transmission of sugarbeet savoy virus by the vector, *Piesma cinereum* (Say). *Plant Dis R* **48**, 843–5.

Simons, J. N. (1962). The pseudo-curly top disease in south Florida. *J Econ Entomol* **55**, 358–63.

Simons, J. N. (1980). Membracids. In *Vectors of plant pathogens* (ed. Harris, K. F. and Maramorosch, K.) Academic Press: London. pp. 93–6.

Slack, S. A. and Scott, H. A. (1971). Haemolymph as a reservoir for the cowpea strain of southern bean mosaic virus in the bean leaf beetle. *Phyt* **61**, 538–40.

Slykhuis, J. T. (1955). *Aceria tulipae* Keifer (Acarina: Eriophyidae) in relation to the spread of wheat streak mosaic. *Phyt* **45**, 116–28.

Slykhuis, J. T. (1970). Factors determining the development of wheat spindle streak mosaic caused by a soil-borne virus in Ontario. *Phyt* **60**, 319–31.

Slykhuis, J. T. (1972). Transmission of plant viruses by Eriophyid mites. In *Principles and techniques of plant virology*, (ed. Kado, C. I. and Agrawal H. O.) Van Nostrand Reinhold: New York. pp. 204–25.

Slykhuis, J. T. (1976). Wheat spindle streak mosaic virus. CMI/AAB, Descriptions of plant viruses No. 167.

Smith, F. L. and Hewitt, W. B. (1938). Varietal susceptibility to common bean mosaic and transmission through seed. *Calif Agric Exp Stn Bull* **621**, 1–18.

Stevenson, W. R. and Hagedorn, D. J. (1970). Effect of seed size and condition on transmission of pea seed-borne mosaic virus. *Phyt* **60**, 1148–9.

Sylvester, E. S. (1969). Evidence of transovarial passage of sowthistle yellow vein virus in the aphid *Hyperomyzus lactucae Virol* **38**, 440–8.

Sylvester, E. S. and Richardson, J. (1970). Infection of *Hyperomyzus lactucae* by sowthistle yellow vein virus. *Virol* **42**, 1023–42.

Takahashi, Y. and Orlob, G. (1969). Distribution of wheat streak mosaic virus-like particles in *Aceria tulipae*. *Virol* **38**, 230–40.

Tamada, T. (1975). Beet necrotic yellow vein virus. CMI/AAB, Descriptions of plant viruses No. 144.

Taylor, C. E. and Raski, D. J. (1964). On the transmission of grape fan-leaf by *Xiphinema index*. *Nematologica* **10**, 489–95.

Taylor, C. E. and Robertson, W. M. (1969). The location of raspberry ringspot and tomato black ring viruses in the nematode vector *Longidorus elongatus* (de Man). *Ann Appl Biol* **64**, 233–7.

Taylor, C. E. and Robertson, W. M. (1970*a*). Location of tobacco rattle virus in the nematode vector, *Trichodorus pachydermus* Seinhorst. *J Gen Virol* **6**, 179–82.

Taylor, C. E. and Robertson, W. M. (1970*b*). Sites of virus retention in the alimentary tract of the nematode vectors, *Xiphinema diversicaudatum* (Micol.)

and X. index (Thorne and Allen). *Ann Appl Biol* **66**, 375–80.

Teakle, D. S. (1972). Transmission of plant viruses by fungi. In *Principles and techniques in plant virology*, (ed. Kado, C. E. and Agrawal, H. O.). Van Nostrand Reinhold: New York. pp. 248–66.

Teliz, D., Grogan, R. G. and Lownsbery, B. F. (1966). Transmission of tomato ringspot, peach yellow bud mosaic and grape yellow vein viruses by *Xiphinema americanum*. *Phyt* **56**, 658–63.

Temmink, J. H. M., Campbell, R. N. and Smith, P. R. (1970). Specificity and site of *in vitro* acquisition of tobacco necrosis virus by zoospores of *Olpidium brassicae*. *J Gen Virol* **9**, 201–13.

Thomas, C. E. (1969). Transmission of tobacco ringspot virus by *Tetranychus* sp. *Phyt* **59**, 633–6.

Tomlinson, J. A. (1970). Lettuce mosaic virus. CMI/AAB, Descriptions of plant viruses No. 9.

Tomlinson, J. A. and Carter, A. L. (1970). Studies in the seed transmission of cucumber mosaic virus in chickweed (*Stellaria media*) in relation to the ecology of the virus. *Ann Appl Biol* **66**, 381–6.

Van Hoof, H. A. (1970). Some observations on retention of tobacco rattle virus in nematodes. *Neth J Plant Pathol* **76**, 329–30.

Walters, H. J. (1969). Beetle transmission of plant viruses. *Adv Virus Res* **15**, 339–61.

Walters, H. J. and Henry, D. G. (1970). Bean leaf beetle as a vector of the cowpea strain of southern bean mosaic virus. *Phyt* **60**, 177–8.

Watson, M. A. (1972). transmission of plant viruses by aphids. In *Principles and techniques in plant virology* (ed. Kado, C. I. and Agrawal, H. O.). Van Nostrand Reinhold: New York. pp. 131–67.

Whitcomb, R. F. (1972). Transmission of viruses and mycoplasma by *Auchenorrhynchous Homoptera*. In *Principles and techniques in plant virology* (ed. Kado, C. I. and Agrawal, H. O.). Van Nostrand Reinhold: New York. pp. 168–99.

Zitter, T. A. and Tsai, J. H. (1977). Transmission of three potyviruses by the leaf-miner, *Liriomyza sativae* (Diptera: Agromyzidae), *Plant Dis* **61**, 1025–9.

Zitter, T. A. and Tsai, J. H. (1980). Flies In *Vectors of plant pathogens*, (ed. Harris, K. F. and Maramorosch, K.) Academic Press: London. pp. 165–76.

7.11 Further selected reading

Bennett, C. W. (1969). Seed transmission of viruses. *Adv Virus Res* **14**, 221–58.

Bos, L. (1977). Seed-borne viruses. In *Plant health and quarantine in international transfer of genetic resources* (ed. Hewitt, W. B. and Chiarappa, L.) CRC Press Inc., pp. 39–69.

Harris, K. F. (1981). Arthropod and nematode vectors of plant viruses. *Ann R Phyto* **19**, 391–426.

Harris, K. F. and Maramorosch, K. (1977). *Aphids as virus vectors*. Academic Press: London.

Harrison, B. D. (1977). Ecology and control of viruses with soil inhabiting vectors. *Ann R Phyto* **15**, 331–60.

Southey, J. F. (1978). *Plant Nematology* HMSO, London.

Teakle, D. S. (1983). Zoosporic fungi and viruses: double trouble. In *Zoosporic plant pathogens.* (ed. Buczacki, S. T.), Academic Press: New York. pp. 231–48.

8 *Ecology and Epidemiology of Plant Viruses*

8.1 Introduction

Before developing an efficient control method for any specific plant virus disease, the virologist must investigate and understand the complex ecology and epidemiology of the virus. Figure 8.1 illustrates some of the major factors that govern the infection pathways between a virus and its host. These factors may concern the virus itself (e.g. through the pathogenicity of different virus strains), its mode of transmission (or vector) or may affect the host plant.

Many of these factors, such as the importance of aphid and nematode vectors in virus transmission have been discussed earlier (*see* Chapter 7), and others are covered in chapters 9–11. In this chapter selected examples are used to demonstrate the complex and varied nature of plant virus ecology and the factors that cause disease epidemics. More detailed information on this subject has been published by Watson (1967); Thresh (1976, 1980, 1981 and 1982) and Maramorosch and Harris (1981).

8.2 Examples of Ecological Factors that Control Virus Epidemiology

8.2.1 Weeds and other alternative host species

Many viruses have weed or other alternative hosts, that provide a reservoir of virus from which economically important crop plants may become infected. When considering control measures for such viruses, it is important to identify the initial sources (*foci*) of infection from which the virus spreads into or within a crop. Three principle foci of infection are usually recognized (Thresh, 1981). Spread may occur from infected weeds within a crop, from weeds or other plants growing in ground adjacent to the crop, or from infected hosts in areas remote from the crop (*see* Figure 8.2).

The amount and range of spread from any particular focus of

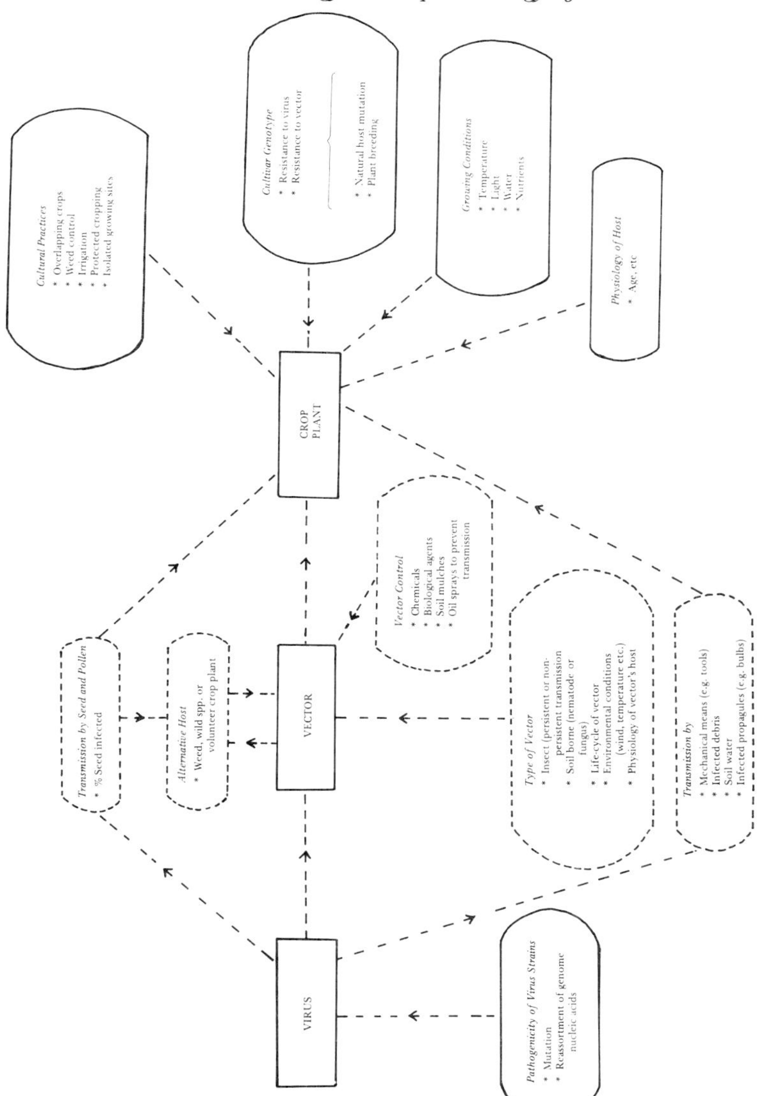

Fig. 8.1 Some of the more important factors influencing the ecology and epidemiology of plant virus infections.

infection, is referred to as the *gradient* of infection. As would be expected, the occurrence of infected plants decreases with increasing distance from the focus of infection. The steepness or shallowness of the infection gradient, depends upon the type of vector or mode of transmission of the virus, and factors that govern the movement and abundance of the vector. Shallow gradients involving spread over considerable distances, result from highly mobile vectors, whereas steep gradients, with spread over short distances, involve less mobile vectors or

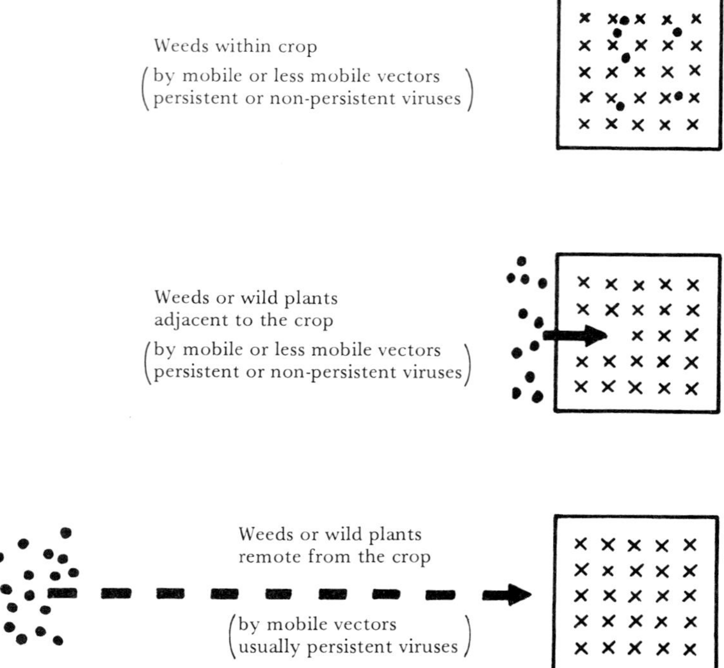

Fig. 8.2 Diagrammatic representation of virus spread from weeds or wild plants growing within, adjacent to, or remote from a cultivated crop (based on Thresh, 1981).

transmission by such means as root or leaf contact between plants. Winged vectors, such as aphids, tend to be carried downwind rather than upwind, consequently downwind gradients are usually shallow and those upwind steep.

The persistent or non-persistent nature of virus transmission by a vector (*see* Section 7.4.2) is also important in governing virus spread. A vector will remain viruliferous over long periods of time and long distances, if it is carrying a persistent virus, but a non-persistent virus will only be retained and spread for short periods of time, over relatively short distances. The many factors governing the nature of infection gradients and virus spread have been described in detail by Thresh (1976).

Spread from weeds within a crop

Virus spread from weeds within a crop may occur through highly mobile vectors such as aphids or other winged insects or through less

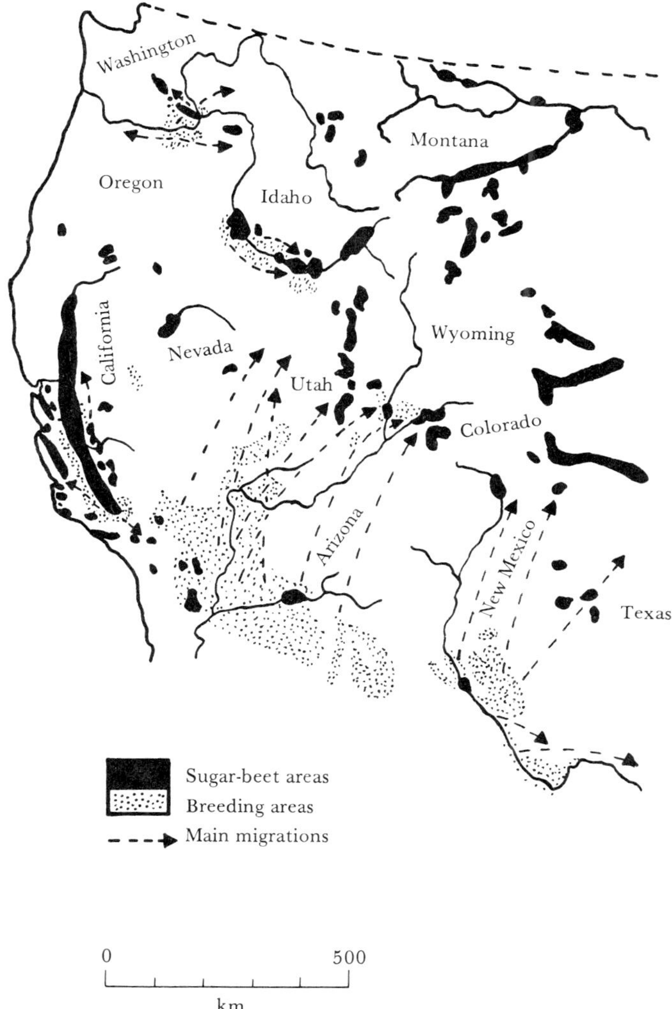

Fig. 8.3 The long distance transmission of beet curly top virus by leafhoppers flying from breeding grounds in Western U.S.A. (based on Douglass and Cook, 1954).

mobile vectors such as wingless beetles or soil-borne nematodes or fungi (*see* Chapter 7). Both persistent and non-persistent transmitted viruses may be involved.

The importance of infected weeds within a crop as a source of infection was demonstrated as early as 1925 in Wisconsin, when cucumber mosaic virus (CMV) infected cucumbers were found in scattered patches around CMV infected perennial mayweed (*Asclepias* syriaca) plants, that had overwintered and regenerated in the spring (Doolittle and Walker, 1925).

More recently, CMV has been shown to overwinter in chickweed (*Stellaria media*) in Britain, and this provides a source of infection for the following season's lettuce crop (Tomlinson *et al.*, 1970). The virus was transmitted from the infected chickweed to lettuce by aphids, and since CMV is seed-borne in chickweed, the weed could provide a source of infection for many years.

Weeds within crops are also important as sources of infection of various nematode transmitted viruses (Murant, 1981). Many common weeds of arable land are the natural hosts of the nematode vectors (McNamara and Flegg, 1981), and the nematode transmitted viruses are also frequently seed-transmitted to a high level in these weeds (Murant and Lister, 1967). Consequently the ecology of these viruses is closely related to that of their vectors, and the longevity of nematode transmitted viruses in weeds, is a major factor in their epidemiology.

Another source of infection within a crop are plants regenerating from a previous season's crop. These *volunteer* plants are common in potato and sugar beet crops for instance, and may frequently be virus infected. Aphid spread of potato virus Y and potato leafroll virus has been shown to result from volunteer tubers in England (Doncaster and Gregory, 1948), and regenerated beet debris is a source of beet mosaic and yellowing viruses in Washington State (Howell and Mink, 1971). Similarly, cotton leaf curl virus, which is transmitted by whiteflies in the Sudan, may spread from infected cotton plants that regenerate from stumps of the previous year's crop (Tarr, 1951). In the Columbia Basin region of Washington, volunteer carrot plants were found to be the overwintering source for both carrot thin leaf and carrot motley dwarf viruses (Howell and Mink, 1977), and onion yellow dwarf virus has been shown to overwinter in volunteer onions, before infecting the new season's crop (Louie, 1968).

These examples emphasize the importance of good weed control within a crop and adequate cultivation to avoid volunteer plants.

Spread from sources adjacent to a crop

Infected trees, shrubs or herbaceous plants growing around crops may provide a reservoir of virus infection (Duffus, 1971; Bos, 1981). Such plants may be the only foci of infection for the crop, or they may be the primary source, with secondary spread occurring from plants that become infected within the crop.

The nature of the infection gradient into a crop will again depend upon the mode of transmission of the virus and the mobility of the vector. Steep gradients are most likely to occur with soil-inhabiting nematode vectors moving from infected plants in adjacent hedgerows

(Murant, 1981) or with non-persistent, insect transmitted viruses. In contrast, shallow gradients are more probable with persistently transmitted viruses.

Care should be taken when interpreting such patterns of infection, for a large number of infected plants near the edge of a field, does not always indicate that the source of infection is nearby. Sometimes, infection of the peripheral plants may result from incoming vectors that have travelled relatively long distances and accumulate on plants at the edge of the crop. This is often the case on the leeward side of a hedge or windbreak, where air turbulence causes the insects to drop suddenly as they migrate inwards (Lewis, 1969).

Often virus spread from adjacent host plants may occur as the initial infected host matures and senesces. Under these circumstances the vector may produce winged migrants on the crowded host plant, which fly off to transmit the virus to the nearby crop. This type of spread has been reported to cause epidemics of barley yellow dwarf virus in California, when aphids move from wilting grass to young cereal crops in adjacent fields (Oswald and Houston, 1953).

The control of weeds and other plants growing along headlands by cultivation, burning, or herbicides, is often an effective way or preventing the spread of such viruses (Stubbs *et al.*, 1963; Adlerz, 1981; Bos, 1981).

Spread from remote sources

The transmission of sugar beet curly top virus by vectors moving long distances from infected weeds to beet, is a classical example of the spread of a virus from areas quite remote from the crop. Both the virus and its leafhopper vector *Circulifer tennellus* overwinter in various weed hosts, such as *Chenopodium* species, in warm areas including California, Arizona and New Mexico (Douglass and Cook, 1954). Large populations of leafhoppers are produced on the weeds during winter and early spring, and these are forced to migrate in the late spring as the weed-hosts mature. If the prevailing winds are suitable, the leafhoppers are carried considerable distances to infect beet crops in distant regions (*see* Figure 8.3), where characteristically the virus suddenly appears over wide areas.

The importance of weeds and other non-crop species as sources of virus infection, has recently been reviewed by various workers. These articles include the epidemiology of nematode transmitted viruses (Murant, 1981), aphid transmitted viruses (Adlerz, 1981), lettuce necrotic yellows virus (Martin and Randles, 1981), hopper transmitted viruses (Conti, 1981), potato viruses (Jones, 1981) and the viruses of amenity trees (Cooper, 1981).

8.2.2 *Introduction of crops into new areas*

Virus epidemics may occur when crops are grown in regions remote from their normal endemic growing area. Cacao swollen shoot disease is such an example. Cacao (*Theobroma cacao*) is an indigenous tree of the Amazon forests of South America where swollen shoot disease is unknown. The cacao tree was introduced to Nigeria and Ghana in West Africa in the early twentieth century, where it rapidly became an important crop (Posnette, 1981).

Swollen shoot disease was first observed in Ghana in 1936 (Steven, 1937) and has since spread throughout the West African growing area. The virus has been shown to be transmitted by a mealy bug (*Pseudococcid*) (Posnette, 1947), and indigenous native tress and shrubs have been shown to be the source of the virus (Posnette *et al.*, 1950; Dale, 1962).

Similarly, the beetle transmitted rice mottle virus occurs in rice crops, only in a remote area of Kenya. This virus is thought to originate from local, indigenous grasses (Bakker, 1970; 1974).

8.2.3 *Spread of a virus into new areas*

Viruses may spread rapidly into new areas, where often the host plant may have little or no natural resistance. Plum pox virus is a striking example. The virus infects *Prunus* spp., and is non-persistently

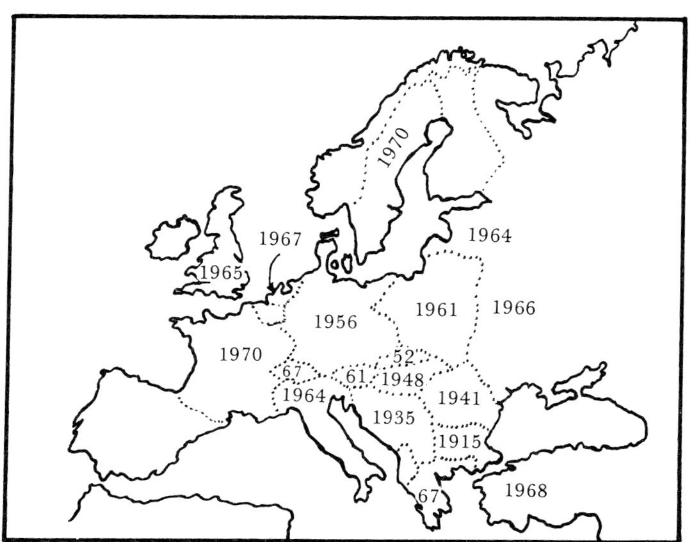

Fig. 8.4 The spread of plum pox virus in Europe and Western Asia. Dates signify the year the virus was first recorded (based on Thresh, 1981).

transmitted by aphids such as *Phorodon humuli* and *Myzus persicae*.

The virus was first reported in Bulgaria in 1915 and has since spread north and west (*see* Figure 8.4). The disease is reported to spread rapidly to new orchards if they are planted adjacent to old infected trees, but the incidence of infection decreases rapidly, with distance from the source of infection (Jordović, 1975). This indicates that aphids, although responsible for the movement within orchards, are not responsible for transmission of the disease to new growing areas and to different countries. It is most probable that the disease spreads from country to country by the distribution of plant material from infected nurseries.

The spread of plum pox in Europe is an important example of how man may inadvertently disseminate a disease quite rapidly over considerable distances. The disease has not yet reached North America, and it is imperative that the import of *Prunus* spp. is strictly controlled to prevent its establishment there.

8.2.4 *Effect of monocropping or short-term rotation on virus occurrence*

Repeated cultivation of a single crop species on the same land may result in a marked increase in the occurrence of diseases, particularly those caused by soil-borne viruses. Wheat spindle-streak virus, for instance, which is transmitted by the root inhabiting fungus, *Polymyxa* graminis, is widespread in fields used regularly to grow wheat in the main cereal growing areas of Ontario in Canada. However the virus is rare or absent in fields where wheat is infrequently grown (Slykhuis, 1970). Studies have shown that in an area new to wheat, the virus will infect only a few plants in the first season, numerous scattered groups of plants during the second season, and almost all plants in subsequent crops if no break in rotation occurs (Slykhuis, 1976). Although wheat is the only host of spindle streak virus, the virus can persist in the fungal resting spores for at least five years, necessitating long periods without wheat if heavy soil infestations are to be controlled.

Similarly, grapevine fanleaf disease is difficult to control, as vines are usually continually cultivated on the same site. The disease is widespread in Europe, Asia, South Africa, South America and North America, where it is transmitted by the soil-inhabiting nematode *Xiphinema index*. Even if the vines are removed, the viruliferous eelworms can survive for some years on the remnants of old roots (Hewitt *et al.*, 1962), and a fallow period of up to ten years may be required for successful eradication (Vuittenez, 1970).

8.2.5 Occurrence of new virus strains

The possibility of new virus strains being produced by mutation, or by recombination of viral nucleic acids among the multicomponent genome viruses, has been described in Chapter 2. Such new strains may arise quite suddenly and cause serious losses in crops, that up to that time have been tolerant to, or unaffected by, the particular virus concerned.

A recent example of a change in virus strain virulence, is seen in the case of maize dwarf virus in the maize crop. For many years the virus was prevalent in sugar cane, where it caused severe losses in many countries until tolerant cultivars of cane were introduced (Klinkowski, 1970). Various strains of the virus were shown to occur in these early studies, but they rarely infected maize unless the crop was planted alongside sugar cane. Recently, however, the situation has dramatically changed in many parts of the U.S.A., and the virus now infects maize, causing heavy losses in many areas, including northern growing sites, well removed from sugar cane crops. The change in disease severity has been caused by an increase in new virus strains.

8.2.6 Viruses in new cultivars

Numerous examples exist in which local virus-resistant or tolerant cultivars have been replaced by new cultivars that prove to be highly susceptible to virus infection. The new cultivars are frequently bred for special agronomic characters, such as high yields and crop uniformity, by breeders working in distant horticultural areas or even in different countries. Often, no consideration is given to the local disease situation in the areas where the new cultivars may be grown.

Such an example occurred with new cultivars of maize. Maize was introduced into Europe from South America in the sixteenth century and has been cultivated for more than 400 years in Italy and other Mediterranean countries. In these areas, maize rough dwarf virus, which is transmitted by the plant hopper (*Laodelphax striatellus*) occurs in wild grasses. The virus caused no problem in maize, however, until new, high yielding hybrid cultivars were introduced into Europe from America after the second world war (Harpaz, 1972). Severe outbreaks of the virus then occurred in Italy and Israel in these highly susceptible hybrids. Maize rough dwarf virus does not occur in North America where the hybrid maize was bred.

A similar example also occurred in new lettuce cultivars introduced in California in 1966. A new crisp-head cultivar called Calmar was produced with downy mildew (*Bremia lactucae*) resistance, but the cultivar was found to be highly susceptible to turnip mosaic virus

(TuMV), which is aphid transmitted in a non-persistent way (Zink and Duffus, 1969). It was later shown that the genes for mildew resistance and TuMV susceptibility were linked (Zink and Duffus, 1970).

The introduction into England of new Hybrid Brussels sprout cultivars bred in Holland, also provides another striking example of unforeseen virus susceptibility. A hybrid cultivar called Fasolt, bred for high yielding, uniform sprouts, was devastated by TuMV and cauliflower mosaic virus as soon as it was widely grown (Tomlinson and Ward, 1981). In contrast, the older, locally popular cultivars, had been highly resistant to these viruses. Further studies showed that both the inbred parent lines used to produce the hybrid Fasolt, were highly susceptible to virus infection.

8.2.7 Effect of favourable climatic conditions on virus spread

The example of favourable wind conditions causing the long distance spread of beet curly top virus from weeds, has already been mentioned in Section 8.2.1. There are also many other examples of favourable climatic conditions being responsible for disease epidemics.

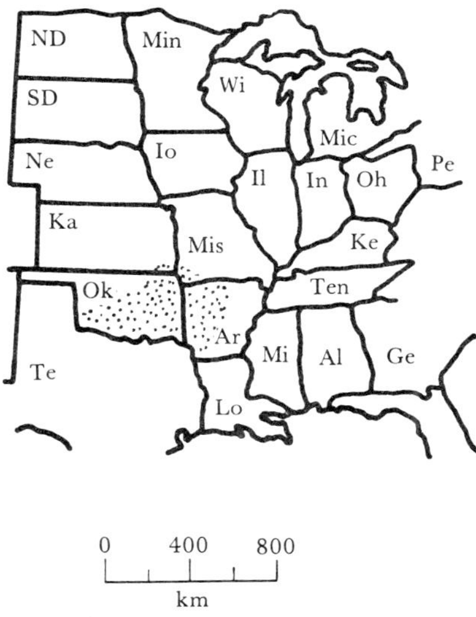

Fig. 8.5 The movement of cereal aphids and the spread of barley yellow dwarf virus in central U.S.A. The stippled area indicates parts of Oklahoma and adjacent states that are major overwintering areas of cereal aphids. These migrate northwards each spring and may spread the virus to Wisconsin and neighbouring areas of the U.S.A. and Canada (based on Medler, 1960).

In Wisconsin and neighbouring areas of the USA and Canada, severe outbreaks of barley yellow dwarf virus may occur as a result of the long distance migration of aphids from maturing cereal crops in Oklahoma and nearby southern states (Medler, 1960) (*see* Figure 8.5). These epidemics are predictable, because they can be associated with favourable winds blowing from the south (Medler and Smith, 1960; Wallin *et al.*, 1967).

Similarly in England, the occurrence of beet yellowing virus in sugar beet crops can be predicted by studying winter weather conditions (Watson *et al.*, 1975). The winter conditions govern the numbers and infectivity of the aphid vector which migrate to beet. These aphids are the source of initial infection in the spring (Hurst, 1965).

Weather conditions have also been shown to be important in the epidemiology of wheat streak mosaic virus, which is transmitted by the eriophyid mite (*Aceria tulipae*). Epidemics in Kansas in 1959 were shown to be related to early sowings of cereals in the autumn of 1958, which resulted in the crop becoming infected at a very young growth stage (King and Sill, 1959). Normally, the mite population is reduced to a low level by winter conditions before the wheat is sown. Unusual weather conditions were also the cause of wheat streak mosaic epidemics in Alberta in 1963 (Atkinson and Slykhuis, 1963).

8.2.8 Cultural practices and virus spread

Various other cultural practices, besides short-term rotations and monocropping (*see* Section 8.2.4) may stimulate virus epidemics. Among the most important of these is the practice of planting new crops alongside a similar crop that has been overwintered. The old mature crop may contain infected plants that act as reservoirs of virus infection.

Overwintered crops can be the main source of beet mosaic and beet yellows virus infection in newly planted sugar-beet crops in California (Shepherd and Hills, 1970; Duffus, 1973). In England, the same viruses may overwinter in mangold or beet seed crops, before infecting the young sugar-beet (Broadbent, 1969). Similarly, in Florida, epidemics of celery mosaic virus may be caused by the overlapping of celery seed-beds with newly planted crops (Zitter, 1977), and in the Netherlands, leek yellow stripe virus may be perpetuated by the overwintering and overlapping of leek crops (Bos, 1982).

The use of irrigation to extend the growing season can also lead to major virus epidemics. In areas around the Mediterranean, irrigated crops may be planted following the natural wet season in March and April. At this time, aphid vectors may migrate in large numbers from the senescing winter crops and natural vegetation, to the young, irrigated crop. Epidemics of cucumber mosaic, maize dwarf mosaic, potato Y and water melon mosaic viruses may then occur on the irrigated crops (Quiot, 1980; Thresh, 1982).

Virus epidemics that would not normally occur in outdoor crops, may occur when the crop is grown under protective glass or polythene. In Northern Italy for example, tobacco mosaic virus is the main disease in protected pepper crops, whereas in the field, the aphid transmitted cucumber mosaic and potato Y viruses are the main problems (Conti and Masenga, 1977). In Israel, the use of plastic tunnels and mulches to extend the growing season of vegetable crops, has resulted in a major increase in the occurrence of viruses transmitted by aphids and whiteflies (Cohen, S unpublished results). Similarly, in Japan, plastic protection has led to an increase in the incidence of rice dwarf and yellow dwarf viruses, which are vectored by leafhoppers (Kiritani, 1979). The rice seedlings, which are raised in nurseries under polythene, can be transplanted earlier in the season than outdoor raised plants. Consequently, the rice fields are now left fallow between crops, to accommodate the earlier crop, whereas a wheat or barley crop used to be grown between rice crops. The vector is able to overwinter in these fallow fields and migrate to the rice seedlings at an early stage of their growth, whereas previously, the vector did not multiply on the wheat or barley crop.

In England, the introduction of nutrient film techniques to grow lettuce in glasshouses, resulted in a serious outbreak of lettuce big-vein disease (Tomlinson and Faithfull, 1979). The zoospores of the fungus vector, *Olpidium brassicae*, were quickly cycled through the nutrient flowing in channels, thus resulting in a decimated crop (*see* Section 9.5.3).

Although increased irrigation and protected cropping may greatly increase the cropping potential of many areas, these examples show that these practices may increase the severity of many existing virus problems and cause the initiation of diseases in new areas. Such problems necessitate the careful management of new cultural practices, and, in some cases, may be a major restraint in their introduction.

8.3 References

Adlerz, W. C. (1981). Weed hosts of aphid-borne viruses of vegetable crops in Florida. In *Pests, pathogens and vegetation* (ed. Thresh, J. M.). Pitman: London. pp. 467–78.

Atkinson, T. G. and Slykhuis, J. T. (1963). Relationship of spring drought, summer rain and high fall temperatures to the wheat streak mosaic epiphytotic in Southern Alberta, 1963. *Can Plant Dis Surv* **43**, 154–9.

Bakker, W. (1970). Rice yellow mottle, a mechanically transmissible virus disease of rice in Kenya. *Neth J Plant Path* **76**, 53–63.

Bakker, W. (1974). Characterization and ecological aspects of rice yellow mottle virus in Kenya. *Agric Res Rep* 829.

Bos, L. (1981). Wild plants in the ecology of virus diseases. In *Plant Diseases and Vectors* (ed. Maramorosch, K. and Harris, K. F.). Academic Press: London. pp. 1–33.

Bos, L. (1982). Viruses and virus diseases of *Allium* species *Acta Hort* **127**, 11–29.

Broadbent, L. (1969). Disease control through vector control. In *Viruses, vectors and vegetation* (ed. Maramorosch, K.). Interscience: New York. pp. 593–630.

Cooper, J. I. (1981). The possible role of amenity trees and shrubs as virus reservoirs in the United Kingdom. In *Pests, pathogens and vegetation* (ed. Thresh, J. M.). Pitman: London. pp. 79–87.

Conti, M. (1981). Wild plants in the ecology of hopper-borne viruses of grasses and cereals. In *Pests, pathogens and vegetation* (ed. Thresh, J. M.). Pitman: London. pp. 109–19.

Conti, M. and Masenga, V. (1977). Identification and prevalence of pepper viruses in northwest Italy. *Phytopathol Z* **90**, 212–22.

Dale, W. T. (1962). Diseases and pests of Cocoa, A. Virus diseases. In *Agriculture and land use in Ghana* (ed. Wills, J. B.). Oxford Univ. Press: London. pp. 286–316.

Doolittle, S. P. and Walker, M. N. (1925). Further studies on the overwintering and dissemination of cucurbit mosaic. *J Agric Res* **31**, 1–58.

Doncaster, J. P. and Gregory, P. H. (1948). The spread of virus diseases in the potato crop. ARC Rep. Ser. No. 7, HMSO: London.

Douglass, J. R. and Cook, W. C. (1954). The beet leafhopper, U.S. Dept. Agr., Washington, Circ. No. 942.

Duffus, J. E. (1971). Role of weeds in the incidence of virus diseases. *Ann Rev Phytopathol* **9**, 319–40.

Duffus, J. E. (1973). The yellowing virus diseases of beet. *Adv Virus Res* **18**, 347–86.

Harpaz, I. (1972). *Maize rough dwarf.* Israel Univ. Press: Jerusalem.

Hewitt, W. B., Goheen, A. C., Raski, D. J. and Gooding, G. V. (1962). Studies on virus diseases of the grapevine in California. *Vitis* **3**, 57–83.

Howell, W. E. and Mink, G. I. (1971). The relationship between agricultural practices and the occurrence of volunteer sugarbeets in Washington. *J Am Soc Sugar beet Technol* **16**, 441–7.

Howell, W. E. and Mink, G. I. (1977). The role of weed hosts, volunteer carrots and overlapping growing seasons in the epidemiology of carrot thin leaf and carrot motley dwarf viruses in Central Washington. *Plant Dis R* **6**, 217–22.

Hurst, G. W. (1965). Forecasting the severity of sugar beet yellows. *Plant Path* **14**, 47–53.

Lewis, T. (1969). Factors affecting primary patterns of infestation. *Ann Appl Biol* **63**, 315–7.

Louie, R. (1968). Epiphytology of onion yellow dwarf. PhD thesis, Cornell Univ.

Jones, R. A. C. (1981). The ecology of viruses infecting wild and cultivated potatoes in the Andean region of South America. In *Pests, pathogens and* vegetation (ed. Thresh, J. M.). Pitman: London. pp. 89–107.

Jordović, M. M. (1975). Study of sharka spread pattern in some plum orchards. *Acta Hort* **44**, 147–54.

King, C. L. and Sill, W. H. (1959). Wheat streak mosaic epiphytotics in Kansas. *Plant Dis R* **43**, 1256.

Klinkowski, M. (1970). Catastrophic plant diseases. *Ann R Phyto* **8**, 37–60.

Kiritani, K. (1979). Pest management in rice. *Ann R Entomol* **24**, 279–312.

McNamara, D. G. and Flegg, J. J. M. (1981). The distribution of virus-vector nematodes in Great Britain in relation to past and present natural vegetation. In *Pests, pathogens and vegetation* (ed. Thresh, J. M.). Pitman: London. pp. 225–35.

Maramorosch, K. and Harris, K. F. (1981). *Plant diseases and vectors; ecology and epidemiology*. Academic Press: London.

Martin, D. K. and Randles, J. W. (1981). Inter-relationships between wild host plant and aphid vector in the epidemiology of lettuce necrotic yellows. In *Pests, pathogens and vegetation* (ed. Thresh, J. M.). Pitman: London. pp. 479–86.

Medler, J. T. (1960). Long range displacement of homoptera in the central United States. *Proc XIth Int Kong für Ento* **3**, 30–5.

Medler, J. T. and Smith P. W. (1960). Greenbug dispersal and distribution of barley yellow dwarf virus in Wisconsin. *J Econ Ent* **53**, 473–4.

Murant, A. F. (1981). The role of wild plants in the ecology of nematode-borne viruses. In *Pests, pathogens and vegetation* (ed. Thresh, J. M.). Pitman: London. pp. 237–48.

Murant, A. F. and Lister, R. M. (1967). Seed-transmission in the ecology of nematode-borne viruses. *Ann Appl Biol* **59**, 63–76.

Oswald, J. W. and Houston, B. R. (1953). Host range and epiphytology of the cereal yellow dwarf disease. *Phyt* **43**, 309–13.

Posnette, A. F. (1947). Virus diseases of cacao in West Africa. 1. Cacao viruses 1A, 1B, 1C, and 1D. *Ann Appl Biol* **34**, 388–402.

Posnette, A. F. (1981). The role of wild hosts in cocoa swollen shoot disease. In *Pests, pathogens and vegetation* (ed. Thresh, J. M.). Pitman: London. pp. 71–8.

Posnette, A. F., Robertson, N. F. and Todd, J. M. (1950). Virus diseases of cacao in West Africa. V. Alternative host plants. *Ann Appl Biol* **37** 229–40.

Quiot, J. B. (1980). Ecology of cucumber mosaic virus in the Rhone Valley of France. *Acta Hort* **88**, 9–21.

Shepherd, R. J. and Hills, F. J. (1970). Dispersal of beet yellows and beet mosaic viruses in the inland valleys of California. *Phyt* **60**, 798–804.

Slykhuis, J. T. (1970). Factors determining the development of wheat spindle streak mosaic caused by a soil-borne virus in Ontario. *Phyt* **60**, 319–31.

Slykhuis, J. T. (1976). Wheat spindle streak mosaic virus. CMI/AAB Descriptions of Plant Viruses No. **167**.

Steven, W. F. (1937). A new disease of cocoa in the Gold Coast. *Trop Agric Trin* **14**, 84.

Stubbs, L. L., Guy, A. D. and Stubbs, K. J. (1963). Control of lettuce necrotic yellow virus disease by the destruction of common sowthistle (*Sonchus oleraceus*). *Aust J Exp Agr Anim Husb* **3**, 215–48.

Tarr, S. A. J. (1951). 'Leaf curl disease of Cotton' CMI: Kew.

Thresh, J. M. (1976). Gradients of plant virus diseases. *Ann Appl Biol* **82**, 381–406.

Thresh, J. M. (1980). The origins and epidemiology of some important plant virus diseases. *Appl Biol* **5**, 1–65.

Thresh, J. M. (1981). The role of weeds and wild plants in the epidemiology of plant virus diseases. In *Pests, pathogens and vegetation* (ed. Thresh, J. M.). Pitman: London. pp. 53–70.

Thresh, J. M. (1982). Cropping practices and virus spread. *Ann Rev Phyto* **20**, 193–218.

Tomlinson, J. A., Carter, A. L., Dale, W. T. and Simpson, C. J. (1970). Weed plants as sources of cucumber mosaic virus. *Ann Appl Biol* **66**, 11–16.

Tomlinson, J. A. and Faithfull, E. M. (1979). Effects of fungicides and surfactants on the zoospores of *Olpidium brassicae*. *Ann Appl Biol* **93**, 13–19.

Tomlinson, J. A. and Ward, C. M. (1981). The reactions of some Brussels Sprout Fl hybrids and inbreds to cauliflower mosaic and turnip mosaic viruses. *Ann Appl Biol* **97**, 205–12.

Vuittenez, A. (1970). Fanleaf of grape. In *Virus diseases of small fruits and grapevines* (ed. Frazier, N. W.). University of California, pp. 217–18.

Wallin, J. R., Peters, D. and Johnson, L. C. (1967). Low level jet winds, early cereal aphid and barley yellow dwarf detection in Iowa. *Plant Dis R* **51**, 527–30.

Watson, M. A. (1967). Epidemiology of aphid-transmitted plant virus diseases. *Outlook Agric* **5**, 155–66.

Watson, M. A., Heathcote, G. D., Lauckner, F. B. and Sowray, P. A. (1975) The use of weather data and counts of aphids in the field to predict the incidence of yellowing viruses of sugarbeet crops in England in relation to the use of insecticides. *Ann Appl Biol* **81**, 181–98.

Zink, F. W. and Duffus, J. E. (1969). Relationship of turnip mosaic virus susceptibility and downy mildew, *Bremia lactucae*, resistance in lettuce. *J Am Soc Hortic Sci* **94**, 403–7.

Zink, F. W. and Duffus, J. E. (1970). Linkage of turnip mosaic virus susceptibility and downy mildew, *Bremia lactucae*, resistance in lettuce. *J Am Soc Hortic Sci* **95**, 420–2.

Zitter, T. A. (1977). Epidemiology of aphid-borne viruses. In *Aphids as virus vectors* (ed. Harris, K. F. and Maramorosch, K.). Academic Press: London. pp. 385–412.

8.4 Further Selected Reading

Maramorosch, K. and Harris, K. F. (1981). *Plant diseases and vectors; ecology and epidemiology*. Academic Press.

Plumb, R. T. and Thresh, J. M. (1983). *Plant virus epidemiology*. Blackwell Scientific Publications: Oxford.

Thresh, J. M. (1976). Gradients of plant virus diseases. *Ann Appl Biol* **82**, 381–406.

Thresh, J. M. (1980). The origins and epidemiology of some important plant virus diseases. *Appl Biol* **5**, 1–65.

Thresh, J. M. (1981). *Pests, pathogens and vegetation*. Pitman: London.

Thresh, J. M. (1982). Cropping practices and virus spread. *Ann Rev Phyto* **20**, 193–218.
zitter, T. A. and Simons, J. N. (1980). Management of viruses by alteration of vector efficiency and by cultural practices. *Ann Rev Phyto* **18**, 289–310.

9 Basic Control Measures

9.1 Introduction

The ultimate aim of the applied plant virologist is to devise measures for the complete or partial control of a disease. To achieve this, the worker must have correctly identified the virus and understood its ecology and epidemiology. Armed with this information the most likely control methods can then be evaluated.

Unlike mycologists or bacteriologists, virologists do not have an array of chemicals with which to attack and control virus diseases. Considerable time and effort has been spent trying to find chemicals, that will directly eliminate or restrict virus multiplication in crop plants. To date, however, for reasons of ineffectiveness, phytotoxicity or economy, no such chemicals have been found. The virologist must, therefore, rely on control methods that prevent or restrict infection. Sometimes it is necessary to use a combination of control measures to combat a particular disease, and this approach is referred to as *integrated control*.

In this chapter various control measures are described, including those dealing with the avoidance of virus infection and the control of virus vectors. Later, in Chapter 10, the control of viruses through the use of resistant plants is discussed and in Chapter 11 procedures for the control of viruses in vegetatively propagated crops are described.

9.2 Elimination of the Source of Infection

9.2.1 Eradication of weeds and other alternative hosts

The importance of virus spread from weeds within or around a crop, has been discussed in the previous chapter. It is self evident, that if these sources can be eradicated by adequate cultivation, or the use of efficient herbicides, these potential reservoirs of virus are eliminated. Studies on the annual weed hosts of cucumber mosaic virus (CMV) occurring within lettuce crops in the United Kingdom (Tomlinson *et al.*, 1970) and the removal of reservoirs of celery mosaic virus, growing around celery crops in Florida (Townsend, 1947), illustrate clearly the importance of adequate weed control.

Virus spread from perennial ornamental hosts, or amenity trees (Cooper, 1981) growing in the vicinity of crop plants, may be more difficult to prevent, as frequently it would be impossible to destroy them. Eradication of these sources is particularly difficult in mixed cropping areas, where commercial crops are grown adjacent to private gardens (Conti and Masenga, 1977).

9.2.2 'Roguing' within the crop

The removal of infected plants from within a crop is another important control measure. Such control may be particularly important during the young stages of a crop's growth, when a few plants may be foci of infection for secondary virus spread. '*Roguing*' as a control measure has long been used in potato 'seed' crop production in the Netherlands, Germany and Scotland, and visual 'roguing' may be supplemented with serological tests to identify infected, symptomless plants (Gibbs and Harrison, 1976).

Removal of infected plants is also used to control cacao swollen shoot disease in West Africa (Kenton and Legg, 1971) and to limit the spread of plum pox virus in Britain.

9.2.3 Eradication of 'volunteer' plants

Plants surviving from a previous season's crop (volunteers) may be a potential reservoir of virus infection within a new crop. Such plants are common in potato crops where 'volunteer' tubers may be infected with the aphid transmitted potato virus Y and leaf roll viruses (Doncaster and Gregory, 1948).

Similarly, 'volunteer' sugar-beet plants, regenerating from root debris of previous seasons, may be a source of beet mosaic and yellowing viruses (Howell and Mink, 1971).

Adequate soil cultivation will prevent the spread of virus from these sources.

9.3 Avoidance of the Source of Infection

9.3.1 Modification of cropping procedures

Continued year round cultivation of the same crop, particularly in a tropical or sub-tropical climate, is a potential cause of virus disease epidemics. In California and Florida the overlapping of celery crops caused serious problems with aphid-transmitted celery mosaic virus, but the disease was readily controlled when a compulsory celery-free period was introduced (Zitter, 1977).

Similarly, overlapping of onion and shallot crops in New Zealand caused outbreaks of onion yellow dwarf virus (Chamberlain and Bayliss, 1948) and more recently in the Netherlands, continual cropping of leeks has led to outbreaks of leek yellow stripe virus (Bos, 1982). Both may be controlled by a break period when the crop is not grown. Such crop-free periods may be difficult to organize, particularly in tropical developing countries when a major food crop is involved. Problems were encountered, for instance, in the Solomon Islands when a control break was required in the taro (*Colocasia esculenta*) crop. This basic crop is traditionally grown throughout the year, in continuous overlapping cycles (Gollifer *et al.*, 1978).

Cropping practices may also be modified to prevent virus spread to new crops from the debris of old crops that remain in the soil. Such transmission can readily occur with highly stable viruses such as tomato mosaic virus (TMV) in the tomato crop. Although general hygiene (*see* Section 9.3.3) is very important in tomato crops (Broadbent, 1976), it may be impossible to eliminate all infected debris. In this case, growing tomato plants in compost contained within plastic bags has been found to be an efficient control measure (Wall, 1973). The polythene prevents the roots of the young plants from penetrating the old, infected soil. More recently, the use of trickle irrigation systems, with the plants growing in straw bales or rock-wood, and hydroponic systems such as the nutrient film culture technique, has eliminated the possibility of soil contamination altogether.

9.3.2 Cultivation in isolated areas

If the source of the virus cannot be eliminated, infection may be avoided by growing the crop in an area distant from the sources of infection. This measure is frequently used in certification schemes for the production of virus-free potato tubers to be used as 'seed'. In Britain (Todd, 1961) and Canada (Wright, N.S, personal communication) for example, potato 'seed' crops are grown in areas remote from the commercial 'ware' crop, and legislation may be used to control these distances and prevent the cultivation of potatoes in local gardens.

Biennial seed crops, such as brassicas and sugar-beet may also provide reservoirs of virus infection for a new season's commercial crop planted near by. The studies of Broadbent (1957) clearly demonstrated the value of isolating brassica seed-beds from commercial crops; similarly, the isolation of beet seed (*steckling*) – beds is an important control measure for beet mosaic and yellowing viruses in England

(Jepson and Green, 1983), and East Germany (Fritzsche *et al.*, 1972).

In the Sacramento Valley of California, isolation has been used to prevent the spread of beet viruses in the commercial root crop (Shepherd and Hills, 1970). In this case the problem of overlapping beet crops was solved, by separating crops by large distances.

9.3.3 Crop hygiene

Hygiene is particularly important with viruses such as tobacco mosaic (TMV) in tomato crops, where the highly stable virus can remain infectious for long periods (Broadbent, 1976; Lanter *et al.*, 1982). For efficient control, even when soil is not used as the culture medium (*see* Section 9.3.1), debris must be carefully removed from the glasshouse. Contamination of clothes, hands and tools must be avoided, and a 3 to 10% trisodium phosphate solution may be used as a disinfectant.

The plants should be handled as little as possible, and if some plants do become infected, one should never visit houses containing infected plants before entering houses with healthy plants.

9.3.4 Use of virus-free seed

Virus dissemination through seed can be an important source of infection in some crops. Seed infection is particularly significant if the virus is also aphid transmitted. This is the case with lettuce mosaic virus (LMV) in lettuce, for even a low rate of seed transmission can result in scattered foci of infection within the crop, and subsequent rapid spread of the disease by aphids (Grogan *et al.*, 1952). Studies have shown that the rate of LMV transmission in lettuce seed must be below 0.1% if adequate control is to be obtained (Zink *et al.*, 1956; Tomlinson, 1962). Lettuce seed producers now go to considerable expense to ensure that commercial seed meets this requirement. Seed crops are grown in isolation and seed lots are tested before marketing to ensure that they are virus-free.

The value of using virus-free lettuce seed, is illustrated by the increase in crop yields in the Salinas Valley of California, following the introduction of a seed certification programme. Prior to the use of virus-free seed, yields were 353 cartons per acre, but in the 5 years following its introduction, yields were increased to 478 cartons per acre (Kimble *et al.*, 1975).

Bean common mosaic virus (BCMV) is another virus that is both seed and aphid transmitted. Sanitation in the commercial production of seed of dwarf beans (*Phaseolus vulgaris*) is good, and the virus rarely occurs in field crops in the United Kingdom. Serious problems can occur, however, in dwarf bean accessions used in bean breeding programmes. If these accessions are obtained from world-wide sources, some will invariably be carrying BCMV, and when these are planted in field plots the virus can be spread rapidly, to invalidate the results of selection programmes (Walkey and Innes, 1979). Control in this case can be achieved by first growing all imported accessions in a glasshouse, and selecting seed only from virus-free plants. The occurrence of seed-borne viruses in breeding germplasm of many other crops, has been reported by Hampton *et al.* (1982).

Pea seedborne mosaic virus is a further example of a virus that can be transmitted in commercial seed lots. It was first discovered in the United States in 1968, and the destruction of infected seed lots controlled the spread of the virus until 1974 (Hampton *et al.*, 1976). It was then re-introduced into the United States in breeding lines raised in Canada, and has since been found in commercial seed batches grown in Europe. Careful management of pea seed production is now required to eradicate the disease.

The examples of seed-borne viruses so far described are all viruses that are transmitted in the embryo of the seed, but some viruses, such as TMV in tomato, may be seed transmitted by contamination of the seed coat (*see* Section 7.7). In this case, TMV spread occurs because the virus contaminates the cotyledons during seed germination, and this virus is then inoculated into the plant when the seedling is pricked out (Broadbent, 1976). This type of seed transmission may be controlled by sterilizing the seed coat in hydrochloric acid, trisodium orthophosphate or sodium hypochlorite (Gooding, 1975), or by sowing the seed directly into soil to avoid handling during transplanting (Broadbent, 1976).

9.3.5 Use of virus-free planting material

Vegetatively propagated crops present a special problem in respect of virus-disease control. Once a clone becomes infected, unless special measures are taken (*see* Chapter 11), all future crops of that clone will be diseased. This will probably result in a reduction of both yield and crop quality, while the infected clone can be a source of virus infection for other crops.

Control can be achieved by producing virus-free plants of the infected clone by meristem-tip culture or heat therapy (or a combination of both) as described in Chapter 11. This is followed by multiplication and distribution of the virus-free material to the grower through nuclear-stock schemes (Hollings, 1965; Walkey, 1980).

Infected planting material also provides a means by which virus diseases may be spread internationally, over great distances. The spread of plum pox virus in Europe and grapevine fan leaf virus throughout the world, in infected rootstocks (*see* Chapter 8), are vivid examples of this. Most agriculturally advanced countries have import and quarantine regulations aimed at controlling the entry of plant diseases and pests, and many countries have regulations for excluding specific virus diseases and/or their vectors.

Often these schemes are expensive to establish and administer, and their effectiveness may be restricted by economic and political considerations. Nevertheless, the enforcement of such regulations is essential, if the spread of virus diseases is to be controlled on a world-wide basis. It is possible that virus-free plantlets, raised by tissue culture (*see* Chapter 11), may provide the most satisfactory means for the world-wide distribution of vegetatively propagated planting material (Button, 1977).

9.4 Avoidance of the Vector

9.4.1 Cropping in vector-free areas

In the United Kingdom, potato 'seed' crops are grown in areas not only isolated from commercial potato fields (*see* Section 9.3.2), but also in areas where the aphid vectors are absent, or occur only in relatively low numbers (Todd, 1961). For this reason, the potato 'seed' crop is grown mainly in Scotland in relatively, cool, windy regions, where if aphids do occur, they arrive late in the season and fly infrequently.

In some countries, such as the Netherlands, areas completely free of aphid vectors do not occur. In these circumstances legislation may require the potato 'seed' crop to be harvested before a certain date, to avoid flights of the aphid vector. The precise date is determined by vector trapping data, and if the tubers are not lifted by that date, the mature haulms must be sprayed with herbicide to destroy the aerial parts of the plant, and so prevent infection (De Bokx, 1972).

If crops are particularly valuable, as is the case with virus-free mother plants of vegetatively propagated crops such as strawberries, bulbs or carnations, a nucleus of stock plants is grown in insect-free glass or gauze-houses. The vectors are excluded from such houses by fine mesh covers on all ventilators, and a double-door porch system is used to enter the house.

If nematode vectors are the problem, valuable crops which might be susceptible, can be grown in soil which has been tested and shown to be vector-free. When insect-free glass or gauze-houses are used for propagating nuclear-stock material, it is usual to grow the plants in sterilized soil, and to sterilize the soil between crops.

9.4.2 Changes in cropping practices

Airborne virus vectors can sometimes be avoided by altering the sowing or transplanting date of the crop. The optimal time for planting will depend on the normal migration times of the vector, so late planting will avoid early vector flights and vice versa. In the Sudan for example, the occurrence of broad bean mosaic virus which is transmitted by an aphid vector in field bean (*Vicia faba*) crops, is markedly influenced by sowing dates (Abu Salih *et al.*, 1973).

Another interesting example of control by the alteration of planting dates, is that of maize rough dwarf virus disease in maize crops in Israel. The virus is transmitted by a planthopper, *Laodelphax striatella*, in which the virus mltiplies. Studies have shown that the virus is incapable of multiplying in the vector after early June, because of high summer temperatures. Consequently, if the planting of the maize crop is delayed from its normal time in April, to late May, the incidence of virus is reduced from 45% to 3% (Harpaz, 1982).

An important consideration that must not be overlooked when planting dates are changed, is the effect of the change upon yield. Broadbent *et al.* (1957) showed for instance, that there was a greatly reduced virus incidence when potatoes were planted early, but yields were reduced to an uneconomic level.

Significant reductions in virus incidence may also be obtained if plants are grown at high densities. A'Brook demonstrated that the occurrence of the aphid transmitted, rosette virus of groundnuts, varied considerably at different planting densities (1964, 1968). Virus incidence was greatest in plants grown at wide spacings, and was associated with higher aphid populations on these plants. The advantages of the lower incidence of rosette infection in plants grown at high densities, must however, be carefully balanced against the lower yields that resulted from increased competition. In this situation it was essential to use a spacing that would achieve maximum ground cover, with the minimum yield reduction due to plant competition.

9.5 Chemical Control of Vectors

9.5.1 Airborne insect vectors

In general, insecticides are more effective against viruses that are transmitted in a persistant, rather than a non-persistent manner. The vectors of non-persistent viruses will eventually be killed after feeding on a plant sprayed with a systemic insecticide. However, because the virus may be transmitted within seconds of the insect starting to feed (*see* Section 7.4) many plants may become infected before the insect dies. In fact, sprays may agitate the insect and encourage it to move around and probe a greater number of plants than normal, with a resulting increase in transmission.

In the case of persistent viruses, however, where the vector requires many hours or even days to acquire and transmit the virus, systemic insecticidal sprays can be a very effective control measure. In the potato crop for instance, insecticidal treatments have been shown to reduce greatly the incidence of the persistently transmitted potato leaf roll virus. These measures have no effect however upon transmission of the non-persistent potato virus Y (Burt *et al.*, 1964; Webley and Stone, 1972).

Besides the effectiveness of spraying to control insect vectors, the cost of spraying and its environmental effects must also be considered. The cost of sprays and their application is expensive, tractor-wheel damage to the crop may be extensive, spray drift damage to adjacent crops can occur, and vector resistance to the insecticide may develop.

To some extent, these problems can be minimized if the number of sprays applied is reduced by, for example, careful timing of the application during the growing season. A warning system based on date records of aphid numbers is operated for sugar-beet crops in England. Under this system, a grower only sprays when virus spread by aphids is forecast (Heathcote, 1973). Similar warning systems have been used to protect potatoes and barley.

The disadvantages of sprays may be overcome to some extent, if insecticides, particularly systemics, are applied in granular form at planting. In the potato crop, granules fed through applicators into the furrows from the planting machine, have been shown to control aphids and hence transmission of the persistent potato leaf roll virus (Smith *et al.*, 1964*a*; Close, 1967). Hull and Selman (1965) reported, however, that granular insecticides had little effect on the incidence of the non-persistent pea mosaic virus, and the persistent pea enation mosaic virus in sweet pea crops.

Granular systemic insecticides should be formulated so that the active ingredients are released slowly to maximize the period over which they will protect the plant. The insecticides disulfton and phorate are particularly suitable in this respect.

Chemical control of the vector may sometimes be achieved by destroying its weed hosts, or overwintering hosts. This is particularly effective in controlling lettuce necrotic yellows virus in Australia (Stubbs *et al.*, 1963), for the aphid vector *Hyperomyzus lactucae* multiplies on sowthistle (*Sonchus oleraceus*) which is also a weed source of the virus.

9.5.2 Nematode vectors

Nematode-transmitted viruses may persist for relatively long periods in their nematode hosts, and may also be acquired by the eelworm from weeds or root fragments that persist in the soil (Thomason and McKenry, 1975; Martelli, 1978). Control by fallow periods, or other cultural procedures is therefore difficult, so chemical treatment to kill the vector is probably the most effective means of control. The difficulty with most nematicide treatments is achieving a 100% kill of the vector, because nematodes quite frequently occur at considerable depths which may be beyond, or on the fringe of effective nematicide penetration. Harrison *et al.* (1963) demonstrated that DD (dichloropropane-dichloropropene) or methyl bromide were effective nematicides. They controlled the infection of strawberry crops with arabis mosaic virus by killing over 99% of the eelworm vector, *Xiphinema diversicaudatum*, with pre-planting soil treatments. Treatment with DD or quintozene (pentachloronitrobenzene) has also been shown to reduce the incidence of tomato blackring and raspberry ringspot viruses in strawberry which are transmitted by the nematode *Longidorus elongatus*.

If soil temperatures are higher than 25°C then EBD (ethylenedibromide) has been found to be a more efficient nematicide than DD (Lamberti and Basile, 1982). Other non-fumigant chemicals may also be used as nematicides, and these are usually applied in a granular form. These include Aldicarb, which has been used to reduce the incidence of tobacco rattle virus (spraing disease), transmitted by trichodorid nematodes in potatoes (Alphey *et al.*, 1975). Another chemical Oxamyl, which may be applied as granules or as a foliar spray, has been shown to reduce the spread of nematode transmitted viruses by inhibiting their feeding and preventing the eelworm from acquiring the virus (Alphey, 1978).

9.5.3 Fungal vectors

Chemicals may be used to kill the resting spores or motile zoospore stages of fungal vectors. Various soil fumigants, such as methyl bromide, may be effective, but large-scale soil sterilization is probably uneconomic for many crops. It would, for instance, be too costly to sterilize the large areas of ground, which would be necessary to control the fungal vectors of soil-borne cereal mosaic viruses (Harrison, 1977). If, however, the ground area to be treated is relatively small, and a high value crop is involved, then soil treatment may be worthwhile. Van Slogteren (1970) for example, successfully controlled Augusta disease in tulips by killing the resting spores of its *Olpidium* vector with Dazomet.

Injection of methyl bromide into contaminated field soils has been successfully used to control big-vein disease of lettuce, which is transmitted by *Olpidium brassicae* (White, 1980 and 1983; Campbell *et al.*, 1980). The disease may also be controlled by incorporating carbendazim (methyl-2yl-benzimidazole carbamate) into peat blocks used for lettuce transplants, prior to transplanting in big-vein contaminated soil (White, 1983).

Carbendazim has also been used to control big-vein disease in lettuce crops grown in nutrient film culture (Tomlinson and Faithfull, 1979). In this type of cultivation, the lettuce are grown in concrete channels along which the nutrient medium is allowed to flow. Such a system provides an ideal situation for the multiplication and dissemination of the motile zoospores of the *Olpidium* vector, enabling the disease to spread rapidly throughout the crop. Control has also been obtained by killing the zoospores with surfactants, such as Agral, which are slowly released into the feeder tanks which contain the nutrient medium (Tomlinson and Faithfull, 1979).

The incidence of *Spongospora subterranea*, the fungal vector of potato mop-top virus has also been controlled by chemical treatment. Cooper *et al.* (1976) reported that if the pH of infected soil was lowered to 5.0 by the application of sulphur, the occurrence of mop-top disease was significantly reduced, although neither the vector nor the virus was eradicated from the soil.

9.6 Non-chemical Control of Insect Vectors

9.6.1 Barriers and reflective mulches

Barrier crops have been reported to be useful in controlling aphid transmitted viruses. Broadbent (1957) demonstrated that a barley barrier planted around a cauliflower seed-bed, reduced the incidence of cauliflower mosaic virus in the bed by 80%. The incoming aphids were thought to land on the barley and probe briefly, causing them to loose the non-persistently transmitted virus they were carrying.

More recently, sticky, yellow polythene sheets erected vertically on the windward side of pepper fields, have been shown to reduce the incidence of potato virus Y (PVY) and cucumber mosaic virus (CMV) in the crop (Cohen and Marco, 1973). The aphids are attracted to the yellow colour and are caught on the sticky polythene. The control obtained was so successful that the method has become a standard control procedure in pepper crops in Israel (Shoham, 1977). Similar traps have also been used to protect 'seed' potato crops, against potato leaf roll virus (Zimmerman-Gries, 1979).

Reflective surfaces (mulches) laid on the soil around the crop plant, have also been found to be highly effective in controlling aphid vectors. Aluminium strips, or grey or white plastic sheets, may be used as the mulch and have successfully protected peppers against CMV and PVY in Israel (Loebenstein *et al.*, 1975) and summer squash (*Cucurbita pepo*) against water melon mosaic virus in the Imperial Valley of California (Wyman *et al.*, 1979).

The mulches are thought to act as a repellent by reflecting UV light as the aphid comes into land (Smith *et al.*, 1964*b*). The disadvantages of mulch protection is that they are expensive and their efficiency tends to decrease as the leaves of the plant cover the mulch. They are, therefore, only economic for high value crops.

Straw mulches have been successfully used to control the whitefly transmitted tomato yellow leaf curl virus in tomato crops in Israel (Cohen *et al.*, 1974). Cohen (1982) believes that the colour of the straw attracts the whiteflies and they are subsequently killed by the reflective heat. The disadvantage with straw mulches is that they eventually lose their yellow colour, but prolonged control may be obtained if straw is replaced by yellow polythene sheets (Cohen and Melamed-Madjar, 1978).

9.6.2 Oil-sprays

If oils, such as paraffin (Bradley, 1963) or the mineral oil, albolineum (Asjes, 1974 and 1975), are sprayed on to the leaf surfaces of plants, aphid transmission of viruses may be prevented. Oil sprays have been effective in controlling the spread of a number of non-persistently transmitted viruses (Loebenstein *et al.*, 1970; Mowat and Woodford, 1976; and Vanderveken, 1977), but their effect against persistent viruses has been variable. Studies with pea enation mosaic virus (Peters and Lebbink, 1973) and potato leaf roll virus (PLRV) (Hein, 1971), both persistently transmitted viruses, showed oil sprays to have little or no effect. In contrast, however, oil was found to reduce the transmission of the persistent tomato yellows disease (probably caused by a strain of PLRV) in tomatoes in Florida (Zitter and Everett, 1979). Oil sprays have also been reported to reduce the incidence of a whitefly transmitted virus, tomato leaf curl virus, in field grown tomatoes in India (Singh *et al.*, 1973).

Problems associated with using oil sprays as a control measure, include phytotoxicity and failure to maintain a continual cover of oil over the whole surface of the plant. Oil sprays as dilute as 1% may cause damage in some plants and the oil layer may be readily removed by rain or irrigation water (De Wijs *et al.*, 1979) and weekly sprays may not be frequent enough if the young expanding leaves are to be protected (Walkey and Dance, 1979).

The mechanism of protection is not fully understood, but since the oil does not directly inactivate the virus, it seems likely that it may interfere with transmission as the aphid stylets probe the leaf.

9.6.3 Biological control by predators

Although some workers have attempted to control virus vectors by means of predators, parasites or pathogens, such measures have met with little success (Harpaz, 1982). Recently, however, attempts have been made in Chile to control the rapid spread of the aphid vectors of barley yellow dwarf virus, by using natural predators such as the insect *Aphidius ervi* (Van den Bosch, 1976).

It is thought that the low incidence of barley yellow dwarf virus in Israel is due to the presence of natural predators of the aphid vectors and that the introduction of such predators to Chile may be successful.

Control by predators on their own is unlikely to be completely successful, but used as an integrated control measure with selective insecticides, they may be effective (Harpaz, 1982).

9.7 Control by Cross-Protection

The phenomenon of cross-protection has already been described in Section 6.4 and may be used as a control measure. If a plant is deliberately infected with a mild strain of a virus, it may be protected against later infection by a more severe strain of the same virus. Some yield loss may be expected from infection by the mild strain, but severe yield losses are avoided.

Commercially, the technique has been used in the tomato industry to protect crops against severe strains of TMV. Rast (1972, 1975) developed mild TMV strains by mutagenic action with nitrous acid, and these strains were widely used in the Netherlands and the United Kingdom by growers in the early and mid-1970s (Fletcher, 1978). In the Netherlands the average yield from early glasshouse crops was increased by 15% and in the United Kingdom by 7%, using this method (Upstone, 1974). In recent years it has been replaced by the cultivation of TMV resistant tomato cultivars.

Other examples of protection by mild virus strains have been reported in citrus crops. In Florida, mild isolates of citrus tristeza virus are widespread in citrus trees, in which they cause symptomless infection. This infection is thought to protect the trees against more severe damage from more virulent strains of the virus (Cohen, 1976). Similar protection is also reported to have been obtained, by inoculating grapefruit, orange and lime crops with mild strains of tristeza in Brazil (Muller and Costa, 1977).

9.8 Control of Disease Symptoms by Chemicals

As mentioned previously, there are no chemicals that can be used commercially to cure crops of virus infection. Recently, however, some systemic fungicides have been shown to suppress virus symptoms when applied to diseased plants, without necessarily reducing the concentration of virus within the plant. Tomlinson *et al.* (1976) demonstrated that fungicides such as Benlate and Bavistin, which contain methyl benzimidazole-2yl-carbamate (MBC or carbendazim), reduced the severity of mosaic symptoms caused by TMV in tobacco, and yellowing symptoms produced by beet western yellows virus in lettuce. The chemicals were applied to the plants as a soil drench.

It is thought that these fungicides may have a cytokinin-like activity, in that they delayed the breakdown of the chloroplasts by the virus, thereby suppressing disease symptoms. The possibility that these chemicals may have a widespread, commercial application against other virus diseases, particularly of the yellowing type, remains to be assessed.

9.9 References

A'Brook, J. (1964). The effect of planting date and spacing on the incidence of groundnut rosette disease and of the vector *Aphis craccivora* Koch, at Mokawa, Northern Nigeria. *Ann Appl Biol* **54**, 199–208.

A'Brook, J. (1968). The effect of plant spacing on the numbers of aphids trapped over the groundnut crop. *Ann Appl Biol* **61**, 289–94.

Abu Salih, H. S., Ishag, H. M. and Siddig, S. A. (1973). Effect of sowing date on incidence of Sudanese broad bean mosaic virus in, and yield of, *Vicia faba*. *Ann Appl Biol* **74**, 371–8.

Alphey, T. J. W. (1978). Oxamyl sprays for the control of potato spraing disease caused by nematode-transmitted tobacco rattle virus. *Ann Appl Biol* **88**, 75–80.

Alphey, T. J. W., Cooper, J. I. and Harrison, B. D. (1975). Systemic nematicides for the control of trichodorid nematodes and of spraing disease caused by tobacco rattle virus. *Plant Path* **24**, 117–21.

Asjes, C. J. (1974). Control of the spread of the brown ring formation virus disease in the lily Mid-Century hybrid Enchantment by mineral-oil sprays. *Acta Hort* **36**, 85–91.

Asjes, C. J. (1975). Control of the spread of tulip breaking virus in tulips with mineral-oil sprays. *Neth J Plant Pathol* **81**, 64–70.

Bos, L. (1982). Viruses and virus diseases of *Allium* species. *Acta Hort* **127**, 11–29.

Bradley, R. H. E. (1963). Some ways in which a paraffin oil impedes aphid transmission of potato virus Y. *Can J Microbiol* **9**, 369–80.

Broadbent, L. (1957). Investigations of virus diseases of Brassica crops. A.R.C. Rep. Ser. No. 14. Cambridge Univ. Press.

Broadbent, L. (1976). Epidemiology and control of tomato mosaic virus. *Ann Rev Phyto* **14**, 75–96.

Broadbent, L., Heathcote, G. D., McDermott, N. and Taylor, C. E. (1957). The effect of date of planting and of harvesting potatoes on virus infection and on yield. *Ann Appl Biol* **45**, 603–22.

Burt, P. E., Heathcote, G. D. and Broadbent, L. (1964). The use of insecticides to find when leaf roll and virus Y, spread within potato crops. *Ann Appl Biol* **54**, 13–22.

Button, J. (1977). International exchange of disease free citrus clones by means of tissue culture. *Outlook Agric* **9**, 155–9.

Campbell, R. N., Greathead, A. S. and Westerlund, F. V. (1980). Big-vein of lettuce infection and methods of control. *Phyt* **70**, 741–6.

Chamberlain, E. E. and Bayliss, G. T. S. (1948). Onion yellow dwarf. Successful eradication. *NZ J Sci Technol* **A29**, 300–1.

Close, R. C. (1967). Granular insecticides for aphid control. *Proc NZ Weed Pest Control. Conf.* **20**, 222–6.

Cohen, M. (1976). A comparison of some tristeza isolates and a cross-protection trial in Florida. *Proc. 7th Conf. Int. Org.* Citrus Virol. Gainesville, Fla., pp. 50–4.

Cohen, S. (1982). Control of whitefly vectors of viruses by colour mulches. In *Pathogens, vectors and plant diseases, approaches to control* (ed. Harris, K. F. and Maramorosch, K.). Academic Press: London. pp. 45–56.

Cohen, S. and Marco, S. (1973). Reducing the spread of aphid-transmitted viruses in peppers by trapping the aphids on sticky yellow polythene sheets. *Phyt* **63**, 1207–9.

Cohen, S. and Melamed-Madjar, V. (1978). Prevention by soil mulching of the spread of tomato yellow leaf curl virus transmitted by *Bemisia tabaci* (Gennadius) in Israel. *Bull Entomol Res* **68**, 465–70.

Cohen, S., Melamed-Madjar, V. and Hameiri, J. (1974). Prevention of the spread of tomato yellow leaf curl virus transmitted by *Bemisia tabaci* in Israel. *Bull Entomol Res* **64**, 193–7.

Conti, M. and Masenga, V. (1977). Identification and prevalence of pepper viruses in Northwest Italy. *Phytopathol Z* **90**, 212–22.

Cooper, J. I. (1981). The possible role of amenity trees and shrubs as virus reservoirs in the United Kingdom. In *Pests, Pathogens and Vegetation* (ed. Thresh, J. M.). Pitman: London. pp. 79–87.

Cooper, J. I., Jones, R. A. C. and Harrison, B. D. (1976). Field and glasshouse experiments on the control of potato mop-top virus. *Ann Appl Biol* **83**, 215–30.

De Bokx, J. A. (1972). Viruses of potatoes and seed potato production. Pudoc, Wageningen.

De Wijs, J. J., Sturm, E. and Schwinn, F. J. (1979). The viscosity of mineral oils in relation to their ability to inhibit the transmission of stylet-borne viruses. *Neth J Plant Pathol* **85**, 19–22.

Doncaster, J. P. and Gregory, P. H. (1948). The spread of visus diseases in the potato crop. *G B Agric Res Counc R Ser* **7**, 1–189.

Fletcher, J. T. (1978). The use of avirulent strains to protect plants against the effects of virulent strains. *Ann Appl Biol* **89**, 110–14.

Fritzsche, R., Karl, E., Lehmann, W. and Proeseler, G. (1972). Tierische Vektoren pflanzenpathogener Viren. Fischer, Jena.

Gibbs, A. J. and Harrison, B. D. (1976). *Plant virology; the principles.* Edward Arnold: London.

Gollifer, D. E., Jackson, G. V. H., Dabek, A. J. and Plumb, R. T. (1978). Incidence and effects on yield, of virus diseases of taro (*Colocasia esculenta*) in the Solomon Islands. *Ann Appl Biol* **88**, 131–5.

Gooding, G. V. (1975). Inactivation of tobacco mosaic virus on tomato seed with trisodium orthophosphate and sodium hypochlorite. *Plant Dis R* **59**, 770–2.

Grogan, R. G., Welch, J. E., and Bardin, R. (1952). Common lettuce mosaic and its control by the use of disease free seed. *Phyt* **42**, 573–8.

Hampton, R. O., Mink, Hamilton, R. I., G. I., Kraft, J. M. and Meuhlbauer, F. J. (1976). Occurrence of pea seedborne mosaic virus in North American pea breeding lines, and procedures for its elimination. *Plant Dis R* **60**, 455–9.

Hampton, R. O., Waterworth, H., Goodman, R. M., and Lee, R. (1982). Importance of seedborne viruses in crop germplasm. *Plant Dis R* **66**, 977–8.

Harpaz, I. (1982). Nonpesticidal control of vector-borne diseases. In *Pathogens, vectors and plant diseases: approaches to control* (ed. Harris, K. F. and Maramorosch, K.). Academic Press: London. pp. 1–21.

Harrison, B. D. (1977). Ecology and control of viruses with soil-inhabiting vectors. *Ann R Phyto* **15**, 331–60.

Harrison, B. D., Peachey, J. E. and Winslow, R. D. (1963). The use of nematicides to control the spread of arabis mosaic virus by *Xiphinema diversicaudatum* (Micol.). *Ann Appl Biol* **52**, 243–55.

Heathcote, G. D. (1973). Beet mosaic – a declining disease in England. *Plant Path* **22**, 42–5.

Hein, A. (1971). Zur Wirkung von Ol ouf die Virusubertragung durch Blattlause. *Phytopathol Z* **71**, 42–8.

Hollings, M. (1965). Disease control through virus-free stock. *Ann Rev Phyto* **3**, 367–96.

Howell, W. E. and Mink G. I. (1971). The relationship between volunteer sugarbeets and occurrence of beet mosaic and beet western yellows viruses in Washington beet fields. *Plant Dis R* **55**, 676–8.

Hull, R. and Selman, I. W. (1965). The incidence and spread of viruses in sweet peas (*Lathyrus odoratus L.*) in relation to variety and the use of systemic insecticides. *Ann Appl Biol* **55**, 39–50.

Jepson, P. C. and Green, R. E. (1983). Prospects for improving control strategies for sugar-beet pests in England. *Adv Appl Biol* **7**, 175–243.

Kenton, R. H. and Legg, J. T. (1971). Varietal resistance of cocoa to swollen shoot disease in West Africa. *Plant Proc Bull FA O* **19**, 1–11.

Kimble, K. A., Grogan, R. G., Greathead, A. S., Paulus, A. O. and House, J. K. (1975). Development, application, and comparison of methods for indexing lettuce seed for mosaic virus in California. *Plant Dis R* **59**, 461–4.

Lamberti, F., and Basile, M. (1982). Chemical control of nematode vectors. In

Pathogens, vectors and plant disease – approaches to control (ed. Harris, K. F. and Maramorosch, K.). Academic Press: London. pp. 57–69.

Lanter, J. M., McQuire, J. M. and Goode, M. J. (1982). Persistence of tomato mosaic virus in tomato debris and soil under field conditions. *Plant Dis R* **66**, 552–5.

Loebenstein, G., Alper, M. and Levy, S. (1970). Field tests with oil sprays for the prevention of aphid spread viruses in peppers. *Phyt* **60**, 212–15.

Loebenstein, G., Alper, M., Levy, S., Palevitch, D. and Menagem E. (1975). Protecting peppers from aphid-borne viruses with aluminium foil and plastic mulch. *Phytoparasitica* **3**, 43–53.

Martelli, G. P. (1978). Nematode-borne viruses of grapevine, their epidemiology and control. *Nematol Mediterr* **6**, 1–27.

Mowat, W. P. and Woodford, J. A. T. (1976). Control of the spread of two non-persistent aphid-borne viruses in lilies. *Acta Hort* **59**, 27–8.

Muller, G. W. and Costa, A. S. (1977). Tresteza control in Brazil by pre-immunization with mild strains. *Proc Int Soc Citriculture* **3**, 368–72.

Peters, D. and Lebbink, G. (1973). The effect of oil on the transmission of pea enation mosaic virus during short inoculation probes. *Ent Exp Appl* **16**, 185–90.

Rast, A. T. B. (1972). M11–16, an artificial symptomless mutant of tobacco mosaic virus for seedling inoculation of tomato crops. *Neth J Plant Pathol* **78**, 110–12.

Rast, A. T. B. (1975). Variability of tobacco mosaic virus in relation to control of tomato mosaic in glasshouse tomato crops by resistance breeding and cross protection. *Agric Res R (Neth)*, **834**, 1–76.

Shepherd, R. J. and Hills, F. J. (1970). Dispersal of beet yellows and beet mosaic viruses in the inland valleys of California. *Phyt* **60**, 798–804.

Shoham, C. (1977). Recommendations for the control of pests in vegetable crops. 1977/78. Israel Min.of Agr., Extension Service, Tel Aviv.

Singh, S. J., Sastry, K. S. M. and Sastry, K. S. (1973). Effect of oil spray on the control of tomato leaf-curl virus in the field. *Indian J Agric Sci* **43**, 669–72.

Smith, H. C., Close, R. C. and Rough, B. F. A. (1964*a*). The efficiency of granular insecticides in controlling virus diseases of crops. *Proc NZ Weed Pest Control Conf* **17**, 68–74.

Smith, F. F., Johnson, G. V., Kahn, R. P. and Bing, A. (1964*b*). Repellancy of reflective aluminium to transient aphid virus-vectors. *Phyt* **54**, 748.

Stubbs, L. L., Guy, A. D., and Stubbs, K. J. (1963). Control of lettuce necrotic yellows virus disease by the destruction of common sowthistle (*Sonchus oleraceus*). *Aust J Exp Agr Anim Husb* **3**, 215–18.

Thomason, I. J., and McKenry, M. V. (1975). Chemical control of nematode vectors of plant viruses. In *Nematode vectors of plant viruses* (ed. Lamberti, F., Taylor, C. E. and Seinhorst, J. W.). Plenum Press: New York. pp. 423–39.

Todd, J. M. (1961). The incidence and control of aphid-borne potato virus diseases in Scotland. *Eur Potato J* **4**, 316–29.

Tomlinson, J. A. (1962). Control of lettuce mosaic by the use of healthy seed. *Plant Path* **11**, 61–4.

Tomlinson, J. A., and Faithful, E. M. (1979). Effects of fungicides and surfactants on the zoospores of *Olpidium brassicae*. *Ann Appl Biol* **93**, 13–19.

Tomlinson, J. A., Carter, A. L., Dale, W. T. and Simpson, C. J. (1970). Weed plants as sources of cucumber mosaic virus. *Ann Appl Biol* **66**, 11–16.

Tomlinson, J. A., Faithful, E. M. and Ward, C. M. (1976). Chemical suppression of the symptoms of two virus diseases. *Ann Appl Biol* **84**, 31–41.

Townsend, G. R. (1947). Celery mosaic in the Everglades. *Plant Dis R* **31**, 118–19.

Upstone, M. E. (1974). Effects of inoculation with the Dutch mutant strain of tobacco mosaic virus on the cropping of commercial tomatoes. *R ADAS*, 1972, pp. 162–5.

Van den Bosch, R. (1976). Report on a second visit to Chile as a consultant to the program of biological and integrated control of cerial aphids (*see* Harpaz, 1982).

Vanderveken, J. J. (1977). Oils and other inhibitors of non-persistent virus transmission. In *Aphids as virus vectors* (ed. Harris, K. F. and Maramorosch, K.). Academic Press: London. pp. 435–54.

Van Slogteren, D. H. M. (1970). Augustaziek in tulpen (tabaksnecrosevirus). Jaarversl. Lab. Bloembollenonderz. Lisse. 1969–70, p. 34.

Walkey, D. G. A. (1980). Production of virus-free plants. *Acta Hort* **88**, 23–31.

Walkey, D. G. A. and Dance, M. C. (1979). The effect of oil sprays on aphid transmission of turnip mosaic, beet yellows, bean common mosaic and bean yellow mosaic viruses. *Plant Dis R* **63**, 877–81.

Walkey, D. G. A. and Innes, N. L. (1979). Resistance to bean common mosaic virus in dwarf beans. *J Agric Sci* **92**, 101–8.

Wall, E. T. (1973). *Isolated growing systems. The UK tomato manual.* Grower Books: London. pp. 94–9.

Webley, D. P. and Stone, L. E. W. (1972). Field experiments on potato aphids and virus spread in South Wales 1966/9. *Ann Appl Biol* **72**, 197–203.

White, J. G. (1980). Control of lettuce big-vein disease by soil sterilisation. *Plant Path* **29**, 124–30.

White, J. G. (1983). The use of methyl bromide and carbendazim for the control of lettuce big-vein disease. *Plant Path* **32**, 151–7.

Wyman, J. A., Toscano, N. C., Kido, K., Johnson, H. and Mayberry, K. S. (1979). Effects of mulching on the spread of aphid-transmitted watermelon mosaic virus to summer squash. *J Econ Entomol* **72**, 139–43.

Zimmerman-Gries, S. (1979). Reducing the spread of potato leaf roll virus, alfalfa mosaic virus and potato virus Y in seed potatoes by trapping aphids on sticky yellow polyethylene sheets. *Potato Res* **22**, 123–31.

Zink, F. W., Grogan, R. G. and Welch, J. E. (1956). The effect of percentage of seed transmission upon subsequent spread of lettuce mosaic virus. *Phyt* **46**, 622–4.

Zitter, T. A. (1977). Epidemiology of aphid-borne viruses. In *Aphids as virus vectors* (ed. Harris, K. F. and Maramorosch, K.). Academic Press: London. pp. 385–412.

Zitter, T. A. and Everett, P. H. (1979). Use of mineral oil sprays to reduce the spread of tomato yellows virus disease in Florida. *Univ Fla Immokalee Res Rep SF* 79–1, 7pp.

9.10 Further Selected reading

Broadbent, L. (1976). Epidemiology and control of tomato mosaic virus. *Ann R Phyto* **14**, 75–96.

Harris, K. F. and Maramorosch, K. (1977). *Aphids as virus vectors*. Academic Press: London.

Harris, K. F. and Maramorosch, K. (1982). *Pathogens, vectors and plant diseases; approaches to control*. Academic Press: London.

Harrison, B. D. (1977). Ecology and control of viruses with soil-inhabiting vectors. *Ann Rev Phyto* **15**, 331–60.

Matthews, R. E. F. (1981). *Plant virology*. Academic Press: London.

Plumb, R. T. and Thresh, J. M. (1983). *Plant virus epidemiology; The spread and control of insect-borne viruses*. Blackwell Scientific Publications: Oxford.

Thresh, J. M. (1981). *Pests, pathogens and vegetation*. Pitman: London.

Zitter, T. A. and Simons, J. N. (1980). Management of viruses by alteration of vector efficiency and by cultural practices. *Ann Rev Phyto* **18**, 289–310.

10 *Control Through Resistant Cultivars*

10.1 Introduction

In addition to the control measures described in Chapter 9, considerable time and effort has been spent selecting and breeding cultivars that are resistant to virus infection. The basis of any selection or breeding programme, is the existence of genetic variation within the species for response to a particular virus. Fortunately, this variation exists and can be exploited for many viruses that are of economic importance.

Genetically controlled, inherited (or *constitutive*) resistance should not be confused with *induced* (or *acquired*) resistance, which occurs when a normally susceptible plant has resistance conferred on it by a predisposing treatment (Ouchi, 1983). Induced resistance may result from prior infection of the host plant by a virus, so that subsequent, younger leaves are resistant to infection by the same virus or a closely related strain of the virus (*see* Section 6.4). The phenomenon, known as *cross-protection*, is the basis of the control procedure used to protect tomato plants, which is described in Section 9.7. Alternatively, induced resistance may be obtained if the host plant is inoculated with certain chemicals, such acetyl salicyclic acid (aspirin) (White, 1979). Following inoculation-treatment with the chemical, later virus infection of the plant may be prevented or restricted. Induced resistance is not inherited and is of no use in resistance breeding programmes.

As far as the grower is concerned, control through the use of resistant cultivars, is probably the cheapest and most effective way of combating virus diseases. The cost of growing a resistant cultivar is likely to be no greater than growing a susceptible one, and savings are made by not having to take other costly measures, such as vector control. In addition, if the use of chemicals is avoided, so are the possibilities of environmental pollution, and the development of vector-resistance to the insecticide. Virus-resistant cultivars are particularly useful in controlling viruses that are transmitted by aphid vectors in a non-persistent manner (Walkey *et al.*, 1982), since these viruses are not effectively controlled by killing the vector with insecticides (*see* Section 9.5.1).

In contrast to the grower's lower costs for disease control, the initial cost of producing a new, resistant cultivar can be high. It may take the breeder many years to combine the required resistance with the necessary agronomic characteristics of the crop species concerned. The value of a resistant cultivar with low yield, or poor crop quality would be questionable, but provided the genes controlling virus resistance and poor agronomic characters are not linked, the breeder, given time, should be able to produce a resistant cultivar which is no less agronomically desirable, than a susceptible one.

When breeding for resistance, it is essential to have a sound knowledge of the virus concerned. The procedures for handling the virus, identifying and working with its various strains, and methods for assessment of field symptoms or virus concentrations, may be complex. Close collaboration between the breeder and a virologist is therefore advisable, if not essential, for a successful programme. A cultivar, bred for virus resistance, must also not be highly susceptible to other diseases. It is important, therefore, for breeders to test virus resistant lines for susceptibility to other important pathogens. In breeding for resistance to bean common mosaic virus (BCMV) in *Phaseolus vulgaris* beans for example, it was necessary to combine the BCMV resistance with resistance to halo-blight (caused by the bacterium *Pseudomonas phaseolicola*) and to anthracnose (caused by the fungus *Colletotrichum lindemuthianum*) (Conway *et al.*, 1982).

The importance of resistance to viruses and other plant pathogens may be overlooked by plant breeders striving to select solely for improved agronomic performance, such as higher yields and increased uniformity. Consequently, new cultivars may be produced which are highly susceptible to virus diseases, even though existing commercial cultivars exhibit adequate resistance. Examples of modern hybrid cultivars with increased susceptibility to virus infection were discussed in Section 8.2.6. In the past, these problems did not normally arise, since most crop species have evolved over years of cultivation during which farmers have selected plants without deleterious features, such as susceptibility to disease (Russell, 1978). Under these selection conditions, only plants with a relatively high level of resistance to the most important diseases survived. These stocks were selected and maintained by individual growers on a local basis over many years, and have come to be referred to as '*land races*'. 'Land races' are usually highly adapted to local conditions and often possess a diverse pool of genetic variability, including resistance to many major diseases. This resistance generally prevents widespread losses due to disease epidemics.

10.2 Definitions and Examples of Host Resistance to Viruses

One of the major problems associated with virus resistance studies, is the multiplicity of meanings that different workers give to the various terms they use. It is, therefore, essential for a worker to define the terms he is using. In this section, the various terms used in resistance breeding are defined in relation to their most common and accepted usage, and examples of each type of resistance response are given. The major terms are listed in Table 10.1, together with their characteristics as expressed by host symptoms and virus multiplication. Examples of virus resistance in some important crop species are presented in Table 10.2.

Further information on resistance terms can be found in *A guide to the use of terms in plant pathology* (Federation B.P.P., 1973) and in an article by Cooper and Jones (1983).

10.2.1 The main responses of the host to virus infection

Susceptibility
A plant is *susceptible* if a virus readily infects and multiples within it (*see* Table 10.1). Susceptible can be considered to be the opposite of *resistant* and low or high levels of host susceptibility can be recognized. High susceptibility and low resistance are considered synonymous and vice versa.

Table 10.1 Resistance terms in relation to host symptoms and virus multiplication

Term	*Host symptoms*	*Virus multiplication*
Susceptibility	+++	+++
Immunity (non-host)	−	−
Resistance (low susceptibility)	±	±
Tolerance	±	++ to +++
Hypersensitivity	local lesions and/or death	+

Immunity
The terms *immune* and *immunity* are often given different meanings by different workers. The consensus of opinion now favours restricting these terms to absolute exemption from infection by a specific pathogen. An immune plant is not attacked at all by the particular virus and can be considered to be a *non-host* of the virus concerned. In fact, most plant species are immune to infection by most viruses, and the plant breeder does not need to breed resistant cultivars of that

Table 10.2 Examples of plant virus resistance

Virus	*Host*	*Resistance gene(s)*	*Reference*
Resistance controlled by a single dominant gene			
Bean common mosaic	*Phaseolus vulgaris*	*I*	Ali (1950)
Cucumber mosaic	Spinach	*	Pound & Cheo (1952)
	Cowpea	*	Sinclair & Walker (1955)
	Lactuca saligna	*	Provvidenti *et al.* (1980)
Potato virus X	*Capsicum pendulum*	*	Nagaich *et al.* (1968)
Tobacco mosaic	*Nicotiana glutinosa*	*N*	Holmes (1938)
	N. tabacum	*N*	Takahashi (1975)
	N. tabacum	*N'*	Valleau (1952)
Tomato mosaic	Tomato	*Tm-2*	Pelham (1972)
Resistance controlled by incompletely dominant or recessive genes			
Barley yellow dwarf	Barley	*	Catherall *et al.* (1970)
Beet mosaic	Sugar beet	*Bm*	Lewellen (1973)
Cucumber mosaic	Melon	*	Karchi *et al.* (1975)
Potato virus Y	*Capsicum annuum*	y^a	Shifriss & Marco (1980)
Tobacco mosaic	C. chinense	L^1, L^3	Boukwma (1980)
Resistance controlled by recessive genes			
Alfalfa mosaic	Alfalfa	*am-1*	Crill *et al.* (1971)
Bean yellow mosaic	*P. vulgaris*	*by-3*	Provvidenti & Schroeder (1973)
Bean common mosaic	*P. vulgaris*	*bc-1, bc-2, bc-3*	Drijfhout (1978)
Tomato spotted wilt	Tomato	sw_{-2} sw_{-3} sw_{-4}	Finlay (1953)

*Gene not named

species, because all cultivars will be non-hosts for the virus concerned. For example, beet yellow virus (BYV) cannot infect barley, and barley yellow dwarf virus (BYDV) cannot infect sugar-beet. Thus barley is immune to BYV and sugar-beet to BYDV. The reasons why plants are immune to infection by a specific virus are not fully understood, and until the factors governing the non-host response are known, non-host immunity cannot be utilized in resistance breeding programmes (Fraser, 1982).

The term 'immunity' has also been frequently used to describe a plant response to virus infection, where a virus has been shown not to infect a particular cultivar of a species, normally susceptible to that virus. In this type of response it may not be possible to detect any virus establishment in the host species concerned, or the virus may have been confined to one or two cells close to the site of inoculation. The term *'extreme'* resistance (Russell, 1978) is probably the most suitable to describe this very high level of resistance response. In some cases, it may be difficult to distinguish between 'extreme' resistance and absolute immunity without detailed histological examination of the cells of the inoculated tissues. It therefore seems logical to restrict the use of the term immunity to the 'non-host' situation.

The term *acquired immunity* is often used to describe a resistance response acquired by the host following a predisposing treatment (*see* Section 10.1). This term is synonymous with *induced* or *acquired resistance*. Although widely used, it is the author's view that it should be avoided, for it is resistance and not absolute immunity that is induced in the host concerned.

Resistance

A host plant is *resistant* if it possesses the ability to suppress or retard the multiplication of a virus or the development of pathogenic symptoms. Resistant is the opposite of susceptible, and may be divided into high (extreme), moderate, or low resistance, depending upon its effectiveness. Essentially a resistant plant shows reduced or no symptom expression and virus multiplication within it is reduced or negligible (*see* Table 10.1).

Several different types of host resistance to viruses are recognized, but in no case is the mechanism fully understood. The host may be resistant to establishment of infection, virus multiplication or virus movement.

The tendency for a host variant to resist infection by a virus, to which the species is usually susceptible, is a characteristic that has been used in breeding for resistance in various crops. A tendency to resist infection has been reported for both mechanically and vector-transmitted viruses. Some tobacco lines have been shown to avoid infection by cucumber mosaic (CMV) (Troutman and Fulton, 1958), tobacco mosaic (TMV) and tobacco necrosis (Holmes, 1961) viruses. The same phenomenon occurs in some tomato cultivars to infection by TMV (Holmes, 1955). Although the mechanism for this type of resistance is not completely understood, it is thought that the thickness of the leaf cuticle and the nature of the epidermal hairs may influence transmission. Thomas and Fulton (1968) reported that resistance in tobacco was correlated with the number of ectodesmata in the epidermal cells, through which entry of mechanically transmitted TMV was thought to occur.

Various examples are known of host cultivars that exhibit resistance to virus multiplication. Such hosts are readily infected by the virus, but the virus does not multiply to the same extent, or as rapidly in them, as it does in others. This type of resistance has been reported to occur in the Ambalema cultivar of tobacco in respect of TMV multiplication (Bancroft and Pound, 1954), and in cucumbers resistant to CMV (Wasuwat and Walker, 1961).

The systemic movement of a virus may be limited or delayed in some plants. This type of resistance has been used in breeding programmes for various crops, including potato and maize. In potato the spread of potato virus Y to the tubers is restricted in certain cultivars (Beemster, 1972), and in some maize cultivars the spread of maize dwarf mosaic virus may be restricted (Jones and Tolin, 1972). This type of restricted movement should not be confused with localization of virus infection that may result from a hypersensitive host reaction, which is described later in this section.

Tolerance

The accepted definition of *tolerance*, is a host response to virus infection that results in negligible or mild symptom expression, but relatively normal levels of virus movement and concentration within the host (*see* Table 10.1). Unfortunately, the term is widely misused by some workers, to describe host responses involving reduced symptom expression due to *resistance* resulting from low levels of infection, or reduced virus multiplication.

Virus tolerance may be heritable and has been bred for in many crops including citrus in respect of tristeza virus (Posnette, 1969), in barley against barley yellow dwarf virus (Catherall *et al.*, 1970), in beans against curly top virus (Thomas and Martin, 1969) and in cacao against swollen shoot disease (Brunt, 1975).

The use of tolerance in breeding programmes has been criticized by some workers, because tolerant plants may be a potential reservoir of virus for infection of nearby susceptible crops (Matthews, 1981). Sometimes, however, tolerance may be the only type of protection that is available to the breeder, and in practice, tolerance has been widely used in breeding programmes to produce cultivars, that have been successfully grown commercially over long periods of time (Russell, 1978). It can well be argued, that in the field, there is no reason why tolerant cultivars should become infected more frequently than susceptible ones. In the case of perennial crops, however, tolerant cultivars should be carefully managed to prevent them acting as long-term symptomless carriers of virus diseases.

Hypersensitivity
Some plants respond to virus infection with a reaction that results in early death of the inoculated tissues. This reaction is often associated with a lack of further virus spread. This type of *hypersensitive* reaction is frequently seen when viruses are mechanically sap transmitted to the leaves of a host plant. The virus may be restricted to the inoculated cells or cells adjacent to the inoculation site, which soon die to form a necrotic *local lesion* (*see* Section 3.2.1 and Plate 4.3). In some cases, the virus is completely restricted to the local lesion site, but in others the virus spreads systemically through the vascular system of the host, causing vascular necrosis and relatively rapid death of the plant. The 'black-root' reaction caused by certain strains of bean common mosaic virus in *Phaseolus* beans, is a classic example of this type of response (*see* Plate 3.7).

The capability to react in a hypersensitive manner may be inherited, and this characteristic is frequently used by the breeder to give protection to crop species. In cultivars carrying hypersensitivity genes, control is achieved either by the virus being restricted to the initial sites of inoculation, or by rapid plant death resulting in reduced secondary spread within the crop. Hypersensitivity has been widely used in breeding programmes, including potatoes against potato viruses Y, C and X (Cockerham, 1943 and 1970) and tobacco against TMV (Apple *et al.*, 1962).

10.2.2 Other terms used in resistance breeding

The terms *horizontal* and *vertical* resistance are widely used to describe host resistance responses, although there is debate over whether resistance can be categorized in this way. 'Horizontal' refers to resistance for which there is no specific interaction with genetic variants of the pathogen (Van der Plank, 1963). It is often quantitative in its effect and results in a reduction in the rate of disease increase. In contrast, 'vertical' refers to resistance in which there is a specific interaction with pathogen strains, resulting in greater resistance to some than to others in a way that cannot be predicted from mean performance. This type of resistance is often explained by a *gene-for-gene* relationship in which corresponding complementary genes for resistance and virulence exist in the host and pathogen respectively (Flor, 1956). An excellent example of resistance conferred by a *gene-for-gene* relationship, has been demonstrated for the complex recessive system of resistance to bean common mosaic virus (BCMV) in *Phaseolus vulgaris* beans (Drijfhout, 1978). This resistance depends upon the combined action of the host gene *bc-u* with one or more other genes *bc-1*, *bc-1*^{2}, *bc-2*, *bc-2*^{2} or *bc-3* (*see* Table 10.3). Various strains of the virus are known (NL1–NL8 series) which carry between them different combinations of virulence genes (*0, 1, 2, 1.1*^{2}, *1.2, 1.1*^{2}*.2, 1.1*^{2}*.2*^{2}), which are capable of overcoming the resistance of the various combinations of resistance genes on a *gene-for-gene* basis as shown in Table 10.3. A BCMV strain carrying a virulence gene to overcome the resistance gene *bc-3* has not yet been found.

Table 10.3 The 'gene for gene' relationship governing resistance to bean common mosaic virus in *Phaseolus vulgaris* beans

				Virus strains and their virulence genes						
Resistance genes				NL1 *0*	NL7 *1*	NL8 *2*	NL6 *1.1*^{2}	NL2 *1.2*	NL3(5) *1.1*^{2}*.2*	NL4 *1.1*^{2}*.2*^{2}
bc-u				S	S	S	S	S	S	S
bc-u	*bc-1*			R	S	R	S	S	S	S
bc-u	*bc-1*^{2}			R	R	R	S	R	S	S
bc-u		*bc-2*		R	R	S	R	S	S	R
bc-u	*bc-1*	*bc-2*		R	R	R	R	S	S	R
bc-u	*bc-1*^{2}	*bc-2*^{2}		R	R	R	R	R	R	S
bc-u		*bc-2*^{2}	*bc-3*	R	R	R	R	R	R	R

Based on the work of Drijfhout (1978), S = susceptible; R = resistant

Under natural field conditions some hosts show *field resistance* to a virus, although the same host may be susceptible to the virus under experimental conditions. Frequently, field resistance results from low inoculum levels under natural conditions, and this type of resistance may be especially sensitive to environmental conditions.

Although it is not essential for a breeder to know exactly how resistance is inherited before a breeding programme is carried out, some knowledge of the genetics of the resistance being used is helpful (Russell, 1978). It is useful to know if the resistance is inherited in a *dominant* or *recessive* manner. Some resistance may be under simple genetic control but may segregate progeny with resistance intermediate between that of the two parents in crosses between resistant and susceptible parents. In this type of inheritance the resistance is said to be *incompletely dominant*. Examples of dominant, incompletely dominant and recessive resistance are given in Table 10.2.

Resistance may be controlled by a single (*monogenic*), a few (*oligogenic*) or many (*polygenic*) genes. These genes may be *major* genes which have a large, observable effect, or *minor* genes which have a small observable effect. Classification into major and minor genes is subjective, however, as the individual breeder must judge the relative size of these effects.

10.3 Procedures Used in Breeding for Virus Resistance

10.3.1 Sources of resistance

The first task in any resistance breeding programme, is to identify plant germplasm possessing a high level of resistance to the virus concerned. Sometimes an adequate source of resistance may already be known in an existing commercial cultivar, and its genetical nature well documented. Such resistance can be used by the breeder who wishes to combine this virus resistance, with resistance to another pathogen or other agronomic character, that is present in another cultivar. Provided the characters concerned are inherited in a straightforward way, and particularly if they are controlled by major genes, development of the new cultivar is likely to be relatively rapid, as both parents are commercially acceptable. Recent examples of such breeding programmes to develop virus resistant cultivars, include the production of dwarf beans (*Phaseolus vulgaris*) possessing the dominant *I* gene for resistance to bean common mosaic virus, halo-blight and anthacnose resistance (Conway *et al*., 1982); and lettuce containing lettuce mosaic virus resistance conferred by the *'Gallega'* gene, combined with downy mildew resistance (Ward and Walkey, 1983).

In other cases, no known source of resistance to a particular virus may be documented. In these circumstances it will be necessary to test (*screen*) available commercial cultivars and breeding lines (referred to as *accessions*) against the virus concerned. Accessions for screening can be sought from commercial seed firms, germplasm gene banks and other breeders working with the same species. If, during this screening programme cultivars possessing the necessary resistance are found, and provided they are agronomically acceptable to the grower, it may suffice simply to recommend that these resistant cultivars are grown instead of susceptible ones. If only one or two resistant cultivars are found, however, it is unlikely that they alone, will provide the grower with the range of commercial types that he requires. In these circumstances it will be necessary to use these resistant cultivars as parents in a breeding programme to generate a range of commercial cultivars possessing virus resistance and the necessary agronomic features (Walkey *et al.*, 1982). Again, the development of these new cultivars should be relatively straightforward, as all the parents will be commercially acceptable.

Occasionally, following extensive screening, all commercial cultivars of a crop species, may be found to be highly susceptible to a virus, although individual plants of some cultivars may possess different

Plate 10.1 Examples of host resistance to virus infection.
(*a*) A resistant marrow plant cv. Cinderella (*left*) following inoculation with cucumber mosaic virus, and an inoculated plant of the susceptible cultivar Goldrush (*right*); (*b*) a susceptible plant of dwarf bean (*Phaseolus vulgaris*) following inoculation with bean common mosaic virus (*left*) and a BCMV inoculated resistant plant (*right*).

levels of resistance to the virus concerned. This situation has been reported in vegetable marrows (*Cucurbita pepo*) in respect of cucumber mosaic virus (CMV) resistance (Walkey and Pink, 1984). Although all cultivars of marrow were found to be susceptible to CMV infection, individual plants of a few cultivars possessed extreme resistance to the virus (*see* Plate 10.1). These resistant individuals were only detected when large numbers of the particular cultivars were screened, and were later used as parents in a resistance breeding programme.

If no resistant cultivars or individual plants can be found within a particular species, the worker must search further afield for resistance. The next step will be to screen related wild and exotic species. Often such species will provide good sources of resistance, as, for example, with wild *Cucurbita* species, which carry high levels of resistance to CMV and water-melon mosaic virus (Provvidenti *et al.*, 1978; Pitrat and Dumas de Vaulx, 1979).

The use of distantly related species as sources of resistance to produce commercial crop cultivars, usually involves an extensive and prolonged breeding programme. Problems may be encountered in obtaining fertile crosses between the crop species and the resistant wild species. This may necessitate the use of tissue culture to produce F_1 plants from the abnormal embryos which result from such crosses (Dumas de Vaulx and Pitrat, 1980). In addition, because many of the characters of the parent wild species are likely to be agronomically unacceptable to the grower, an extensive back-crossing programme will be necessary to confer commercially acceptable features to the resistant progeny. The use of alien germplasm as a source of resistance to disease has been reviewed by Knott and Dvořák (1976).

10.3.2 Screening procedures

In any screening programme it is essential that the plants to be tested should be of a uniform age and development, and that each should be inoculated with standard amounts of inoculum. The inoculum must cause adequate symptoms in susceptible plants. Preliminary experiments are usually necessary to determine the optimum time of inoculation for symptom development (Walkey and Pink, 1984), and optimum concentration of virus inoculum (Kenton and Lockwood, 1977). Too high an inoculum pressure, however, may result in too severe an infection to select for certain forms of resistance. In all screening tests, it is advisable to include a cultivar which is known to be highly susceptible, and in which a uniform symptom response can be expected.

Many successful resistance screening programmes have been carried out in the field, relying upon natural virus infection. In the USA for example, selection of sugar-beet resistance to curly top virus is possible, if the beet is planted in fields adjacent to the foothill scrubland, which is the main breeding ground of the leaf hopper vector. In the spring the viruliferous leafhoppers migrate from the virus-infected weeds to the sugar-beet plots. Early, and uniform, infections occur as the leafhoppers multiply and spread the virus throughout the beet crop. In other circumstances, where the insect vectors of a virus are abundant, it is possible to obtain adequate infection by growing rows of infected plants among the rows of test plants to be screened.

Frequently, however, reliance on spread of the virus by natural vectors in the field can be unreliable, or only reliable in certain seasons. In this situation it is necessary to use artificial methods of virus inoculation. Mechanical sap inoculation is the quickest and most convenient method of infecting large numbers of plants, but insect-vectors may have to be used if a virus is not sap transmissible. Occasionally, aphid-transmission may be more efficient than sap transmission for a specific strain of a virus (Walkey *et al.*, 1983). In these circumstances, a relatively rapid procedure is to use a multiple-aphid transfer technique, as described in Section 12.2.2.

The age of the plant at inoculation is critical for a successful screening procedure. The optimum time of inoculation will vary depending on the host and the virus concerned, but usually the plants must be young if satisfactory infection is to be obtained. When screening for virus resistance in *Phaseolus* beans, for example, it is necessary to inoculate the primary leaves before the trifoliate leaves develop (Walkey and Innes, 1979), and the cotyledons of *Cucurbita* plants should be inoculated before the first true leaves appear (Walkey and Pink, 1984).

With both *Phaseolus* beans, and *Cucurbita* species, it is possible to carry out the complete screening programme in the glasshouse using seedling plants, because the full range of virus symptoms develops in pot-grown plants within four to six weeks (*see* Plate 10.1). In the case of many other host/virus screening programmes, however, it is necessary to inoculate the plants at an early age and then grow them to maturity in the field, in order to make adequate resistance selections. A satisfactory procedure with many species is to inoculate the seedlings in the glasshouse at an early age, harden them off in a cool insect-proof glasshouse or gauzehouse, and then transplant to the field-trial site (Walkey and Neely, 1980). The advantage of this procedure is that the environmental conditions may be controlled before and immediately after virus inoculation, to ensure optimal conditions for uniform infection.

At all times during the trial, it is essential to ensure that the test plants do not become contaminated with any virus or other pathogen, that might confuse the selection process. This is particularly important where the plants might become contaminated with an insect-vectored virus. Often it will be necessary to carry out the screening trials at isolation sites distant from possible sources of virus contamination, and if suitable isolation sites are not available, it may be necessary to carry out the trial in insect-proof gauzehouses.

After a resistant line has been selected or bred, it will probably be necessary to test it against the virus at sites with varying soil, and climatic conditions. This will ensure that the observed resistance is not adversely affected or modified by different environmental conditions.

10.3.3 Assessment of resistance

The main objective of any resistance screening programme, is to distinguish between susceptible and resistant plants. For some host/virus responses the selection of resistant plants is obvious and straightforward. Susceptible hosts develop severe symptoms and resistant plants may show no symptoms. This type of response often occurs when the resistance is controlled by one or only a few major genes, as is the case with the *I* gene for resistance against BCMV in *Phaseolus* beans. The distribution of the resistance response in segregating populations in this case is *discontinuous* and easy to identify. In other cases the distribution of the resistance response may be a *continuous* gradient from severe infection to extreme resistance, and it is necessary to measure or estimate the intensity of the disease, before a reliable selection can be made.

The range of the scoring system used for measuring a continuous response will depend on the host/virus reaction concerned, but often a system based on a 0 (symptomless) to 5 (severe symptoms) scale is satisfactory. The scoring system that has been used successfully in the field to measure the reaction of cabbage to turnip mosaic virus infection is illustrated in Table 10.4 (Walkey and Neely, 1980). Occasionally, if it is necessary to distinguish between mainly mild symptoms, it may be helpful to score on a 0 (symptomless) to 10 (severe symptoms) scale, so that the mild symptom categories at the lower end of the scale are separated.

In the case of a symptomless response to infection, it is often necessary to back-test sap samples from the symptomless plants to a susceptible host, to detect the presence, and indicate the concentration of virus in the resistant plant. An indication of virus concentration in the resistant plant is also essential, to distinguish between tolerance and other forms of resistance.

Enzyme-linked immunosorbent assay (ELISA) tests (*see* Section 6.6.3) may also be used to provide a highly sensitive test for the

Table 10.4 An example of a scoring system used for measuring resistance to turnip mosaic virus in cabbage

Symptom grade	*Description of symptom*
0	No necrosis
1	necrotic lesions visible, only a few leaves affected with >10% of an individual leaf surface affected
2	Necrotic lesions on several leaves with 10–25% of the leaf surface affected
3	Necrotic lesions on many leaves with 25–50% of the leaf surface affected
4	Necrotic lesions on most leaves with 50–75% of the leaf surface affected
5	Severe necrotic lesions on all leaves with <75% of the leaf surface affected, often accompanied by premature leaf fall

Information based on the studies of Walkey and Neely (1980)

detection of virus in a symptomless host, and to provide an accurate quantitative measurement of the virus present (Marco and Cohen, 1979; Ward and Walkey, 1983).

In some screening programmes, particularly if a virus produces only mild visual symptoms in an infected plant, it may be necessary to select resistant plants by measuring the yield, or scoring the quality of trial plants in terms of marketability. Measurements of yield can give an accurate assessment of the relative resistance of different cultivars, and it has been used in many screening programmes, including trials to select spring cabbage cultivars resistant to turnip mosaic virus (Walkey, 1982) and lettuce resistant to CMV (Walkey *et al.*, 1983). Selection based on the marketability of the infected plants is particularly useful in leaf crops such as lettuce, where failure to 'heart' is a common symptom that makes the crop unmarketable.

If selection is to be based on yield measurements, problems will arise over the siting of inoculated and healthy plots. These problems will be considerable if the virus concerned has a highly mobile vector, for the healthy and inoculated plots will have to be widely separated in the field, and site differences can have a considerable effect upon crop yields. In this situation, it is essential to ensure adequate replication and randomization of both healthy and inoculated sites for statistical analysis.

10.3.4 Virus strains

The occurrence of variant strains of a virus (*see* Section 2.3) is a major problem in any virus resistance breeding programme. A knowledge of the behaviour and characteristics of the different strains is important for a successful programme to be carried out. The worker must know which host species is best for the propagation of a particular virus strain, how the strain is best transmitted (Walkey *et al.*, 1983), and how stable the strain is during successive propagation passages in the propagation host.

It is also essential to avoid contamination of the virus strain during the breeding programme. Once a culture of a strain has been established, samples of the strain should be stored by freeze-drying and in liquid nitrogen. Maintenance by repeated sub-culture in a host plant in the glasshouse, is only advisable for relatively short periods, in case of contamination or attenuation. Possible contamination should be regularly checked for by electron microscopy, and by reactions in host range tests. Every few months, or before a major screening experiment, it is advisable to revive the isolate from the stored samples and sub-culture it to optimal inoculum levels in the propagation host, before use in the screening tests.

The choice of virus strain or strains to be used in the resistance breeding programme, is critical in respect of the usefulness of the resistance that will be selected. Sometimes the virulence of various strains of a virus is well documented, as is the case with BCMV (Drijfhout, 1978). If this is the situation, the breeder will be able to select the strain or strains of the virus that are most suitable for his particular purpose. In the case of BCMV, the use of just two strains, (NL3 and NL4, *see* Table 10.3) enables the breeder to screen for resistance against all the known virulence genes of the virus. Consequently, hosts which are resistant to dual inoculation with these two strains, should also be resistant to all other known strains of the virus.

If the genetics of the pathogenic variants are not known, as is the case with most viruses, then the breeder is faced with a difficult decision. One possibility is to take the most virulent strain that occurs in the area where the crop is grown, and use this in the initial screening and subsequent breeding programmes. Often this is the only practical option, but it must be remembered that the selected resistance may only give effective protection in the immediate locality. Alternatively, a large number of virus strains (preferably severe ones) from as wide a geographical area as possible, may be used in the screening programme. This may enable the breeder to identify host lines which are resistant to the maximum number of strains. Such an approach was adopted in screening for resistance to bean yellow mosaic virus (BYMV) in *Phaseolus* beans (Walkey *et al.*, 1983). In this programme, numerous bean accessions were separately screened against seven different BYMV strains. This enabled host germplasm that was resistant to all seven strains to be identified, but it would not be practical to use all the strains in a subsequent breeding programme. Then, it would be necessary to use one or two of the most virulent and representative strains.

If more than one strain is used to test segregating progeny in the breeding part of the programme, they must be inoculated jointly to the test plants, ensuring that equal concentrations of both strains are applied.

If the seedling to be inoculated is bisymmetrical, as in beans, the most satisfactory method of inoculation is to rub one strain into one of the primary bean leaves (Innes and Walkey, 1980), and the second strain on to the other. Alternatively, a 'cocktail' of equal concentrations of the two strains could be used, but in beans this has been shown to be less efficient than inoculating the separate strains to different primary leaves (Walkey, 1983). In these joint inoculation experiments, there was no evidence that one strain protected the seedling against infection from the other strain (*see* Section 10.1).

In the case of beans, the separate inoculation of primary leaves, has also been successfully used to screen against two separate pathogens, BCMV and halo-blight (*Pseudomonas phaseolicola*) (Walkey and Taylor, 1983).

10.3.5 Breeding methods

The breeding procedures used when developing a resident cultivar, are basically the same as when breeding for any other crop character. The method used will depend on whether the host plant is self-pollinated, cross-pollinated or vegetatively propagated, and a list of the breeding systems of some of the world's major crops is shown in Table 10.5.

The basic screening procedures for selecting for virus resistance are the same, irrespective of the breeding system, but the subsequent breeding procedures will differ considerably with different breeding systems. If a crop plant is cross-pollinated, the individual plants selected for resistance from the screening programme, cannot themselves be directly used to produce a new cultivar (unless they can be vegetatively propagated in commercially significant numbers). This is because cross-pollinated plants cannot usually be selfed (i.e. they are self-incompatible), or if they are self-compatible, inbreeding normally causes them to suffer a considerable depression in yield and vigour (referred to as *inbreeding depression*). In the breeding of cross-pollinated crops, therefore, the selected plants must be crossed in suitable combinations. This may be done by such methods as the mass pollination (mass selection) of a population of selected resistant plants, or by F_1 hybrid crosses.

In contrast, inbreeding does not usually cause a significant reduction

Table 10.5 Examples of breeding systems in major crop species

Crop	*Usually self-pollinated*	*Usually cross-pollinated*
Cereals	Barley, millet, oats, rice, sorgham, wheat	Maize, rye
Legumes	Pea, peanut, soybean, sweet clover	Alfalfa, red and white clover, runner beans
Vegetables	Brassicae (broccoli, cauliflower), lettuce, parsnip, tomato	Brassicae (Brussels sprout, cabbage, kohlrabi, radish, turnip), carrot, celery, cucumber, onion, parsley, spinach
Fruit*	Peach, citrus	Apple, banana, blackberry, cherry, date, mango, pear, plum, raspberry, strawberry
Other commodities	Cotton, flax, pepper, tobacco	Castor bean, hemp, potato, sugar-beet, sunflower
Forage grasses	Mountain brome, soft brome	Italian rye, perennial rye, meadow fescue, timothy

*Most fruit crops are vegetatively propagated.

in yield or vigour, in self-pollinated crop plants. Consequently, in such species, resistant cultivars may be produced from an individual plant selected for resistance.

If a crop plant is vegetatively propagated, once the breeder has selected a plant with the necessary level of resistance, it can be maintained and multiplied without further selection, irrespective of the heterozygous nature of its genotype. In the case of a vegetatively propagated crop, therefore, a resistant cultivar may be bred by screening from a mixed population of clones, or by selecting from the progeny following the hybridization of different clones. In the latter situation, because the parent clones are all heterozygous, segregation will occur in the F_1 progeny, and each F_1 plant will be a potential new cultivar.

More detailed information on breeding methods in the production of resistant cultivars is given in a book by Russell (1978).

10.4 Durability of Virus Resistance

The terms *durable* and *durability* are used to describe long-lasting resistance (Johnson and Law, 1975). They do not imply that the resistance is effective against all strains of a virus, but that the resistance may be effective for many years. When the resistance of a particular cultivar is overcome by a new variant of the pathogen the resistance is commonly said to have 'broken down', but it would be more correct to speak of the control as 'breaking down'. The cultivar concerned has not lost its resistance to the original pathogen, but rather it does not possess resistance capable of combating the new variant of the pathogen.

In contrast to some fungal diseases, such as yellow rust (*Puccinia striiformis*) and brown rust (*P.reconita*), which frequently produce new variant strains (races) to overcome host resistance (Russell, 1978), resistance to virus diseases has usually been more durable. Many examples of virus resistance exist, where resistance has been effective for considerable periods. In *Phaseolus* beans, resistance conferred by the dominant *I* gene against BCMV, has been effective in most cultivars of dry and snap beans in the USA for nearly 40 years (Zaumeyer and Meiners, 1975). Although strains of the virus are known that will overcome this resistance (Hubbeling, 1972), these strains have not become prevalent in the field. In Britain potato cultivars such as Epicure and King Edward, have shown field resistance against PVX for more than 50 years (Russell, 1978), and the resistance to curly-top virus in sugar-beet in California has been durable for a similar period of time (Carsner, 1926). Long-term durability of resistance has also been observed for TMV resistance in tobacco (Russell, 1978), although

TMV resistance in some cultivars of tomato, has been overcome by certain strains of TMV (Pelham *et al.*, 1970).

The reasons why resistance by some genes to many virus diseases is so durable is not known, for the capacity of viruses to mutate is considerable (*see* Section 2.3), and strain variations of many viruses are common. In some cases, the durability may result because more than one type of resistance mechanism is involved, with each type of resistance being independently inherited. Resistance in potatoes to PVX and PVY, for example, can involve extreme resistance (controlled by a major gene), hypersensitivity (controlled by a different major gene) and resistance to infection (polygenically controlled) (Russell, 1978).

Such multiple resistance is undoubtedly desirable in respect of durability, and workers breeding for virus resistance should combine different types of resistance wherever possible.

10.5 Host Resistance to the Vector

In addition to screening and breeding for resistance to the virus in a host plant, it is also possible to find resistance to the vector in some crops, which may provide useful control against the virus disease concerned. Most work in this field has been devoted to finding host resistance against insect vectors.

The three main types of vector resistance recognized are *non-preference* for a host, *antibiosis* and *tolerance* (Painter, 1951). In the case of a non-preference host, the vector will land, carry out a feeding probe and quickly move on to another host. Hosts with this type of vector-resistance would be useful in controlling a virus which is transmitted in a persistent manner, but could increase the rate of virus spread of a non-persistently transmitted virus, by increasing the number of probes by feeding vectors (*see* Section 7.4.2). Antibiosis refers to host resistance in which the growth and multiplication of the vector is inhibited. Cultivars possessing this type of vector-resistance could be expected to reduce the spread of both persistently and non-persistently transmitted viruses, by reducing the vector population. The third type of vector-resistance, tolerance, refers to the ability of a host-plant to withstand insect attack, without the plant suffering severe damage. This type of vector-resistance is of no use in controlling virus spread.

Examples of virus control by vector-resistance involving non-preferred hosts, has been reported for several aphid transmitted viruses. These include a reduction of the incidence of rosette disease transmitted by *Aphis craccivora* in groundnut (Evans, 1954), several viruses transmitted by *Amphorophora rubi* in raspberry (Jones, 1976), and CMV transmitted by *Aphis gossypii* in melon (Pitrat and Lecoq,

1980). Feeding preferences by leafhoppers can also reduce the level of tomato curly top virus in certain tomato cultivars (Thomas and Martin, 1971).

Besides resistance to insect-vectors, certain cultivars of wheat are reported to have resistance to the fungal vector (*Polymyxa graminis*) of soil-borne wheat mosaic virus (Palmer and Brakke, 1975).

10.6 Production of Resistant Plants by Cell Manipulation

In addition to the conventional methods for selecting and breeding resistant plants that have already been described in this chapter, exciting new possibilities for releasing and utilizing variation for response to viruses have arisen in the last few years. These new techniques involve *cell manipulation* (sometimes referred to as *genetic engineering*) in tissue culture (Day, 1980; Ingram, 1983; Scowcroft *et al.*, 1983). They include *somaclonal variation*, somatic hybridization through *protoplast fusion* and *transformation* by the insertion of foreign DNA into the cell genome.

The procedures for these techniques are now being established and although the full extent of their usefulness remains to be evaluated, results to date suggest that they may well revolutionize the production of resistant cultivars by the end of the 1980s.

10.6.1 Somaclonal variation

Recent research has shown that tissue cultures of callus, single cells or protoplasts, that have been derived from a genetically stable parent plant, may be differentiated into new plants that are genetically variable. The clones produced from these cultured plants have been called *somaclones* and the variation exposed or induced in them is referred to as *somaclonal* (Larkin and Scowcroft, 1981). In some instances the variation induced in the somaclones has produced virus or other disease resistance that was not present in the parent plants.

In sugar cane some clones produced from single cells derived from callus cultures, have shown resistance to mosaic virus (Nickell and Heinz, 1973), and to Fiji disease virus (Heinz *et al.*, 1977). Similarly, clones regenerated from single-leaf cell protoplasts of the genetically stable potato cultivar *Russet Burbank*, have shown enhanced resistance to the fungal diseases early blight (*Alternaria solani*) and late blight (*Phytophthora infestans*) (Shepard *et al.*, 1980; Shepard, 1981). Research is currently in progress in laboratories in Europe and North America, to test potato somaclones derived from protoplasts, for resistance against various potato viruses (Gunn, 1983).

The reasons why somaclonal variation occurs in offspring of

genetically stable parents is still unclear. Initially, it was thought that the cell population that made up a basically stable, diploid plant, might contain some individual cells with an abnormal number of chromosomes. When separated in tissue culture, these genetically abnormal cells might be expected to regenerate into genetically variable plants. More recent thinking, however, favours the possibility that the variation is caused during tissue culture, by cultural effects upon the chromosomes of individual cells.

10.6.2 Somatic hybridization and DNA transformation

The techniques of somatic hybridization by protoplast fusion (Carlson *et al.*, 1972; Melchers *et al.*, 1978; Power *et al.*, 1980), and the insertion of foreign DNA into the genome of another cell (*transformation*) (Drummond *et al.*, 1977; Drummond, 1979), have been successfully demonstrated in recent years.

Somatic hybridization through protoplast fusion, involves the isolation of individual cells, the removal of the cell walls, fusion of the protoplasts and regeneration of the somatic hybrid into a plant by tissue culture. This procedure has the obvious potential for the transfer of virus resistance and the production of new resistant plant cultivars. At the present time, however, the use of the technique to produce virus resistant plants has not been reported. Undoubtedly rapid progress will be made towards this end in the next few years, but, to date, the production of somatic hybrids by protoplast fusion has been restricted to a relatively small number of species belonging to quite closely related genera.

The insertion of the DNA of one species into the genome of another (*DNA transformation*), has been carried out using the *Ti* plasmid, of the crown-gall disease bacterium *Agrobacterium tumefaciens* as a vector. Theoretically, using this procedure, it should be possible to transfer a chosen DNA sequence, including ones for virus resistance, into the genome of any dicotyledonous plant that is a host of *A.tumefaciens* (Drummond, 1979; Van Montagu and Schell, 1982). At present, however, our genetical knowledge has not advanced far enough to enable us to identify and isolate specific, individual genes that could be transferred in this way. In the future, however, increased genetical knowledge will almost certainly allow us to use this or other DNA transformation procedures, to introduce virus resistance into susceptible hosts.

10.7 References

Ali, M. A. (1950). Genetics of resistance to common bean mosaic virus (bean virus 1) in the bean (*Phaseolus vulgaris* L.). *Phyt* **40**, 69–79.

Apple, J. L. Chaplin, J. F. and Mann, T. J. (1962). Single dominant gene from *N.glutinosa* gives resistance to TMV in tobacco. *Phyt* **52**, 722.

Bancroft, J. B. and Pound, G. S. (1954). Effect of air temperature on multiplication of tobacco mosaic virus in tobacco. *Phyt* **44**, 481–2.

Beemster, A. B. R. (1972). Virus translocation in potato plants and mature-plant resistance. In *Viruses of potatoes and seed-potato production* (ed. de Bokx J. A.), *Cen Agric Pub Doc*, Wageningen.

Boukwma, I. W. (1980). Allelism of genes controlling resistance to TMV in *Capsicum L. Euphytica* **29**, 433–9.

Brunt, A. A. (1975). The effects of cocoa swollen-shoot virus on the growth and yield of Amelonado and Amazon cocoa (*Theobroma cacoa*) in Ghana. *Ann Appl Biol* **80**, 169–80.

Catherall, P. L., Jones, A. T. and Hayes, J. D. (1970). Inheritance and effectiveness of genes in barley that condition tolerance to barley yellow dwarf virus. *Ann Appl Biol* **65**, 153–61.

Carlson, P. S., Smith, H. H. and Dearing, R. D. (1972). Parasexual interspecific plant hybridization. *Proc Nat Acad Sci USA* **69**, 2292–4.

Carsner, E. (1926). Resistance in sugar-beets to curly top. *USDA Cir* 388.

Cockerham, G. (1943). The reaction of potato varieties to viruses X, A, B and C. *Ann Appl Biol* **30**, 338.

Cockerham, G. (1970). Genetical studies on resistance to potato viruses X and Y. *Heredity* **25**, 309.

Conway, J., Hardwick, R. C., Innes, N. L., Taylor, J. D. and Walkey D. G. A. (1982). White seeded beans (*Phaseolus vulgaris*) resistant to halo blight (*Pseudomonas phaseolicola*), to bean common mosaic virus, and to anthracnose (*Colletotrichum lindemuthianum*). *J Agric Sci* **99**, 555–60.

Cooper, J. I. and Jones, A. T. (1983). Responses of plants to viruses; proposals for the use of terms. *Phyt* **73**, 127–8.

Crill, P., Hanson, E. W. and Hagedorn, D. J. (1971). Resistance and tolerance to alfalfa mosaic virus in alfalfa. *Phyt* **61**, 369–71.

Day P. R. (1980). Tissue culture methods in plant breeding. In *Tissue culture methods for plant pathologists* (ed. Ingram, D. S. and Helgeson, J. P.). Blackwell Scientific Publications: Oxford. pp. 223–31.

Drijfhout, E. (1978). Genetic interaction between *Phaseolus vulgaris* and bean common mosaic virus with implications for strain identification and breeding for resistance. *Agric Res Rept Cen Agr Pub Doc*, Wageningen, pp. 1–90.

Drummond, M. H., Gordon, M. P., Nester, E. W. and Chilton, M. D. (1977). Foreign DNA of bacterial plasmid origin is transcribed in crown gall tumours. *Nature* **269**, 535–6.

Drummond, M. H. (1979). Crown gall disease. *Nature* **281**, 343–7.

Dumas de Vaulx, R. and Pitrat, M. (1980). Realization of the interspecific hybridization (F_1 and BC_1) between *Cucurbita pepo* and *C.ecuadorensis*. *Cuc Gen Coop R*, p. 42.

Evans, A. C. (1954). Groundnut rosette disease in Tanganyika. 1. Field studies. *Ann Appl Biol* **41**, 189–206.

Federation of British Plant Pathologists (1973). A guide to the use of terms in plant pathology. Phytopath. Papers No. 17, CMI: Kew.

Finlay, K. W. (1953). Inheritance of spotted wilt resistance in the tomato. II. Five genes controlling spotted wilt resistance in four tomato types. *Aust J Biol Sci* **6**, 153–63.

Flor, H. H. (1956). The complementary genetic systems in flax and flax rust. *Adv Gen* **8**, 29–54.

Fraser, R. S. S. (1982). Biochemical aspects of plant resistance to virus disease: A review. *Acta Hort* **127**, 101–16.

Gunn, R. E. (1983). Somaclonal variation in potato breeding. AFRC Conf. on Gen. man. of Crop Plants. p. 79.

Heinz, D. J., Krishnamurthi, M., Nickell, I. G. and Maretzki, A. (1977). Cell, tissue and organ culture in sugarcane improvement. In *Applied and fundamental aspects of plant cell, tissue and organ cutlure* (ed. Reinert, J. and Bajaj, Y. P. S.). Springer: Berlin. pp. 3–17.

Holmes, F. O. (1938). Inheritance of resistance to tobacco-mosaic disease in tobacco. *Phyt* **28**, 553–61.

Holmes, F. O. (1955). Additive resistance to specific viral diseases in plants. *Ann Appl Biol* **42**, 129.

Holmes, F. O. (1961). Concomitant inheritance of resistance to several viral diseases of tobacco. *Virol* **13**, 409–13.

Hubbeling, N. (1972). Resistance in beans to strains of bean common mosaic virus. *Med Fak Land weten Gent* **37**, 458–66.

Ingram, D. S. (1983). Challenges for the future. In *Use of tissue cultures and protoplasts in plant pathology* (ed. Helgeson, J. P. and Deverall, B. J.). Academic Press: London.

Innes, N. L. and Walkey D. G. A. (1980). The genetics of resistance to two strains of bean common mosaic virus in three cultivars of *Phaseolus vulgaris* L. *J Agric Sci* **95**, 619–30.

Johnson, R. and Law, C. N. (1975). Genetic control of durable resistance to yellow rust (*Puccinia striiformis*) in the wheat cultivar Hybride de Bersée. *Ann Appl Biol* **81**, 385–91.

Jones, A. T. (1976). The effect of resistance to *Amphorophora rubi* in raspberry (*Rubus idaeus*) on the spread of aphid borne viruses. *Ann Appl Biol* **82**, 503–10.

Jones, R. K. and Tolin, S. A. (1972). Concentration of maize dwarf mosaic virus in susceptible and resistant corn hybrids. *Phyt* **62**, 640–4.

Karchi, Z., Cohen, S. and Govers, A. (1975). Inheritance of resistance to cucumber mosaic virus in melons. *Phyt* **65**, 479–81.

Kenton, R. H. and Lockwood, G. (1977). Studies on the possibility of increasing resistance to cocoa swollen-shoot virus by breeding. *Ann Appl Biol* **85**, 71–8.

Knott, D. R. and Dvořák, J. (1976). Alien germplasm as a source of resistance to diease. *Ann Rev Phyt* **14**, 211–35.

Larkin, P. J. and Scowcroft, W. R. (1981). Somaclonal variation – a novel source of variability from cell cultures for plant improvement. *Theor Appl Genet* **60**, 197–214.

Lewellen, R. T. (1973). Inheritance of beet mosaic resistance in sugarbeet. *Phyt* **63**, 877–81.

Marco, S. and Cohen, S. (1979). Rapid detection and titre evaluation of viruses in pepper by enzyme-linked immunosorbent assay. *Phyt* **69**, 1259–62.

Matthews, R. E. F. (1981). *Plant Virology*. Academic Press: London.
Melchers, G., Sacristan, M. D. and Holder, A. A. (1978). Somatic hybrid plants of potato and tomato regenerated from fused protoplasts. *Carlsberg Res Com* **43**, 203–18.
Nagaich, B. B., Upadhya, M. D., Prakash O and Singh, S. J. (1968). Cytoplasmically determined expression of symptoms of potato virus X crosses between species of *Capsicum*. *Nature* **220**, 1341–2.
Nickell, L. G. and Heinz, D. J. (1973). Potential of cell and tissue culture techniques as aids in economic plant improvement. In *Genes, enzymes and populations* (ed. Srb, A. M.). Plenum: New York. pp. 109–28.
Ouchi, S. (1983). Induction of resistance or susceptibility. *Ann Rev Phyt* **21**, 289–315.
Painter, R. H. (1951). *Insect resistance in crop plants*. Macmillan: New York.
Palmer, L. T., and Brakke, M. K. (1975). Yield reduction in winter wheat infected with soilborne wheat mosaic virus. *Plant Dis R* **59**, 469–71.
Pelham, J. (1972). Strain-genotype interaction of tobacco mosaic virus in tomato. *Ann Appl Biol* **71**, 219–28.
Pelham, J., Fletcher, J. T. and Hawkins, J. H. (1970). The establishment of a new strain of tobacco mosaic virus resulting from the use of resistant varieties of tomato. *Ann Appl Biol* **65**, 293–7.
Pitrat, M. and Dumas de Vaulx, R. (1979). Recherche de géniteurs de résistance à l'Oidium et aux virus de la mosaique du concombre et de la mosaique de la pasteque chez cucurbita sp. *Ann Amelior Plantes*, **29**, 439–45.
Pitrat, M. and Lecoq, H. (1980). Inheritance of resistance to cucumber mosaic virus transmission by *Aphis gossypii* in *cucumis melo*. *Phyt* **70**, 958–61.
Posnette, A. F. (1969). Tolerance of virus infection in crop plants. *Rev Appl Mycol* **48**, 113.
Pound, G. S. and Cheo, P. C. (1952). Studies on resistance to cucumber virus 1 in spinach. *Phyt* **42**, 301–6.
Power, J. B., Berry, S. F., Chapman, J. V. and Cocking, E. C. (1980). Somatic hybridisation of sexually incompatible petunias: *Petunia parodii*, *Petunia parviflora*. *Theor Appl Gen* **57**, 1–4.
Provvidenti, R., Robinson, R. W. and Munger, H. M. (1978). Resistance in feral species to six viruses infecting *Cucurbita*. *Plant Dis R* **62**, 326–9.
Provvidenti, R., Robinson, R. W. and Shail, J. W. (1980). A souce of resistance to a strain of cucumber mosaic virus in *Lactuca saligna L*. *Hortic Sci* **15**, 528–9.
Provvidenti, R and Schroeder, W. T. (1973). Resistance in *Phaseolus vulgaris* to the severe strain of bean yellow mosaic virus. *Phyt* **63**, 196–7.
Russell, G. E. (1978). *Plant breeding for pest and disease resistance*. Butterworths: London.
Scowcroft, W. R., Larkin, P. J. and Brettell, R. I. S. (1983). Genetic variation from tissue culture. In *Use of tissue cultures and protoplasts in plant pathology* (ed. Helgeson, J. P. and Deverall, B. J.). Academic Press: London.
Shepard, J. F. (1981). Protoplasts as sources of disease resistance in plants. *Ann Rev Phyt* **19**, 145–66.
Shepard, J. F., Bidney, D. and Shahin, E. (1980). Potato protoplasts in crop improvement. *Sci* **28**, 17–24.
Shifriss, C. and Marco, S. (1980). Partial dominance of resistance to potato

virus Y in *Capsicum*. *Plant Dis* **64**, 57–9.

Sinclair, J. B. and Walker, J. C. (1955). Inheritance of resistance to cucumber mosaic virus in cowpea. *Phyt* **45**, 563–4.

Takahashi, T. (1975). Studies on viral pathogenesis in plant hosts. VIII. Systemic virus invasion and localization of infection in 'Samsun-NN' tobacco plants resulting from tobacco mosaic virus infection. *Phytopathol Z* **84**, 75–87.

Thomas, P. E. and Fulton, R. W. (1968). Correlation of ectodesmata number with non-specific resistance to initial virus infection. *Virol* **34**, 459–69.

Thomas, P. E. and Martin, M. W. (1969). Association of recovery from curly top in tomatoes with susceptibility. *Phyt* **59**, 1864.

Thomas, P. E. and Martin, M. W. (1971). Vector preference, a factor of resistance to curly top virus in certain tomato cultivars. *Phyt* **61**, 1257–60.

Troutman, J. L. and Fulton, R. W. (1958). Resistance in tobacco to cucumber mosaic virus. *Virol* **6**, 303.

Valleau, W. D. (1952). The evolution of susceptibility to tobacco mosaic in *Nicotiana* and the origin of the tobacco mosaic virus. *Phyt* **42**, 40–2.

Van der Plank, J. E. (1963). *Plant diseases: Epidemics and Control.* Academic Press: New York.

Van Montagu, M. and Schell, J. (1982). The Ti plasmids of *Agrobacterium. Curr-Top Microbiol Immunol* **96**, 237–54.

Walkey, D. G. A. (1982). Reactions of spring cabbage cultivars to infection by turnip mosaic virus. *J Nat Inst Agric Bot* **16**, 114–25.

Walkey, D. G. A. (1983). Unpublished results.

Walkey, D. G. A. and Innes, N. L. (1979). Resistance to bean common mosaic virus in dwarf beans (*Phaseolus vulgaris L.*) *J Agric Sci*, **92**, 101–8.

Walkey, D. G. A., Innes, N. L. and Miller, A. (1983). Resistance to bean yellow mosaic virus in *Phaseolus vulgaris*. *J Agric Sci* **100**, 643–50.

Walkey, D. G. A. and Neely, H. A. (1980). Resistance in white cabbage to necrosic caused by turnip and cauliflower mosaic viruses and pepper-spot. *J Agric Sci* **95**, 703–13.

Walkey, D. G. A. and Pink, D. A. C. (1984). Resistance in vegetable marrow and other *Cucurbita* spp. to two British strains of cucumber mosaic virus. *J Agric Sci* **102**, 197–205.

Walkey, D. G. A. and Taylor, J. D. (1983). Unpublished results.

Walkey, D. G. A., Tomlinson, J. A., Innes, N. L. and Pink, D. A. C. (1982). Breeding for virus resistance in British vegetable crops. *Acta Hort* **127**, 125–35.

Walkey, D. G. A., Ward, C. M., Bolland, C. J. and Miller, A. (1983). Cucumber mosaic and beet western yellows virus. *Nat Veg Res Stn Rept 1983*, pp. 86–7.

Ward, C. M. and Walkey, D. G. A. (1983). Lettuce mosaic virus. *Nat Veg Res Stn Rept 1982*, p. 79.

Wasuwat, S. L. and Walker, J. C. (1961). Inheritance of resistance in cucumber to cucumber mosaic virus. *Phyt* **51**, 423–6.

White, R. F. (1979). Acetyl salicylic acid (aspirin) induces resistance to TMV in tobacco. *Virol* **99**, 410–12.

Zaumeyer, W. J. and Meiners, J. P. (1975). Disease resistance in beans. *Ann Rev Phyt* **13**, 313–34.

10.8 Further Selected Reading

Federation of British Plant Pathologists (1973). 'A guide to the use of terms in plant pathology.' Phytopath. Papers No. 17, CMI: Kew.

Helgeson, J. P. and Deverall, B. J. (1983). *Use of tissue cultures and protoplasts in plant pathology*. Academic Press: London.

Russell, G. E. (1978). *Plant breeding for pest and disease resistance*. Butterworths: London.

11 *The Production of Virus-Free Plants*

11.1 Introduction

The various control measures by which viruses may be prevented from infecting or causing severe losses in crop plants are discussed in Chapters 9 and 10. In this chapter, the techniques by which viruses may be eradicated from plants that are already infected are considered.

Most, if not all, crop plants are likely to become infected with viruses at some time or another. In the case of annual or biennial crops infection may result in reduced yields or even loss of the infected crop, but provided appropriate preventative measures are taken, a new crop that will mature in a healthy condition may be grown from seed in the following seasons. Many other economically important crops are, however, vegetatively propagated in order to maintain the desirable horticultural characteristics of particular clones and cultivars. Virus infection of such vegetatively propagated clones may have serious consequences, for once infection occurs, the virus will automatically be transmitted in the vegetative propagule (e.g. bulb, tuber, etc.) to most, if not all offspring (*see* Section 7.8). Frequently, the distribution of infected propagules may result in a clone becoming totally virus infected, and many old cultivars that have been propagated vegetatively for decades or even a century or more, may be infected with numerous viruses (Tomlinson and Walkey, 1967). Infected clones, such as the rhubarb cultivar Timperley Early, that have survived commercially for many years, undoubtedly show a high level of tolerance to multiple virus infection (Walkey *et al.*, 1982), but others show severely reduced yields and loss of vigour (Hollings, 1965) (*see* Plate 11.1). In the past, the susceptibility of complete clones to virus infection has frequently led to their commercial extinction, for unlike fungal (Bent, 1969) or bacterial (Taylor and Dudley, 1977) pathogens, viruses cannot be controlled in infected field crops by chemical treatments.

It is therefore, advantageous and often essential, that viruses be

Plate 11.1 Reduction in yield caused by virus infection of a vegetatively propagated crop.
(*a*) An infected clone of rhubarb (cv. Timperley Early); (*b*) plant of the same clone from which the virus has been eradicated by meristem-tip culture.

eradicated from infected clones if the clones are to continue in commercial production. Since 1950, when Kassanis first used a high temperature treatment to eradicate potato leaf roll virus from potato tubers, techniques involving thermotherapy or tissue culture, and frequently a combination of both, have been developed and successfully used to eradicate viruses from infected plant tissues (Walkey, 1980).

Many attempts have also been made to use chemotherapeutic treatments to produce virus-free plants, but most have been unsuccessful for practical purposes (Tomlinson, 1982), although recently a few chemicals have given promising results in laboratory experiments (*see* Section 11.4).

The use of thermotherapy, tissue culture and chemotherapy in eradicating viruses from infected plant tissues is described in the following sections. At the onset of any virus eradication programme however, and before these various techniques are used, it is important to select a suitable parent clone for treatment. Individual clones, although they may be infected with the same virus, often vary considerably in their vigour and capacity to be propagated. In some cases, this variation may be so pronounced, that plants freed of virus may be less vigorous than plants of the same cultivar that remain

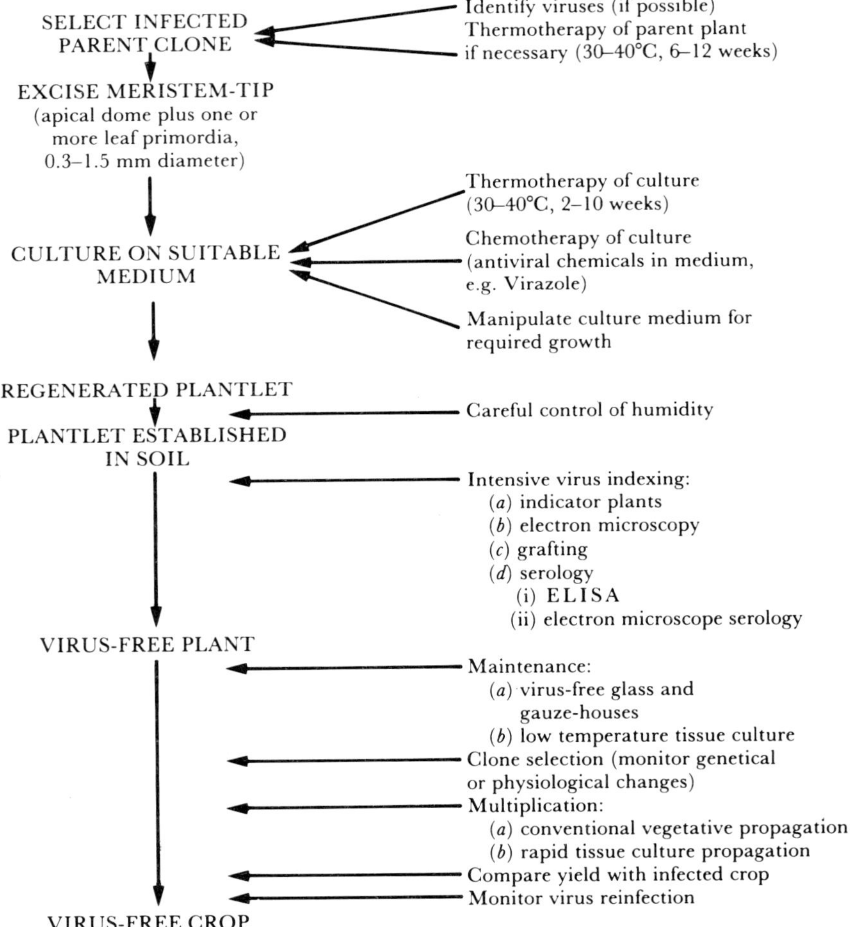

Fig. 11.1 A scheme for virus-free plant production by meristem-tip culture (based on Walkey, 1980).

infected. Therefore, it is advisable to select clones for treatment that are known to be high yielding and to propagate freely. Having selected the clone, it is also advisable, if not essential, to identify the virus or viruses with which it is infected, so that the plants eventually produced may be accurately indexed for the viruses concerned (*see* Figure 11.1).

11.2 Thermotherapy

11.2.1 Introduction

High temperature treatment has been widely used in the production of virus-free plants (*see* Table 11.1) and has been reviewed by Nyland and Goheen (1969). Such treatments usually involve the infected parent plant, or organ of the plant, being grown in hot air in a temperature controlled cabinet at 30° to 40°C for periods of six to twelve weeks. Although virus may be eradicated from a whole organ of a plant, such as potato tuber, by heat treatment (Kassanis, 1950), it is generally impossible to eradicate virus from a complete plant without severely damaging or killing it. Usually a temperature differential is established in the plant under treatment, between the exposed leaves and the soil-embedded roots. Consequently, the leaves may be exposed to temperatures several degrees higher than the roots, so that the virus is inactivated in the leaves and shoots, but not in the base of the stem and roots (Pennazio *et al.*, 1976). It is necessary, therefore, to remove portions of potentially virus-free shoots from the heat treated plant and to grow these as macro (cuttings or bud grafts) or micro (meristem-tips, *see* Section 11.3.2) explants to produce a healthy plant.

When considering temperature inactivation of plant viruses, it is important to distinguish between *in vivo* thermotherapeutic treatments involving the use of temperatures between 30° and 40°C (which are discussed in this section) and the *in vitro* high temperature treatments used to determine the thermal inactivation temperature of a virus (as described in Section 6.3.1). The thermal inactivation temperature of a virus, is the temperature at which the virus is actually killed in sap homogenates and may vary between 40° and 90°C, depending upon the particular virus concerned, whereas the 30° to 40°C treatments described here are considerably lower than the thermal inactivation temperature of most viruses.

11.2.2 Mechanism of virus eradication by high temperature

It has been postulated that within an infected plant, virus synthesis and virus degradation occur simultaneously, and at high temperatures virus synthesis stops, but degradation continues (Kassanis, 1957). This explanation seems highly plausible, and would explain why viruses are eventually eradicated from infected plants maintained at temperatures between 30° and 40°C, even though a much higher temperature is required actually to kill the virus *in vitro*.

Thus, when an infected plant is heat treated at between 30° and 40°C, virus replication is halted, but the young shoots continue to grow and these will be free of the virus that is still present in the older parts of

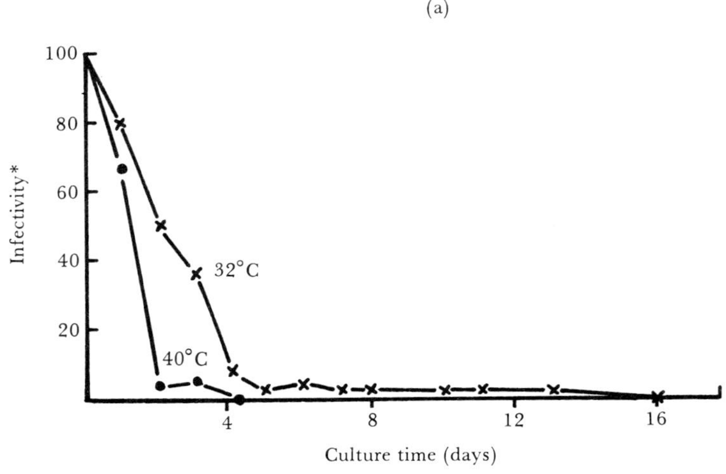

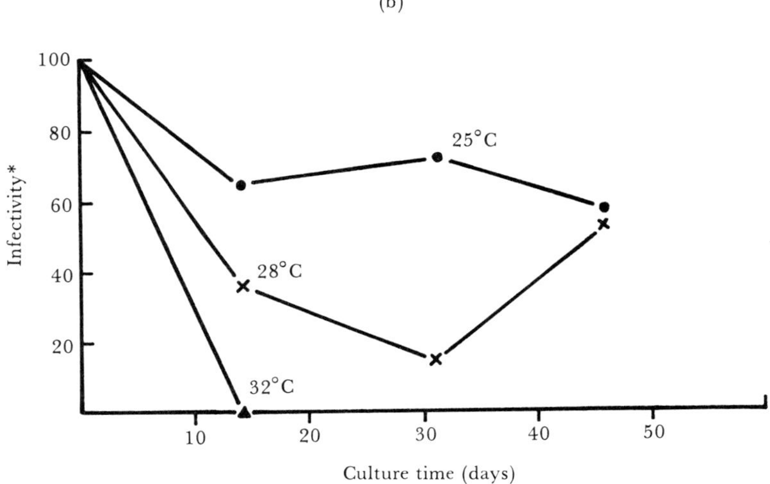

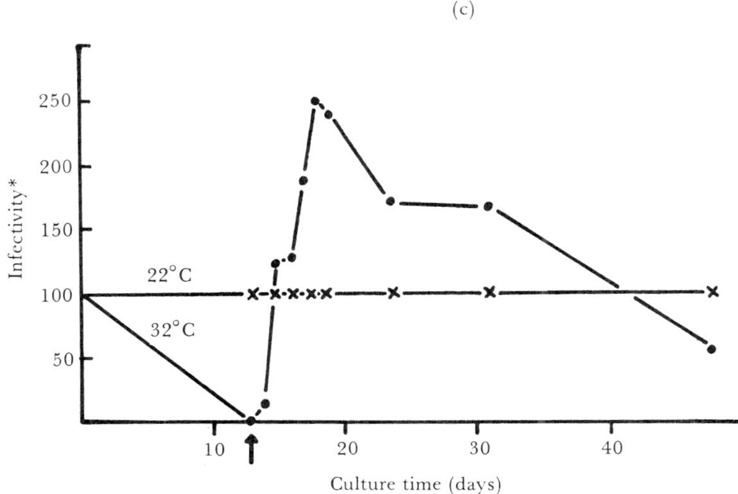

Fig. 11.2 The inactivation of cucumber mosaic virus in *Nicotiana rustica* cultures incubated at high temperature (based on Walkey, 1976). (*a*) The inactivation of CMV at 32° and 40°C; (*b*) the inactivation of CMV at 25°, 28° and 32°C; (*c*) the recovery of CMV infectivity in cultures grown at 32°C and then transferred to 22°C as indicated by the arrow. The control cultures were grown at 22°C.
* Infectivity is expressed as a relative percentage of that of cultures grown at 22°C.

the plant. If these shoots, or buds from them, are removed and grown into plants, the resulting plants will be virus-free. If it is necessary to eradicate virus from a complete organ, such as a potato tuber, it is likely that a longer period of high temperature treatment may be required, for in this situation, virus replication must not only be stopped, but time must be allowed for the existing virus to degrade. The method by which existing virus degrades is unknown, but presumably the virus breaks down and is utilized by the cell during its normal metabolic processes.

As an alternative to thermotherapy of the infected parent plant, it is possible to eradicate virus, such as cucumber mosaic (CMV), by growing virus infected meristematic tissues in tissue culture at 30° to 40°C (Walkey, 1976). Using proliferating cultures of *Nicotiana rustica* systemically infected with CMV, it was possible to study thermotherapeutic effects on uniform material in successive experiments. These experiments showed that sixteen to eighteen days at 32°C, or five days treatment at 40°C, were required to reduce the virus concentration in infected tissues to a level at which it could not be detected by sap assay (*see* Figure 11.2*a*). A temperature of 32°C was critical for the eradication of CMV, for if the tissues were incubated at 25° to 28°C, a reduction in virus concentration occurred, but the virus was not eradicated (*see* Figure 11.2*b*).

Although CMV infectivity was not detected in cultures after sixteen to eighteen days treatment at 32°C, the virus was still present in the tissues in very low concentrations and at least a further thirty days treatment was required at this temperature for complete virus eradication. At 40°C CMV was eradicated from cultures after nine days treatment. If the cultures were removed from the high temperature (32°C) and grown at a lower temperature (22°C) at a time when the virus concentration was too low to be detected, but before complete eradication had occurred, a rapid increase in virus concentration resulted (*see* Figure 11.2*c*). After such treatment virus infectivity was as much as two-and-half times greater than in the infected control tissues grown continually at 22°C, although it gradually fell back to the level of the control material over a three to four week period at the lower temperature. This result suggested that high temperature, in addition to stopping virus synthesis, also inactivated a resistance factor in the host plant. Consequently, if virus inactivation is incomplete following heat treatment, and the plant or tissues are transferred from a restrictive to a non-restrictive temperature for virus replication, a rapid resurgence in virus concentration can be expected because of the removal of the restraining influence of the host's resistance factor (Walkey, 1976). This hypothesis may explain reports that the proportion of plants 'cured' by thermotherapy was sometimes less with longer, than with shorter treatment periods (Mellor and Stace-Smith, 1970; Johnstone and Wade, 1974*a*).

These experiments illustrate the importance of the duration of thermotherapeutic treatments whether treating whole plants or cultured tissues.

11.2.3 Methods of applying the high temperature treatment

Exposure of complete plants or cultured explants to high temperatures for prolonged treatment periods, usually causes some deterioration of the tissues of the treated material. Studies have been made, therefore, on optimal methods of applying the high temperature treatment, in order to minimize damage to the plant tissues being treated. It has been shown that preconditioning treatment periods at temperatures between 27° and 35°C, prior to treatment at 35° to 40°C, may increase the plant's survival capacity (Fulton, 1954; Welsh and Nyland, 1965). Other workers have demonstrated that diurnal alternating periods at high and low temperatures, as an alternative to continuous high

Table 11.1 Species which have been freed of virus by the use of heat therapy combined with tissue culture, or by tissue culture alone

Host	*Viruses eradicated*	*Temperature treatment*	*Reference*
Allium sativum (garlic)	Mosaic	–†	*
Asparagus officinalis (asparagus)	Unspecified	–	Yang and Clore, 1976
Armoracia lapathifolia (horse radish)	Cauliflower mosaic (CaMV) turnip mosaic (TuMV)	–	Hickman and Varma, 1968
Ananas sativus (pineapple)	Unspecified	–	*
Brassica oleracea (cauliflower)	CaMV, TuMV	–	Walkey *et al.*, 1974
Caladium hortulanum (aroid)	Dasheen mosaic	–	*
Chrysanthemum morifolium	Virus B, vein mottle, greenflower, aspermy, stunt.	35–8°C (4–37 wk)	*
Colocasia esculenta (taro)	Dasheen mosaic	–	*
Cymbidium spp.	Mosaic	–	*
Dactylis glomerata (cocksfoot grass)	Cocksfoot mild mottle, mottle and streak	–	Dale, 1979
Dahlia spp.	Mosaic	–	Mullin and Schlegel, 1978
Daphne odora	Daphne virus S	–	Cohen, 1977
Dianthus barbatus	Ringspot, mottle, latent, vein-mottle	–	Stone, 1968
Dianthus caryophyllus (carnation)	Ringspot, mottle, latent, streak, vein-mottle, etched-ring	35–40°C (3–15 wk)	Stone, 1968
Dioscorea alata (yam)	Unspecified rods	36°C (14 d)	Mantell *et al.*, 1980
Forsythia X intermedia	Unspecified	–	Duron, 1977
Fragaria chiloensis (strawberry)	Latent A and C, crinkle, yellow edge, vein chlorosis, vein banding.	33–40°C (4–7 wk)	*

Host	*Viruses eradicated*	*Temperature treatment*	*Reference*
Freesia spp.	Mosaic, bean yellow mosaic	–	*
Geranium spp.	Tomato ringspot	–	Pillai and Hildebrandt, 1968
Gladiolus spp.	Unspecified	–	*
Hippeastrum spp. (amaryllis)	Mosaic	–	*
Humulus lupulus (hop)	Mosaic, latent, prunus necrotic ringspot	40°C (10–28 d)	Schmidt, 1974
Hydrangea hortensia	Ringspot	–	Preil *et al.*, 1978
Hyacinthus orientalis (hyacinth)	Mosaic, lily symptomless	–	*
Ipomoea batatus (sweet potato)	Internal cork, mosaic, mottle	27–8°C	Alconero *et al.*, 1975
		–	Stone *et al.*, 1978‡
Iris spp.	Latent, mosaic	–	*
Lavendula spp. (lavender)	Dieback	–	Maia *et al.*, 1973
Lilium spp.	Cucumber mosaic (CMV), hyacinth mosaic, lily symptomless	–	*
Lolium multiflorum	Rye-grass mosaic	–	Dale, 1977
Manihot utilissima (cassava)	Mosaic, leaf distortion mosaic, brown streak	35°C (4–5 wk)	*
		37°C (30–6 d)	Kaiser and Teemba, 1979
Musa sapientum (banana)	CMV, unspecified	35–43°C (14–15 wk)	*
Narcissus tazetta (daffodil)	Arabis mosaic (AMV), degeneration	–	*

Nasturtium officinale (watercress)	CMV, CaMV, TuMV	–	Walkey and Thompson, 1978
Nerine spp.	Latent, unspecified	–	*
Nicotiana rustica (tobacco)	AMV, cherry leaf roll (CLRV), tobacco ring-spot, alfalfa mosaic	32–40°C	Walkey and Cooper, 1972, 1975
Nicotiana tabacum (tobacco)	Tobacco mosaic (TMV)	–	Martin *et al.*, 1967
Pelagonium spp.	Unspecified	–	*
Petunia hybrida	TMV, tobacco necrosis	–	*
Ranunculus asiaticus	Tobacco rattle, CMV	–	*
Rheum rhaponticum (rhubarb)	TuMV, CMV, CLRV, AMV, strawberry latent ringspot	–	Walkey, 1968
Ribes grossularia (gooseberry)	Vein banding	35°C (2 wk)	*
Rubus ideaus (raspberry)	Mosaic	–	*
Saccharum officinarum (sugar cane)	Mosaic	–	*
Solanum tuberosum (potato)	Potato viruses A, G, M, S, X, Y, paracrinkle,	33–8°C (4–18 wk)	*
Ullucus tuberosum	Unspecified	–	Stone, 1982‡
Xanthomosa brasiliense (cocoyam)	Unspecified	–	*

*See review by Quak, 1977.
†No heat treatment used.
‡Chemotherapeutic treatment combined with meristem-tip culture.

temperature treatment, may be advantageous. Potato leaf roll virus was successfully eradicated from potato tuber 'eye' pieces following diurnal treatments of 40°C (four hours) plus 16–20°C (twenty hours), whereas continuous treatment at 40°C killed the tissues (Hamid and Locke, 1961). Similarly, diurnal periods were found to be better than continuous high temperature treatments in eradicating chrysanthemum virus B from chrysanthemum (Larsen, 1966) and various apple viruses from apple rootstocks (Larsen, 1974).

More detailed studies, using cultured tissues of *Nicotiana rustica* infected with CMV, demonstrated that diurnal treatments of 40°C (eight hours) plus 22°C (sixteen hours), 40°C (sixteen hours) plus 22°C (eight hours), or 36°C (twenty hours) plus 22°C (four hours), were all preferable to continuous high temperature treatment at 40°C for host survival (Walkey and Freeman, 1977).

In addition to the use of high temperature to eradicate viruses from infected plants, it has also been shown that low temperatures can be used in a similar way. Temperatures of between 5° and 15°C followed by meristem-tip culture, (*see* Section 11.3.2) have been used to eradicate potato viruses A and Y from infected potatoes (Moskovets *et al.*, 1973). In this treatment, it seems likely that the low temperature stops virus synthesis, allowing virus-free explants to be taken and tissue cultured into healthy plants.

11.3 Virus Eradication by Tissue Culture

11.3.1 Introduction

The importance of the combined use of tissue culture and thermotherapy in producing virus-free plants has already been mentioned in Section 11.2, and examples of commercially valuable crop plants that have been freed of virus by these combined treatments are given in Table 11.1.

In addition to this combined treatment, since Morel and Martin first cultured meristem-tips to eradicate virus from dahlias (1952) and potatoes (1955), tissue culture on its own has been extensively used to produce virus-free plants from infected clones of numerous species (*see* Table 11.1). Basically the technique involves the excision of a suitable explant from the infected parent plant, the aseptic culture of the explant into a plantlet on a nutrient medium, and finally the establishment of the plantlet in soil (*see* Figure 11.3).

A technique for aseptically isolating explants, and a suitable nutrient medium for their culture is described in the practical exercises in Section 12.7.1. Most culture media used at present are based on

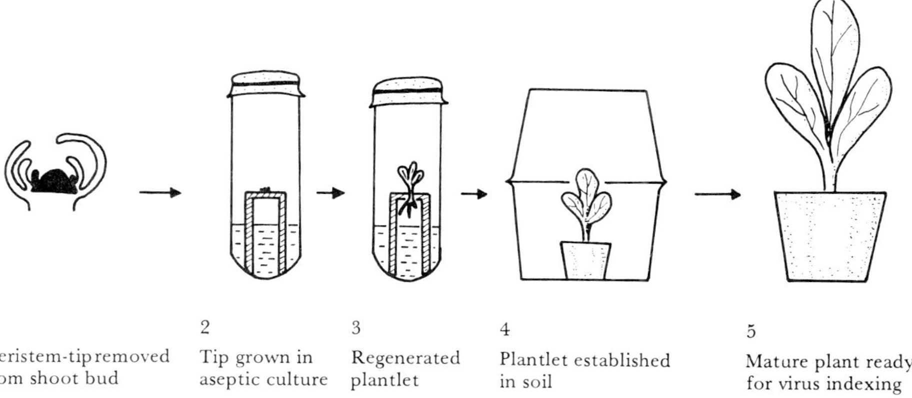

Fig. 11.3 The regeneration of a plant by aseptic tissue culture of a meristem-tip explant.

Murashige and Skoog's medium (1962), which has been successfully used to culture a wide range of plant species (De Fossard, 1976). Cultures may be supported on filter paper bridges in liquid medium, or agar may be incorporated into the medium to solidify it. Details of the various techniques used in tissue culture have been described in recent publications (Gamborg and Wetter, 1975; De Fossard, 1976).

11.3.2 Tissue culture techniques used for virus eradication

Various types of tissues have been cultured to produce virus-free plants from infected parent material, including callus, protoplasts, various reproductive tissues and meristem tips.

Callus culture

Several workers have shown that healthy plantlets can be regenerated from tobacco mosaic virus (TMV) infected tobacco callus (Svobodva, 1965; Hansen and Hildebrant, 1966; Mori, 1977). These virus-free plantlets probably result from 'sectoring' of the cultured callus tissues into healthy and infected cells, before plantlet regeneration. Infectivity tests carried out on individual cells isolated from TMV infected callus showed that only 30–40% of the cells were infected (Hansen and Hildebrant, 1966). It has been suggested that the virus-free areas of tissue arise because virus replication is slower than cell proliferation (Svobodva, 1965). This possibility is supported by evidence which showed, that cell division is accelerated in kinetin-treated callus, with a corresponding increase in the proportion of virus-free plants subsequently regenerated.

Various studies suggest, however, that the use of callus culture to produce virus-free plants should be avoided, if clonal uniformity is required in the regenerated plants. Plants regenerated from callus are frequently genetically different from their parent clone (*see* Section 10.6.1).

Protoplast culture

Shepard (1975) has shown that virus-free plants may be regenerated from protoplasts taken from potato virus X (PVX) infected tobacco leaves. He found that of 4140 plants regenerated, 7½% were virus-free. The reason for this loss of virus, as was the case with callus cultures, appears to be the failure of the virus to infect every cell. Unfortunately, plants regenerated from protoplasts are also likely to be genetically variable (*see* Section 10.6).

The culture of reproductive tissues

A few workers have successfully cultured floral tissues to produce virus-free plants. This method has been particularly useful for citrus species, in which most of the viruses are not seed transmitted. The failure of citrus viruses to enter nucellar and ovular tissues has been used to produce healthy oranges (Bitters *et al.*, 1972; Navarro and Juarez, 1977).

In addition, the culture of floral meristems has been used to produce cauliflowers free of turnip and cauliflower mosaic viruses (Walkey *et al.*, 1974). In this species the primordial floral meristem (the curd) reverts to vegetative growth in tissue culture, enabling many plantlets to be regenerated from a single plant which normally has only a single terminal bud and no axillary buds in the vegetative phase.

Meristem-tip culture

The most important and effective method of tissue culture for the production of virus-free plants, has been meristem-tip culture. Healthy plants of a wide range of economically important crops have been regenerated from meristem-tips (*see* Table 11.1). On a suitable medium, meristem-tips grow more quickly into plantlets than cultured tissues from other sources, but of even greater importance, the regenerated plantlets usually retain the genetic characteristics of the parent plant. The greater genetic stability of plants regenerated from meristem-tips, is probably due to the more uniformly diploid nature of the tips' cells (Murashige, 1974).

Various terms have been used to describe the technique, including bud-tip, axillary-bud, shoot apex, meristem-tip, meristem or simply tip culture. In addition, the orchid industry introduced the phrases meristemming and mericloning. Unfortunately, these terms do not accurately describe the nature of the explant that is taken for culture. The meristem dome of cells alone, cannot usually be successfully cultured into a plantlet, nor is the complete apical or axillary bud normally taken for culture. In practice, for successful culture, the explant must consist of the meristematic dome of cells together with at least one, if not several, leaf primordia (*see* Figure 11.4). The term meristem-tip, has been most frequently used to describe this unit of tissue, and is the one used in this chapter.

In recent years, meristem-tip and other tissue culture methods have been increasingly used for the rapid propagation of clonal plants. It should be emphasized, however, that when meristem-tip culture is used for the production of virus-free plants, in contrast to its use for plant propagation, it is only necessary for one healthy plantlet to be produced from an infected parent, and this can be further propagated by conventional vegetative propagation, or by rapid tissue culture propagation, as required.

A scheme for the production of virus-free plants from meristem-tips is presented in Figure 11.1.

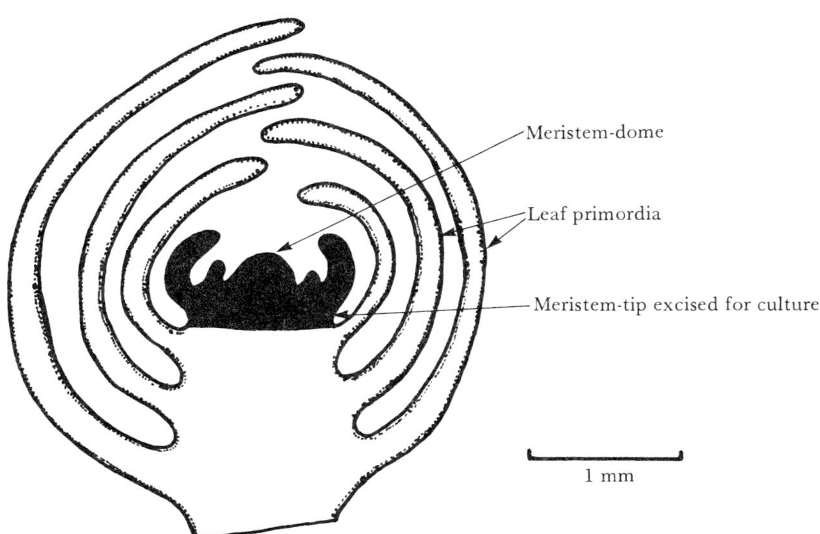

Fig. 11.4 Diagram of a bud showing the meristem-tip region that is usually removed as the explant for tissue culture.

11.3.3 Factors controlling the production of virus-free plants by meristem-tip culture

When meristem-tip culture was first used to produce virus-free plants, the workers assumed that the regenerated plants were healthy because virus did not invade the meristematic cells of the bud. Unfortunately, in some virus/host combinations this assumption was incorrect and some nurserymen propagating orchids by this technique, were found to be inadvertently selling infected plants as virus-free material.

It is now known that some viruses invade the meristem to varying degrees, dependent upon the type of virus and host species involved (Walkey and Webb, 1970; Mori, 1977). Success in obtaining virus-free plants by meristem-tip culture, may therefore, depend upon the initial size of the tip removed for culture, as demonstrated for carnations infected with carnation mottle, vein-mottle and latent viruses (Stone, 1968). Tips varying in size from 0.1 to 2 mm in diameter have been cultured into plantlets. Most workers, however, have cultured tips between 0.3 and 1.5 mm in diameter. Generally, the number of virus-free plantlets produced is inversely proportional to the size of the tip cultured. Thus in some instances, it is impossible to excise a meristem-tip small enough to be free of the virus present in the infected parent and to regenerate it into a healthy plant (*see* Figure 11.5*a*).

In other virus/host combinations viruses may be eradicated from meristem-tips during tissue culture, even when the tips can be shown to be infected at the time of excision (*see* Figure 11.5*b*), as in the

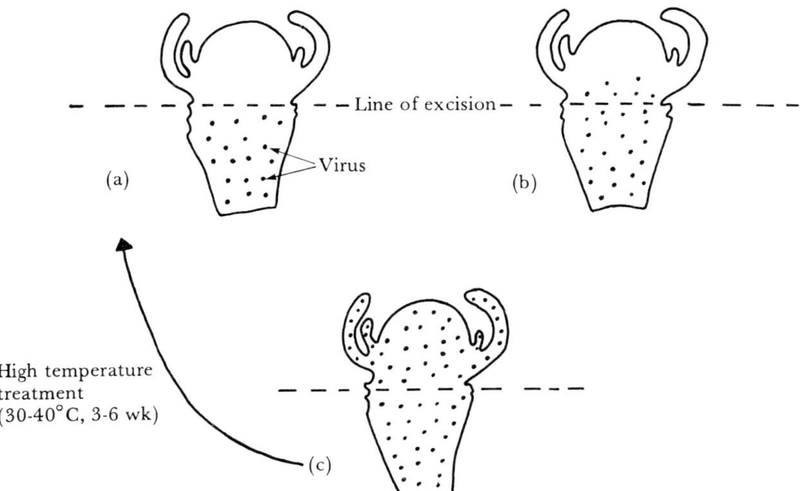

Fig. 11.5 Diagram showing virus invasion of the meristem-tip in relation to virus eradication by tissue culture and thermotherapy.

elimination of carnation mottle virus from carnation (Hollings and Stone, 1964) and cherry leaf roll virus from tobacco (Walkey *et al.*, 1969). It has been suggested by some workers that this '*in-vivo*' virus eradication is caused by metabolic disruption resulting from cell injury during the excision process, the smaller the tip excised, the greater the injury and resultant disruption (Mellor and Stace-Smith, 1977). Others have suggested that auxins and other growth-promoting chemicals in the nutrient solution may stimulate a host tissue's resistance mechanism and assist eradication of virus during tissue culture (Antoniw *et al.*, 1981). To date, the mechanism of such *in vivo* virus inactivation remains unknown, but whatever the explanation, it seems probable that this type of virus eradication is more likely to occur if low, rather than high, concentrations of virus are present in the tip.

In many plant/virus combinations it is impossible to excise a tip, which is either small enough to avoid virus or to allow subsequent *in vivo* eradication, and which is still large enough to be regenerated into a plantlet. In these circumstances larger tips must be taken for culture, and these frequently contain high concentrations of virus (Walkey and Webb, 1970). It is, however, still possible to obtain virus-free plants from such material by using thermotherapeutic treatments combined with meristem-tip culture, as discussed in Section 11.2. The heat treatment is usually applied to the infected parent plant before excision of the tip (*see* Figure 11.5*c*), although thermotherapy of the infected tip during tissue culture is also possible.

Meristem-tips may be taken for culture from either terminal or axillary buds, but the position of the bud on the infected plant may influence its virus content. It has been shown in some virus/host combinations that buds high on the stem may have a lower virus content than buds lower down (Stone, 1981). The reason for this is not known, but it is possible that the auxin gradient in the shoot, which is high at the apex and low at the base, may have a differential influence on bud metabolism at different positions and indirectly affect virus synthesis.

11.4 Chemotherapy

The use of chemicals to suppress virus symptoms and multiplication in infected plants has recently been reviewed by Tomlinson (1982). In general, the evidence suggests that although many chemicals are able to suppress virus symptoms or multiplication, few, if any, eradicate virus. In addition, many proven antiviral compounds have been unsuitable for use on crops because of their côst and phytotoxicity.

It has been suggested that growth-promoting chemicals, such as cytokinins, will eradicate virus during meristem-tip culture (Quak, 1961) and that high cytokinin concentrations will stimulate host, rather than viral protein synthesis (Johnstone and Wade, 1974*b*). There is, however, little or no experimental evidence to support these theories, and recent experiments showed that cucumber mosaic virus was not eradicated from *Nicotiana rustica* tissue, grown in culture media containing a wide range of cytokinin concentrations (Cohen and Walkey, 1978; Simpkins *et al.*, 1981). It is possible, however, that although growth-promoting chemicals may not directly inactivate virus, they may be responsible for stimulating the host plant's resistance (Antoniw *et al.*, 1981) which could in some circumstances, result in a virus's eradication (*see* Section 11.3.3).

Recent chemotherapeutic studies suggest that a more promising approach may be the incorporation of antimetabolite chemicals such as ribavirin (1,2,4,-triazole-3-carboxamide syn. Virazole) in the tissue culture medium (Shepard, 1977; Cassells and Long, 1980). These chemicals in common with high temperature treatments, block virus replication in infected tissues, and presumably, while virus synthesis is stopped, degradation of existing virus continues until eradication has occurred (Simpkins *et al.*, 1981). The cost of these chemicals, and possible phytotoxic effects will prevent them from being used on a field scale, but their use in laboratory treatments combined with meristem-tip culture, may prove advantageous in eradicating certain viruses.

Another antiviral chemical developed originally for medical uses, adenine arabinoside (syn. Vira A, vidarabine), has also been used to increase the efficiency of eradication of viruses from infected meristem-tips grown in tissue culture. The incorporation of Vira-A in the culture medium, enabled two unidentified viruses to be eradicated from sweet potato (Stone *et al.*, 1978) and a complex of viruses from *Ullucus tuberosum* (Stone, 1982). The mechanism of its action on these viruses is not yet known.

11.5 Indexing, Clone Selection and Stock Maintenance

Having regenerated a plantlet by one of the techniques described and established it in soil, it is then essential to ensure that the plantlet is virus-free. Because many viruses have a delayed resurgence period following thermotherapy and tissue culture treatments (Hollings, 1965), considerable emphasis must be placed on virus indexing. Virus tests must be made several times during the first year or so following culture, before the plant can be confidently considered virus-free and used as nuclear stock material for commercial multiplication.

Methods of virus indexing depend on the species concerned and include sap transmission to susceptible hosts, serology and electron microscopic examination of leaf and sap material (*see* Chapter 6). Undoubtedly, the most important developments in virus indexing of nuclear stock material in recent years, has been the introduction of the ELISA (Enzyme Linked Immunosorbent Assay) and immunosorbent electron microscopy techniques (*see* Section 6.6.3), which are considerably more sensitive than other assay methods. If a virus is not sap or vector transmissible and cannot be purified for antiserum production, as is the case with some fruit tree viruses, the treated material may have to be grafted to a susceptible root stock to confirm its freedom from virus (*see* Section 7.3).

From the time the regenerated plantlet is removed from tissue culture and established in soil, it must be maintained under conditions that prevent it from becoming reinfected with virus. The regenerated plantlets and virus-free nuclear stock mother plants, have traditionally been maintained in insect-proof glass or gauze houses and grown in sterilized soil. Recent studies have shown, however, that virus-free stock material may be readily stored for long periods by aseptic tissue culture at low temperatures (Mullin and Schlegel, 1976), a technique that may prove less expensive and less labour intensive than maintaining adult plants.

Although there is considerable evidence to show that plants derived from meristem-tips usually exhibit little or no genetical variation from their virus-infected parents, it is essential to check that no genetic changes have occurred. The regenerated plants should be monitored for any changes in agronomic characteristics over several cropping seasons in the field, and attention should be paid to small differences that might result from physiological changes in the plants' performance due to the absence of virus.

In apple, minor changes in fruit colour and in the time of flowering and fruiting, have been observed in different regenerated, virus-free clones (Campbell, 1974), and in rhubarb, changes in the low temperature requirements to break dormancy have been reported (Case, 1973). Consequently, selection of the most horticulturally desirable clones following plantlet regeneration, should be an important consideration in any virus eradication programme (*see* Figure 11.1).

As soon as a suitable clone has been selected, the plants have to be multiplied to commercial quantities. Nuclear stock associations have been established in many countries to maintain, propagate and distribute virus-free stock plants and in some instances, individual growers are using meristem-tip culture and heat therapy to eradicate virus from their own clones.

It is also important that field trials are conducted to compare the yield and quality of virus-free and virus-infected stocks. Without this information it may prove difficult to persuade some growers, that it is to their economic advantage to replace infected stocks with virus-free clones.

Finally, it is important to monitor rates of virus reinfection of healthy clones grown in various localities. The rate of virus reinfection will be affected by the epidemiology of the virus concerned, and the degree of isolation the healthy crop can be given. In general, it may be expected that aphid-borne viruses will be the first to reinfect, particularly if the healthy crop is planted near virus-infected plants (Adams *et al.*, 1979; Walkey *et al.*, 1982). Information obtained from monitoring virus reinfection rates, will help in determining the economic benefits that are likely to result from the use of virus-free planting material.

11.6 Future Developments in the Use of Virus-free Plants

The use of virus-free plants to replace infected clonal material has already made a considerable impact on horticulture throughout the world, through improvements in yield and quality. It is probable, however, that the demand for virus-free stock plants will increase; first, because of their greater yield potential; and secondly, because of the need to use disease-free stock material, as more and more propagators turn to tissue culture techniques for the rapid multiplication of vegetatively propagated crops.

It is also probable, that as many countries impose higher sanitary standards for the importation of plant material, a guarantee that plants are virus-free will be essential to facilitate their international movement (Button, 1977; Roca *et al.*, 1979).

11.7 References

Adams, A. N., Barbara, D. J., Manwell, W. E. and Thresh, J. M. (1979). Hop mosaic virus (HMV). *East Malling Res Stn Rept 1978*, 103.

Alconero, S., Santiago, A. G., Morales, F. and Rodriquez, F. (1975). Meristem-tip culture and virus indexing of sweet potato. *Phyt* **65**, 769–73.

Antoniw, J. F., Kueh, J. S. H., Walkey, D. G. A. and White, R. F. (1981). The presence of pathogensis-related proteins in callus of Xanthi-nc tobacco. *Phytopathol Z* **101**, 179–84.

Bent, K. J. (1969). Fungicides in perspective. *Endeavour* **27**, 129–34.

Bitters, W. P., Murashige, T., Rangan, T. S. and Nauer, E. (1972). In *Proc 5th Conf Intern Organ Citrus Virologists* (ed. W. C. Price). Univ Fla: Gainsville. pp. 267–71.

Button, J. (1977). International exchange of disease free citrus clones by means of tissue culture. *Outlook Agric* **9**, 155–9.

Campbell, A. I. (1974). Unpublished results.

Case, M. W. (1973). Rhubarb. 15th Ann. Rept. Stockbridge House E.H.S., Yorkshire.

Cassells, A. C. and Long, R. D. (1980). The regeneration of virus-free plants from cucumber mosaic virus and potato virus Y infected tobacco explants cultured in the presence of virazole. *Zeitsch Nature* **35c**, 350–1.

Cohen, D. (1977). Thermotherapy and meristem-tip culture of some ornamental plants. *Acta Hort* **78**, 381–8.
Cohen, D. and Walkey, D. G. A. (1978). Unpublished results.
Dale, P. J. (1977). The elimination of ryegrass mosaic virus from *Lolium multiflorum* by meristem-tip culture. *Ann Appl Biol* **85**, 93–6.
Dale, P. J. (1979). The elimination of cocksfoot streak virus, cocksfoot mild mosaic virus and cocksfoot mottle virus from *Dactylis glomerata* by shoot tip and tiller bud culture. *Ann Appl Biol* **93**, 285–8.
De Fossard, R. A. (1976). *Tissue culture for plant propagators.* University of New England.
Duron, M. (1977). Utilisation de la culture *in vitro* pour l'amelioration de l'état sanitaire de cultivars de Forsythia (Vahl.). *C R Hebd Séances Acad Sci Ser D* **284**, 183–5.
Fulton, J. P. (1954). Heat treatment of virus-infected strawberry plants. *Plant Dis R* **38**, 147–9.
Gamborg, O. L. and Wetter, L. R. (1975). *Plant tissue culture methods.* N.R.C. Canada, Saskatchewan.
Hamid, A. D. and Locke, S. B. (1961). Heat inactivation of leaf roll virus in potato tuber tissues. *Am Potato J* **28**, 304–10.
Hansen, A. J. and Hildebrandt, A. C. (1966). Distribution of tobacco mosaic virus in plant callus cultures. *Virol* **28**, 15–21.
Hickman, A. J. and Varma, A. (1968). Viruses in horse-radish. *Plant Path* **17**, 26–30.
Hollings, M. (1965). Disease control through virus-free stock. *Ann Rev Phyto* **3**, 367–96.
Hollings, M. and Stone, O. M. (1964). Investigations of carnation viruses I. Carnation mottle. *Ann Appl Biol* **53**, 103–18.
Johnstone, G. R. and Wade, G. C. (1974*a*). Therapy of virus-infected plants by heat treatment. I. Some properties of tomato aspermy virus and its inactivation at 36°C. *Aust J Bot* **22** 437–50.
Johnstone, G. R. and Wade, G. C. (1974*b*). Therapy of virus-infected plants by heat treatment. II. Host protein synthesis and multiplication of tomato aspermy virus at 36°C. *Aust J Bot* **22**, 451–60.
Kaiser, W. J. and Teemba, L. R. (1979). Use of tissue culture and thermotherapy to free East African cassava cultivars of African cassava mosaic and cassava brown streak viruses. *Plant Dis R* **63**, 780–4.
Kassanis, B. (1950). Heat inactivation of leaf-roll virus in potato tubers. *Ann Appl Biol* **37** 339–41.
Kassanis, B. (1957). The use of tissue culture to produce virus-free clones from infected potato varieties. *Ann Appl Biol* **45**, 422–7.
Larsen, E. C. (1966). Daily temperature cycles in heat inactivation of viruses in chrysanthemum and apple. Proc 17th Int Hort Cong Abst 104.
Larsen, E. C. (1974). Heat inactivation of viruses in apple. CLMM 109 by use of daily temperature cycles. *Tidss Plant* **78**, 422–8.
Maia, E., Bettachini, B., Beck, D., Vénard, P. and Maia, N. (1973). Contribution à l'amélioration de l'état sanitaire du lavandim, clone 'abrial'. *Ann Phytopathol* **5**, 115–24.
Mantell, S. H., Haque, S. Q. and Whitehall, A. P. (1980). Apical meristem-tip

culture for eradication of flexuous rod viruses in yams (*Dioscorea alata*). *Trop Pest Man* **26**, 170–9.

Martin, C., Dulieu, H. and Carré, M. (1967). Sur la possibilité de rendre des plantes virosés indemnés de virus par la culture de méristèms inflorescentiels et floraux. *C R Hebd Séances Acad Sci* **264**, 1994–6.

Mellor, F. C. and Stace-Smith, R. (1970). Virus strain differences in eradication of potato viruses X and S. *Phyt* **60**, 1587–90.

Mellor, F. C. and Stace-Smith, R. (1977). Virus free potatoes by tissue-culture. In *Plant cell, tissue and organ culture* (ed. Reinert, J. and Bajaj, Y. P. S.). Springer-Verlag: Berlin.

Morel, G. M. and Martin, C. (1952). Guérison de dahlias atteints d'une maladie à virus. *C R Hebd Séances Acad Sci* **235**, 1324–5.

Morel, G. M. and Martin, C. (1955). Guérison de pommes de terre atteints de maladies à virus. *C R Hebd Séances Acad Agric* **41**, 472–5.

Mori, K. (1977). Localisation of viruses in apical meristems and production of virus-free plants by means of meristem and tissue culture. *Acta Hort* **78**, 389–96.

Moskovets, S. N., Gorbarenko, N. I. and Zhuk, I. P. (1973). The use of the method of the culture of apical meristems in the combination with low temperature for the sanitation of potato against mosaic virus. *Sel' skokhozyaistvennaya Biol* **8**, 271–5.

Mullin, R. H. and Schlegel, D. E. (1976). Cold storage maintenance of strawberry meristem plants. *Hortic Sci* **11**, 1004.

Mullin, R. H. and Schlegel, D. E. (1978). Meristem-tip culture of dahlia infected with dahlia mosaic virus. *Plant Dis R* **62**, 565–6.

Murashige, T. (1974). Plant propagation through tissue culture. *Ann Rev Plant Physiol* **25**, 135–66.

Murashige, T. and Skoog, F. (1962). A revised medium for rapid growth and bioassays with tobacco tissue culture. *Physiologia Plantarum*, **15**, 473–97.

Navarro, L. and Juarez, J. (1977). Tissue culture techniques used in Spain to recover virus-free citrus plants. *Acta Hort* **78**, 425–35.

Nyland, G. and Goheen, A. C. (1969). Heat therapy of virus diseases of perennial plants. *Ann Rev Phyto* **7**, 331–54.

Pennazio, S., Redolfi, P., Vecchiati, M. and Cantisani, A. (1976). Tobacco plant temperatures during thermotherapy of potato vitrus X. *Phytopathol Mediterr* **15**, 106–9.

Pillai, S. K. and Hildebrandt, A. C. (1968). Geranium plants differentiated in vitro from stem tips and callus cultures. *Plant Dis R* **52**, 600–1.

Preil, W., Kuhne, H. and Hoffmann, M. (1978). Hydrangea ringspot virus-free Hydrangea hortensia from meristem culture. *Deut Pflanzenschutzd* **30**, 88–90.

Quak, F. (1961). Heat treatment and substances inhibiting virus multiplication in meristem culture to obtain virus-free plants. *Adv Hort Sci Appl* **1**, 144–8.

Quak, F. (1977). Meristem culture and virus-free plants. In *Plant cell, tissue and organ culture* (ed. Reinert, J. and Bajaj, Y. P. S.) Springer-Verlag: Berlin. pp. 598–615.

Roca, W. M., Bryan, J. E. and Roca, M. R. (1979). Tissue culture for the international transfer of potato genetic resources. *Am Potato J* **56**, 1–10.

Schmidt, H. E. (1974). Investigation on the therapy of virus infected hop (*Humulus lupulus* L.) by heat treatment of cuttings and shoot tips. *Zentralbl Bakt Hyg Abt* **11**, 259–70.

Shepard, J. F. (1975). Regeneration of plants from protoplasts of potato virus X infected tobacco leaves. *Virol* **66**, 492–501.

Shepard, J. F. (1977). Regeneration of plants from protoplasts of potato virus X infected tobacco leaves. II. Influence of virazole in the frequency of infection. *Virol* **78**, 261–6.

Simpkins, I., Walkey, D. G. A. and Neely, H. A. (1981). Chemical suppression of virus in cultured plant tissues. *Ann Appl Biol* **99**, 161–9.

Stone, O. M. (1968). The elimination of four viruses from carnation and sweet william by meristem-tip culture. *Ann Appl Biol* **62**, 119–22.

Stone, O. M. (1981). Unpublished results.

Stone, O. M. (1982). The elimination of four viruses from *Ullucus tuberosus* by meristem-tip culture and chemotherapy. *Ann Appl Biol* **101**, 79–83.

Stone, O. M., Hollings, M. and Pawley, R. R. (1978). Virology. *Ann R Glasshouse Crops Res Inst.*

Svobodva, J. (1965). Elimination of viruses by means of callus tissue cultures. In *Viruses of plants* (ed. Beemster, A. B. R. and Dijkstra, J.). Elsevier/North Holland: London. pp. 48–53.

Taylor, J. D. and Dudley, C. L. (1977). Seed treatment for the control of halo blight of beans (*Pseudomonas phaseolicola*). *Ann Appl Biol* **85**, 223–32.

Tomlinson, J. A. (1982). Chemotherapy of plant viruses and virus diseases. In *Pathogens, vectors and plant diseases* (ed. Harris, K. F. and Maramorosch, K.). Academic Press: London. pp. 23–44.

Tomlinson, J. A. and Walkey, D. G. A. (1967). The isolation and identification of rhubarb viruses occurring in Britain. *Ann Appl Biol* **59**, 415–27.

Walkey, D. G. A. (1968). The production of virus-free rhubarb plants by apical tip-culture. *J Hort Sci* **43**, 283–7.

Walkey, D. G. A. (1976). High temperature inactivation of cucumber and alfalfa mosaic viruses in *Nicotiana rustica* cultures. *Ann Appl Biol* **84**, 183–92.

Walkey, D. G. A. (1980). Production of virus-free plants by tissue culture. In *Tissue culture methods for plant pathologists* (ed. Ingram, D. A. and Helgeson, J. P.). Blackwell: Oxford. pp. 109–17.

Walkey, D. G. A. and Cooper, V. C. (1972). Some factors affecting the behaviour of plant viruses in tissue culture. *Physiol Plant Path* **2**, 259–64.

Walkey, D. G. A. and Cooper, V. C. (1975). Effect of temperature on virus eradication and growth of infected tissue cultures. *Ann Appl Biol* **80**, 185–90.

Walkey, D. G. A., Cooper, V. C. and Crisp, P. (1974). The production of virus-free cauliflowers by tissue culture. *J Hort Sci* **49**, 273–5.

Walkey, D. G. A., Creed, C., Delaney, H. and Whitwell, J. D. (1982). Studies on the reinfection and yield of virus-tested and commercial stocks of rhubarb cv. Timperley Early. *Plant Path* **31**, 253–61.

Walkey, D. G. A., Fitzpatrick, J. and Woolfitt, J. M. G. (1969). The inactivation of virus in cultured tips of *Nicotiana rustica* L. *J Gen Virol* **5**, 237–41.

Walkey, D. G. A. and Freeman, G. H. (1977). Inactivation of cucumber

mosaic virus in cultured tissues of *Nicotiana rustica* L. by diurnal alternating periods of high and low temperature. *Ann Appl Biol* **87**, 375–82.
Walkey, D. G. A. and Thompson, A. (1978). Virus inactivation studies – watercress. Ann Rept Nat Veg Res Stn 1977.
Walkey, D. G. A. and Webb, M. J. W. (1970). Tubular inclusion bodies in plants infected with viruses of the Nepo type. *J Gen Virol* **7**, 159–66.
Welsh, M. F. and Nyland, G. (1965). Elimination and separation of viruses in apple clones by exposure to dry heat. *Can J Plant Sci* **45**, 443–54.
Yang, H. J. and Clore, W. J. (1976). Obtaining virus-free plantlets of *Asparagus officinalis* L. by culturing shoot tips and apical meristems. *Hortic Sci* **11**, 474–5.

11.8 Further Selected Reading

De Fossard, R. A. (1976). Tissue culture for plant propagators. University of New England.
Gamborg, O. L. and Wetter, L. R. (1975). Plant tissue culture methods. N.R.C. Canada, Saskatchewan.
Nyland, G. and Goheen, A. C. (1969). Heat therapy of virus diseases of perennial plants. *Ann Rev Phyto Path* **7**, 331–54.
Quak, F. (1977). Meristem culture and virus-free plants. In *Plant cell, tissue and organ culture* (ed. Reinert, J. and Bajaj, Y. P. S.). Springer-Verlag: Berlin. pp. 598–615.

12 Practical Information and Introductory Exercises

12.1 Introduction

The information outlined in this chapter should enable the newcomer to applied plant virology, to carry out some of the basic procedures for virus isolation, transmission and identification. Details are also given of meristem-tip culture procedures that may be used to eradicate viruses from infected parent clones.

The procedures described in this chapter are based on techniques used every day in an applied research laboratory, and should be of value to the student or post-graduate research worker. Wherever possible, practical class exercises are suggested that could be carried out by students.

12.2 Virus Isolation and Transmission

12.2.1 Mechanical sap transmission

Solutions for preparing inoculum

(*a*) 1% di-potassium hydrogen orthophosphate (K_2HPO_4)+0.1% sodium sulphite (Na_2SO_3).

: prepared by dissolving 1 g K_2HPO_4 and 0.1 g Na_2SO_3 in 100 ml of distilled water (use ice-cold).

(*b*) Alternatively use 0.1 M potassium phosphate buffer at pH 7.5

: prepared by dissolving 17.4 g di-potassium hydrogen orthophosphate (K_2HPO_4) in 1 l distilled water (0.1 M solution), and add potassium di-hydrogen orthophosphate (KH_2PO_4) (3.4 g dissolved in 250 ml distilled water (0.1 M solution), to achieve a pH 7.5.

Sap transmission procedure

(*a*) Take infected leaf material and place in a pre-cooled mortar. Add ice-cold K_2HPO_4/Na_2SO_3 solution in the ratio of 1.5 ml solution to 1 g leaf and grind with a pestle until a fine sap-homogenate is obtained.

(*b*) Filter homogenate through a square of butter-muslin into a 1 cm diameter test-tube. Keep the test-tube containing the inoculum in an ice-bucket until used.

(*c*) Dust the test plant with a fine covering of *carborundum 300*, using a throat-spray (*see* Plate 4.2). Mark the leaves to be inoculated by piercing them with a pencil point. Label the pot with the date and identity of the treatment being applied.

(*d*) Take a folded square of muslin, moisten thoroughly with the sap inoculum, and squeeze gently to remove excess sap. Gently, but firmly stroke the upper leaf surface with the moist pad, supporting the leaf with the free hand (*see* Plate 4.2). Ensure that the complete upper surface of the leaf is covered.

A finger dipped in the sap-inoculum may be used to rub the leaf instead of the moist pad. Wash hands thoroughly using sodium triphosphate soap when inoculation is completed.

(*e*) Rinse the surface of inoculated leaves under a trickle of cold tap-water, and place the plant in a glasshouse compartment at approximately 22° to 25°C.

(*f*) Inoculate a carborundum dusted plant with the phosphate solution alone, for control purposes.

(*g*) Observe the test plants daily for the development of symptoms, especially from 4 to 5 days after inoculation.

Class exercises

(*a*) Sap transmit cucumber mosaic virus (CMV) to *Chenopodium quinoa* to demonstrate the production of primary local lesion symptoms, without a further systemic reaction; or tobacco mosaic virus to *Nicotiana glutinosa* to demonstrate necrotic local lesions.

(*b*) Sap transmit CMV to marrow or cucumber seedlings by inoculating the seedling cotyledons as soon as they unfold. Chlorotic local lesions may develop on the inoculated cotyledons (depending on the cultivar used), and the virus will spread systemically to give a mosaic reaction.

Source of virus

For class purposes, virus isolates will probably have to be obtained from another plant virus laboratory. These may be sent in a dried form (*see* Section 4.5), or if the postal services are efficient, by placing a piece of fresh infected leaf between moist filter paper and sealing it in a polythene bag.

Host Plants

Examples of standard laboratory host plants are given in Table 4.1. Seed of some species, such as tobacco, may be obtained from a commercial seed company, but in most cases it will be necessary to approach another plant virus laboratory. Consequently, only a small amount of seed is likely to be available, and workers must expect to multiply seed for their own requirements.

A regular supply of test plants will be necessary, and attention must be given to the time required from seed sowing to the plant's availability for inoculation. This is particularly important in temperate regions during the winter months, when supplementary lighting will be essential to ensure an adequate supply of plants. Table 12.1 illustrates the approximate time required for the more common test species to mature from sowing to inoculation size, at different times of the year (information based on Pawley, 1973).

Table 12.1 *Maturation of some common virus test plants*: number of days from sowing seed until the plants are ready for inoculation*

	Plant species				
Ready for use in	*Chenopodium amaranticolor*	*Chenopodium quinoa*	*Nicotiana clevelandii*	*Nicotiana glutinosa*	*Nicotiana tabacum*
Jan.	65	62	76	72	61
Feb.	73	59	71	74	64
Mar.	62	47	67	66	53
Apr.	51	42	57	61	46
May	45	34	50	48	39
June	39	31	42	42	36
July	37	31	41	39	34
Aug.	43	34	44	43	36
Sept.	47	38	48	46	41
Oct.	49	42	54	49	44
Nov.	55	46	63	57	51
Dec.	60	53	66	63	58

*Information based on Pawley (1973).

Maintenance of virus isolates

Virus isolates can be maintained for long periods by sub-culturing in their respective hosts, but may become contaminated or attenuated during repeated sub-culturing (*see* Section 4.5). It is advisable, therefore, to store samples of the original isolates as soon as possible after they have been acquired. When a fresh isolate is obtained, its concentration should be built up by sub-culturing once or twice in a suitable host plant. Then it should be preserved by one of the following methods.

(*a*) As dried powder

(i) Place infected leaves over anhydrous calcium chloride ($CaCl_2$) in an airtight, screw-topped bottle for several months.

(ii) Grind the dry leaves to a powder with a pestle and mortar, and store in a tightly sealed capsule.

(*b*) In liquid nitrogen

(i) Grind infected leaves in K_2HPO_4/Na_2SO_3 solution as if preparing inoculum for sap-transmission.

(ii) Place 0.5 ml aliquots of sap in sealed capsules and store under liquid nitrogen in a commercial storage flask.

(*c*) By lyophilization

(i) Take a small piece of infected leaf (approximately 2 cm square) and place in a glass ampoule suitable for the freeze-drying machine to be used. Add a label to identify and date the isolate. Constrict the tube and evacuate the air. Seal the tube and store at room temperature.

(ii) Alternatively, grind infected leaf in a minimal volume of K_2HPO_4/Na_2SO_3 solution. Take 10 ml of the filtered sap and add 0.7 g D-glucose+0.7 g peptone. Shake the mixture to dissolve the additives. Pipette 0.25 ml aliquots of the mixture into the freeze-drying ampoule. Add a small plug of non-absorbent cotton wool, followed by a suitable paper label. Proceed with lyophilization as described above. Lyophilized virus should be reactivated from the storage ampoule by resuspension in a minimal volume of K_2HPO_4/Na_2SO_3 solution and inoculation to the host plant.

12.2.2 Aphid transmission

Workers in applied aspects of plant virology find it necessary to maintain colonies of aphids for transmission studies. For many purposes the potato-peach aphid, *Myzus persicae*, is suitable, but other aphid species may be required for specific viruses.

A healthy colony of *M.persicae* may be cultured on *Brassica juncea* cv. Tendergreen Mustard (available from the Ferry Morse Seed Co., California, U.S.A.). The colony should be maintained at between 18° and 23°C, with a 16 h daylength, and at a relative humidity of between 75 and 80%. The aphids will require sub-culturing to fresh host plants every 7 to 14 days.

Experimental procedure for transmission
The procedure used will depend on whether the virus is transmitted in a non-persistent or persistent manner (*see* Section 7.4.2).

(*a*) Non-persistent virus

(i) Leaves carrying large numbers of non-viruliferous aphids are removed from the healthy culture plant, and lightly tapped over a sheet of white paper. Aphids which are not feeding will drop off and can be gently funnelled into a plastic box, with a securely fitting lid. Removal of the aphids from the leaf is generally easier, if the leaf is warmed over an electric light bulb to agitate the aphids into withdrawing their stylets from the leaf tissues. Under no circumstances should the aphids be brushed off the leaf or removed by violent tapping of the leaf.

(ii) The aphids removed from the leaf should be starved for 1 to 2 hours in the sealed container, which should be kept in the dark.

(iii) After starving, the aphids should be removed from the box using a fine paintbrush, and placed upon the surface of a detached leaf infected with the virus to be transmitted. This leaf should also be contained within a sealed plastic box.

(iv) Using a binocular stereomicroscope, check that the aphids have inserted their stylets into the leaf to feed. Allow them to feed for a minimum of 2 to 3 min (referred to as the *acquisition feeding time*), and then start to transfer them individually with the paintbrush to the healthy test plant to be infected.

(v) When transferring the aphid, it is essential to ensure that the aphid has withdrawn its stylets before it is removed. Often it is necessary to 'tickle' the aphid with the hairs of the brush, until it stops feeding and starts to walk.

(vi) When the required number of aphids have been transferred to the test plants (5 to 10 aphids per plant is often an adequate number), at least several minutes feeding time (the *inoculation feeding period*) should be allowed for them to transmit the non-persistent virus. In practice, however, it is often more convenient to allow them to remain on the plant overnight and then kill them by watering the plant with a systemic aphicide, such as *Metasystox* (demeton-S-methyl), the following morning.

(vii) During the inoculation feeding period the test plants should be kept in a cage or plastic propagation chamber, to prevent the aphids from spreading to other plants in the glasshouse.

The transfer of aphids singly to a test plant, is very time-consuming, and may be impractical if large numbers of plants have to be infected, for example, in a resistance screening programme. An alternative method is to transfer aphids *en masse* by following the above procedure up to the end of the *acquisition feeding period*, but then removing portions of infected leaf each carrying at least 10 feeding aphids. One leaf-portion is then pinned on to a leaf of each test plant to be infected. The test plants are then left for 48 hours in an insect-proof cage. During this period the aphids will walk off the infected leaf as it wilts and start feeding on the test plant. After the 48 hour period the aphids may be killed with an aphicide.

Class exercises

Carry out the above procedures to transmit turnip mosaic virus from infected to healthy Tendergreen Mustard plants. Symptoms will appear 12 to 16 days after inoculation. Alternatively, transmit cucumber mosaic virus from infected to healthy marrow plants. Symptoms will appear 7 to 10 days after inoculation.

(*b*) Persistent virus

Remove non-viruliferous aphids from the culture host as described above. Allow the aphids at least a 48 hour acquisition feeding period on the infected plant (shorter or longer periods may be required depending on the virus). Transfer the aphids to the test plant and allow at least a 48 hour inoculation feeding period (longer periods may be necessary). Kill the aphids with an aphicide.

12.3 *In vitro* Properties in Crude Sap

Three simple tests can be carried out on crude sap from an infected plant, that will give an indication of a virus' stability and concentration (*see* Section 6.3). These tests are used to determine a virus' *thermal inactivation temperature* (TIP), its *dilution end-point* (DEP) and its *longevity 'in-vitro'* (LIV).

12.3.1 Experimental procedures

Thermal inactivation temperature

(i) Grind infected leaf material in 0.01 M phosphate buffer (1 g leaf to 1.5 ml buffer) using a pestle and mortar. Filter the homogenate through muslin.

(ii) Pipette 0.5 ml aliquots of the infected sap into thin-walled glass tubes and store in crushed ice.

(iii) The samples should be heated to temperatures between 45° and 70°C at intervals of 5°C. In the case of a few viruses, such as tobacco mosaic virus, treatment at 90°C will be required for inactivation.

(iv) Preheat a waterbath to the required temperature and then treat each sample for 10 min. Immediately the treatment period is completed, rapidly cool the contents by plunging the tube into ice-cold water.
(v) Inoculate the sample immediately to the test host.
(vi) The temperature of the waterbath can be easily controlled if it is heated by a gas burner and a paddle stirrer is used to stir the water. Treatments should start with the highest temperature, as it is easier to control the temperature as it falls than to raise it.
(vii) The method used for assaying the treated samples will depend on whether the virus concerned has a local lesion host. If a local lesion host is available, then a half-leaf replicate method of assay can be used (*see* Section 4.4). If not, each sample must be inoculated to one or more individual test plants.

Dilution end-point

(i) Prepare a homogenate of infected sap as described above.
(ii) Dilute the sap in a ten-fold dilution series to 10^{-6} using distilled water (D.W.) or 0.01 M phosphate buffer.

e.g. 1.0 ml sap	1.0 ml, 10^{-1}	1.0 ml, 10^{-2}	1.0 ml 10^{-5}
+	+	+ etc.	+
9.0 ml D.W.	9.0 ml D.W.	9.0 ml D.W.	9.0 ml D.W.
10 ml, 10^{-1}	10 ml, 10^{-2}	10 ml, 10^{-3}	10 ml, 10^{-6}

(iii) Ensure that a clean pipette is used to remove each sample for dilution.
(iv) Inoculate the various dilutions to a suitable test host, using half-leaf replicates, or whole assay plants, as necessary.

Longevity ***in vitro***

(i) Prepare a crude sap homogenate as described above and divide it into a number of 0.5 ml aliquots in glass tubes.
(ii) Store samples at room temperature and assay individual aliquots after increasing periods of time e.g. 1 day, 2 days, 4 days, 8 days, 16 days, etc.
(iii) Longevity tests may also be carried out on samples stored at 0° to 2°C.

Class exercises

All three tests are highly suitable for student experiments. Use viruses with significantly different inactivation properties. For example, tobacco mosaic virus, which can be assayed on its local lesion host *Nicotiana glutinosa*, has a TIP of 85° to 90°, a DEP of 10^{-6} and can survive for months at 20°C, and years at 0° to 2°C. In contrast, lettuce mosaic virus, which may be assayed by its systemic reaction on *Chenopodium quinoa*, has a TIP of 55–60°C, a DEP of 10^{-2} and survives for only 1 to 2 days at 20°C.

12.4 Virus Purification

12.4.1 Preparation of buffers for virus purification

Potassium phosphate

Usually used between 0.01 and 0.5 M, and between pH 7.0 and 8.0. To prepare 0.5 phosphate buffer at pH 7.5.

(i) Dissolve 174 g di-potassium hydrogen orthophosphate (K_2HPO_4) in 2 l of distilled water: solution a.
(ii) Dissolve 17 g of potassium di-hydrogen orthophosphate (KH_2PO_4) in 250 ml of distilled water: solution b.
(iii) Add solution b to solution a until the pH reaches 7.5.

Sodium borate

Usually used between 0.05 and 0.5 M and between pH 7.5 and 8.5. To prepare 0.5 M borate buffer at pH 7.5.

(i) Dissolve 61.8 g of boric acid (H_3BO_3) in 2 l of distilled water. Add a small volume of 10% sodium hydroxide (NaOH) solution to help dissolve the boric acid crystals.
(ii) Prepare N/l sodium hydroxide solution by dissolving 40 g of NaOH pellets in 1 l of distilled water.
(iii) Add the N/l sodium hydroxide solution to the boric acid solution until pH 7.5 is reached.

Sodium citrate

Usually used between 0.1 and 0.5 M and between pH 6.0 and 7.4. To prepare 0.5 M citrate buffer at pH 6.5.

(i) Dissolve 105 g of citric acid in 1 l of N/l NaOH.
(ii) Add 0.5 M NaOH to the above solution until pH 6.5 is reached (an almost equal volume of NaOH will be required to reach the correct pH).

Tris-HCl

Usually used between 0.1 and 0.5 M and between pH 7.2 and 8.4. To prepare 0.5 M tris/HCl at pH 8.0.

(i) Dissolve 30.3 g of tris(hydroxymethyl)aminomethane ($NH_2.C(CH_2OH)_3$) in distilled water (approximately 400 ml).
(ii) Titrate 10% hydrochloric acid (HCl) (approximately 100 ml) against tris solution until a pH 8.0 is reached. Make up total volume of solution to 500 ml with distilled water.

12.4.2 Procedure for purifying cucumber mosaic virus (CMV)

Virus propagation

Inoculate the cotyledons of 100 marrow seedlings, cv. Goldrush, with CMV, a few days after the development of the first true leaf. Harvest shoots for purification 14 to 18 days after inoculation.

Purification procedure

(i) Take 200 g of infected leaf and place it in the glass jar of a kitchen liquidizer.

(ii) Add 400 ml of 0.5 M sodium citrate buffer at pH 6.5, containing 0.4 ml thioglycollic acid, and 400 ml of chloroform.

All additives must be precooled to 3°C. All operations should be carried out in a cold room at 3°C and in a refrigerated centrifuge.

(iii) Homogenize the mixture for several minutes until a fine homogenate is produced. Centrifuge the homogenate at low speed (500×*g*) in a bench centrifuge for 15 min. Pipette off the aqueous supernatant and retain; discard the pellet and the chloroform.

(iv) Add 10% (W/V) polyethylene glycol (mol. wt. 6000) to the supernatant and shake until it is dissolved and leave to stand for 30 min.

(v) Centifuge at 8000×*g* for 20 min. Discard supernatant and retain pellet. Resuspend the pellet in 0.05 M citrate buffer at pH 7.0, containing 2% Triton X-100 and using 3.5 ml of buffer to each 35 ml capacity centrifuge tube. Leave overnight.

(vi) Centrifuge at 15 000×*g* for 20 min in an ultra centrifuge. Retain the supernatant and discard the pellet.

(vii) Centrifuge at 75 000×*g* for 150 min in an ultra-centrifuge. Discard the supernatant and retain the pellet. Resuspend the pellet in 1 ml of 0.05 M citrate buffer at pH 7.0. Leave for several hours.

(viii) Centrifuge the resuspended pellet at 5 000×*g* for 10 min. Retain the supernatant and discard the pellet. The supernatant contains the partially purified virus which may be subjected to further sucrose gradient separation if required.

Class exercises

Purify CMV as detailed above.

(i) Examine the partially purified preparation in an electron microscope (*see* Section 12.6).

(ii) Use the preparation in double gel-diffusion tests (*see* Section 12.5), to confirm the identity of the virus against CMV antiserum.

(iii) Test the infectivity of the partially purified preparation and determine its dilution end-point by a series of 10 fold dilutions (10^{-1} to 10^{-5}) in 0.01 M phosphate buffer. Assay each dilution on half-leaf replicates of *Chenopodium quinoa*, and count the local lesions that develop 6 to 8 days after inoculation.

12.4.3 Sucrose gradient separation

(i) Preparation of a 10 to 40% (W/V) gradient.

(*a*) Prepare sucrose solutions by dissolving sucrose in distilled water or a weak buffer (0.01 to 0.1 M) suitable for the virus concerned.

5 g sucrose in 50 ml solution (10%)
10 g sucrose in 50 ml solution (20%)
15 g sucrose in 50 ml solution (30%)
20 g sucrose in 50 ml solution (40%)

(*b*) Add the sucrose solutions to a 9 cm×2.5 cm cellulose nitrate centrifuge tube using a drawn-out 10 ml pipette or a hypodermic syringe. First add 7 ml of the 10% solution. Then slowly and carefully, with the tip of the pipette in contact with the bottom of the tube, allow 7 ml of the 20%

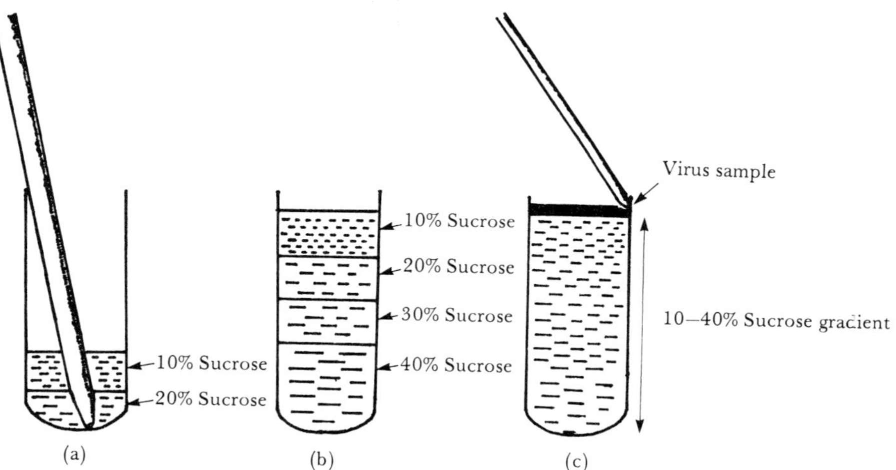

Fig. 12.1 Preparation of a sucrose gradient for ultra-centrifugation. (*a*) The lowest concentration is added to the tube first and the higher concentrations are added, in order, at the bottom of the tube; (*b*) the 10–40% gradient is allowed to stand overnight; (*c*) the virus sample is layered onto the surface of the gradient.

solution to run out below the 10% solution. Then add successively, 7 ml of the 30% and 15 ml of the 40% solution (*see* Figure 12.1). Allow the tube to stand overnight at 3°C to enable the solutions to diffuse to form the gradient.

(ii) Carefully layer 0.5 ml of the partially purified virus into the top of the gradient, by slowly running the virus down the side of the tube above the surface of the sucrose (*see* Figure 12.1).

(iii) Ensure that all the tubes used in the rotor are perfectly balanced against each other. Centrifuge the gradient in a swingout-bucket rotor for 1 to 1.5 hour at the speed recommended for the rotor concerned.

(iv) Following centrifugation, the virus layer may be observed in the tube as an opalescent band, when a narrow beam of light is shone vertically down through the tube. The virus band may be removed with a hypodermic syringe.

(v) The virus may be separated from the remaining sucrose solution by dialysis against 0.01 M citrate buffer overnight at 3°C, or by dilution with the buffer. The diluted virus is then concentrated by centrifugation at 75 000 *g* for 120 min and the resulting pellet resuspended in a small volume of distilled water.

The purified virus should then be suitable for antiserum production, or biochemical analysis.

12.5 Serology Tests

12.5.1 Antiserum production

The production of antiserum is an essential requirement for most laboratories. Rabbits are usually used for antiserum production, but whatever animal is used, it is necessary and often mandatory for the worker to receive training in the handling, injection and bleeding of the animal concerned. In Britain, for example, a government licence must be obtained before experiments can be carried out with a living animal and such licences are granted only under the most stringent conditions.

Virus injection

The following procedure can be used to produce an antiserum suitable for most purposes. It must be emphasized, however, that the purified virus preparation must be as free as possible of host protein contaminants.

(i) Take 1 ml of purified virus and thoroughly emulsify with 1 ml of *Freund's incomplete adjuvant* (a substance which allows slow release of the antigen within the animal). Inject mixture into the thigh muscle.

(ii) Repeat the procedure 3 to 4 days later, injecting into the other thigh. Six to eight such injections should be made over a 4 week period.

(iii) Blood can be taken from the rabbit (by drip bleeding approximately 20 to 25 ml from the outer vein of the ear) 10 to 14 days after the final injection, and then at weekly intervals. Usually 4 or 5 bleeds will provide sufficient antiserum.

(iv) A further booster injection(s) 4 to 6 weeks after the last of the first series of injections, will often provide a high-titred antiserum.

(v) When the blood has been removed from the rabbit, it can be left to coagulate in a sealed tube at room temperature for 24 h. Then the clear serum containing the antibodies, is carefully separated from the blood-clot using a Pasteur pipette. If necessary, the serum can be further clarified (to remove red corpuscles) by centrifugation in a bench-centrifuge at 500 *g* for 5 min.

(vi) The antiserum may then be stored by mixing with glycerol (1 vol. glycerol: 1 vol. serum) or by freeze-drying.

12.5.2 Determination of antiserum titre

The *titre* of an antiserum is the highest dilution of the antiserum that will react with its own homologous virus, and is normally determined in *precipitation-tube* tests (*see* Section 6.6.3).

(i) The tests are carried out in thin-walled glass tubes (75×9 mm). A dilution series of the antiserum (diluted with 0.9% sodium chloride (NaCl) is mixed with a constant dilution of the virus (see Table 12.2). A good result is often obtained if the partially purified virus is used at a dilution of ⅕ (with 0.9% NaCl), although several preliminary tests may be required to determine the optimal dilution for the virus.

(ii) Add 0.5 ml of virus to 0.5 ml of each dilution of the antiserum. Shake each tube for a few seconds to mix the reactants thoroughly, and then incubate in a waterbath at 38°C (*see* Plate 6.2).

(iii) Observe each tube at 5 to 10 min intervals against a black background and record the intensity of the precipitation reactions by a pre-determined scoring system (−: no precipitation to ++++: dense precipitation) is satisfactory, *see* Table 12.2). The reaction is usually complete after 1 h.

(iv) It is essential to include healthy, purified host antigen (prepared in the same way as the purified virus) and normal serum (removed from the rabbit before virus injection) controls in the experiment.

Table 12.2 An example of the determination of the titre of an antiserum by precipitation tube tests

Antigen† dilution	Incubation time (min)	Antiserum dilution 1/4	1/64	1/256	1/512	1/1024	1/2048	Normal serum dilution 1/4	1/64	1/256	1/512	1/1024	1/2048
	0	–*	–	–	–	–	–	–	–	–	–	–	–
5	+	+	+	+	–	–	–	–	–	–	–	–	
Virus	15	++	++	++	+	+	–	–	–	–	–	–	–
1/5	30	+++	+++	+++	++	++	–	–	–	–	–	–	–
	40	++++	++++	++++	+++	+++	–	–	–	–	–	–	–
	50	++++	++++	++++	++++	+++	–	–	–	–	–	–	–

*Precipitation measured on a – (no precipitation) to ++++ (dense precipitation) scale

†Antigen of similar dilution, prepared from healthy host sap by the same purification procedure as the virus, should also be used as a control for each antiserum reaction.

(v) In respect of the results in Table 12.2, the highest dilution at which precipitation occurred is $\frac{1}{1024}$, and if the antiserum has been stored in 50% glycerol, this would represent a titre of $\frac{1}{2048}$.

12.5.3 Gel double-diffusion test

The gel double-diffusion (Ouchterlony) test may be carried out in plastic petri-dishes or if economy of materials is important, on glass microscope slides.

Preparation of agar

Agar at a concentration of 0.75% will provide a satisfactory medium for diffusion.

(i) Dissolve 7.5 g of purified agar (specifically recommended by the manufacturers for immunodiffusion tests), 9.0 g sodium chloride (NaCl) and 0.4 g sodium azide (NaN_3) in 1 l of distilled water.

For certain viruses such as cucumber mosaic virus, better results may be obtained if the agar is dissolved in 0.05 M di-potassium hydrogen orthophosphate (K_2HPO_4) at pH 7.8, containing 0.005 M sodium EDTA (ethylenediaminetetra-acetic acid-disodium salt) and 0.02% NaN_3. To 1 l of distilled water, add 8.7 g K_2HPO_4, 1.86 g Na EDTA, 0.2 g NaN_3 and 7.5 g agar.

If petri dishes are used they should be filled to a depth of 5 mm

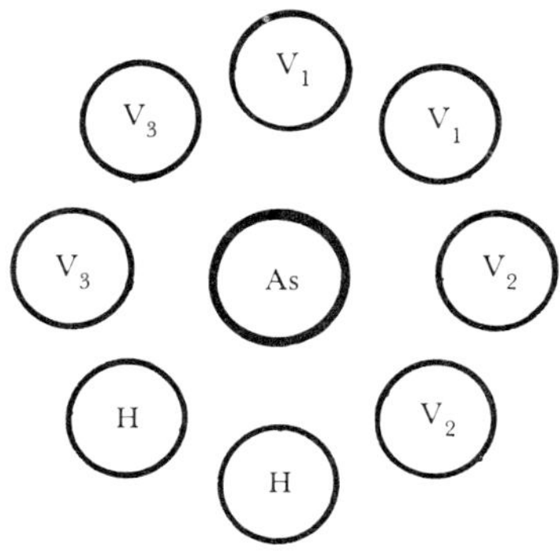

Fig. 12.2 A common arrangement of antigen and antiserum wells for gel double-diffusion tests.
As. = antiserum; V_1, V_2 and V_3 = virus samples 1, 2 and 3; H. = healthy sap for control purposes.

with molten agar, and 5 ml of agar should be pipetted on to the surface of the slide, if glass microscope slides are used. Allow the agar to set, and then cut wells in the agar using hollow steel tubes or cork borers of the required size. Remove the agar plug with a Pasteur pipette attached to a suction pump.

(ii) If glass slides are used, a 5 mm diameter antiserum well and 4 mm diameter antigen wells are suitable. The position of the wells may be varied to suit the experiment concerned, but the design shown in Figure 12.2 is commonly used. In this design, the antiserum is placed in the central well and the antigens in the outside wells. The distance between the edge of the antiserum well and the closest edge of the antigen wells is 2.5 mm. Most workers make a perspex or steel template to accurately reproduce this design.

Preparation of antiserum and antigen

(i) The antiserum can be used undiluted, or diluted with 0.9% NaCl. The dilution required to give optimum precipitation lines will depend on the titre of the antiserum concerned. Antiserum used undiluted, or diluted ⅕ often gives satisfactory results.

(ii) Purified virus or infected crude sap may be used as the antigen, but the latter will only give satisfactory results if the virus concentration is high. If purified virus is used it may be used undiluted, or optimum results may be obtained if it is diluted up to 1/10 with 0.9% NaCl. If crude sap is used, the infected leaf should be homogenized in 1.8% NaCl and the sap filtered before use.

(iii) Carefully add the antiserum and antigens to their respective wells using a fine-pointed Pasteur pipette. Ensure that the wells

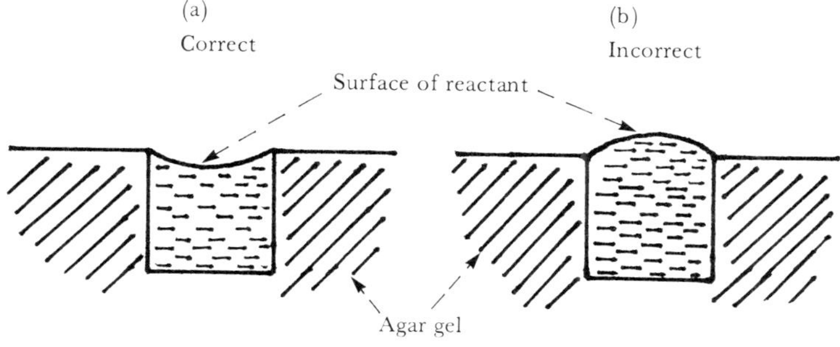

Fig. 12.3 Diagram of how wells should be filled for gel double-diffusion tests.

are filled to the surface, but not overfull (*see* Figure 12.3).

(iv) Control wells, containing healthy plant sap, must be used in all experiments.

Sonication of filamentous virus particles

Long, filamentous particles will not diffuse through the agar unless they are first broken into small fragments. This may be done by sonication (*see* Section 6.6.3).

(i) Place a minimum of 2 to 3 ml of partially purified virus in a thick-walled glass sonication tube. Lower the ultrasonic probe until its end is 1 mm below the surface of the virus preparation.
(ii) Surround the tube with a beaker of crushed ice and water.
(iii) Switch on the sonicator and adjust its frequency until the sound reaches its highest pitch. Maintain this frequency for 5 to 8 min.
(iv) Check that the particles have been broken by electron microscopy.

Incubation and examination

(i) Place the slides or petri dishes in a sealed plastic container on moist wadding to prevent the agar from drying out. Leave for 24 h at room temperature before examining for precipitation lines using an indirect light source beneath the agar.
(ii) If it is necessary to photograph the precipitation lines they can be stained with 0.1 M DOPA (3–4 di-hydroxyphenyl-DL alanine) (*see* Section 6.6.3).

The DOPA is prepared by dissolving 1.97 g DOPA in 100 ml of 0.1 M potassium phosphate buffer at pH 7.4. The stain is applied to the surface of the agar-gel above the precipitation line, by saturating a piece of chromatography paper. Stain diffuses into the gel from the paper. Allow the lines to stain for 6 to 12 h. During this period they will over-stain, and then they may be destained to the required colour, by successive washings in distilled water.

Class exercises

Prepare agar slide or petri dish gel double-diffusion tests as described above. Good precipitation lines are usually obtained with nepoviruses, such as arabis mosaic or tobacco ringspot viruses, propagated in *Chenopodium amaranticolor*.

(i) Remove the systemically infected apex of the plant 14 days after inoculation. Grind tissues in 1.8% NaCl as described above and filter the homogenate through muslin.

(ii) Add the infected sap, and similarly prepared healthy sap, to the outer ring of antigen wells as illustrated in Figure 12.2.
(iii) Add the antiserum to the central well.
(iv) Determine the titre of the antiserum in gel tests (i.e. the lowest dilution of antiserum that will give a precipitation line), by using series of antiserum dilutions. Use antiserum undiluted and diluted with 0.9% NaCl to ½, ¼, ⅛, 1/16, and 1/32.

12.6 Electron Microscopy

It is beyond the scope of this book to describe the complete procedures involved in examining viruses in the electron microscope (E.M.) in detail. The procedures involved in coating E.M. grids with carbon film are highly skilled and require specialized equipment. Frequently, neither an electron microscope nor such specialist equipment are available. However, virologists should know how to prepare and stain a virus preparation ready for examination, when provided with an E.M. grid pre-coated with carbon. A worker may then send the prepared grid to an electron microscopist for examination. This procedure is particularly useful to virologists working in isolated areas, or in undeveloped countries.

12.6.1 Preparation of negative stain

Table 12.3 lists the various salts that can be used for negative staining and indicates viruses for which they might not be suitable.

(i) Prepare a 2% (wt/vol) solution of the stain by dissolving the salt in double-distilled water. The water should be boiled to remove CO_2, just before the salt is added.

Table 12.3 Negative stains that are commonly used in electron microscopy*

Stain†	*pH*	*Application*
Ammonium molybdate	6–10	Excellent for most viruses, moderate contrast
Methylamine tungstate	Not adjusted	Excellent for most viruses, high contrast
Potassium phosphotungstate	7	May disrupt some viruses
Sodium phosphotungstate	7	May disrupt some viruses
Sodium silicotungstate	7	May disrupt some viruses
Uranyl acetate	Not adjusted	Very high contrast, may distrupt some viruses
Uranyl formate‡	Not adjusted	Very high contrast, may distrupt some viruses

*Information supplied by M. J. W. Webb.
†All stains can be prepared as a 2% (wt./vol) concentration in double distilled water.
‡Must be prepared immediately before use.

(ii) Filter the solution and store in a dark-coloured bottle at 4° to 5°C.
(iii) Prepare fresh solutions regularly. Do not use if a precipitate appears.

12.6.2 Preparation of virus specimen

Method One

(i) Take infected leaf material and remove a 2 mm square fragment of epidermis or other leaf tissue with forceps. Place tissue on a clean glass microscope slide. Drop 0.1 ml of stain, containing 0.01% of a wetter such as bacitracin, on to the leaf fragment.
(ii) Grind the fragment to a fine homogenate using a glass rod.
(iii) Pick up a small volume of the sap in a capillary tube (a Pasteur pipette or hyperdermic syringe may equally well be used), and holding an E.M. grid by its edge with forceps, carefully add a drop of sap to the carbon-film covered surface of the grid. Leave for at least 30 seconds. Remove excess sap with a Pasteur pipette and a piece of filter paper applied to the edge of the grid to prevent damage to the film. Allow the surface of the grid to dry thoroughly in air for several minutes.
(iv) The specimen is now ready for examination in the E.M. and may be stored for postage in a gelatine capsule or other similar container. If grids are to be stored for long periods they must be stored at a low humidity (i.e. over calcum chloride or silica gel).

If a purified virus preparation is to be examined, it should have been resuspended in distilled water following ultra-centrifugation (*see* Section 12.4.2).

(*a*) Place 0.1 ml of the virus preparation on a glass slide, add stain and wetter and mix well.
(*b*) Transfer mixture to the surface of a grid, as described above. Method one should not be used with uranium salt stains.

Method Two

(i) Grind a leaf fragment on a glass slide in *distilled water only*.
(ii) Using a capillary tube, pick up the sap homogenate and place it on parafilm or a waxed slide to form a droplet.
(iii) Place an E.M. grid with its carbon-film downwards, on the surface of the droplet, ensuring that the grid's surface becomes wet. *Do not use any wetter in this process*. Leave the grid for 10 min.
(iv) Pick up the grid by its edge with forceps and wash the sap from the filmed surface, using at least 2 ml of double distilled water applied 'dropwise'.
(v) Carefully add 2 ml of uranyl acetate, drop by drop, to the filmed surface of the grid.

(vi) Remove excess stain after the last drop with a strip of filter paper. Allow the surface of the grid to dry thoroughly in air before E.M. examination.

This method is also suitable for examining a purified virus preparation that contains sucrose (following gradient separation) or strong buffer. Place the virus preparation directly on to the parafilm, and place the grid on the droplet as described above. The sucrose or buffer will be removed when the grid is washed with distilled water. Stain and air-dry the grid as described above.

Method two must be used if uranium salts are used as a stain, but can also be used with other stains. In order to withstand the washing treatment, this method requires strong grids freshly coated with carbon.

Class exercises
Prepare various virus specimens on grids pre-coated with carbon. Examine the grids in the E.M.

Table 12.4 Culture medium† for meristem-tip culture

Mineral salts‡	*(mg/l)*	*Organic ingredients*	
Mn_4NO_3	1650		
KNO_3	1900	Sucrose	30 g/l
$CaCl_2\ 2H_2O$	440	Indoleacetic acid	8 mg/l
$MgSO_4\ 7H_2O$	370	Kinetin	2.56 mg/l
KH_2PO_4	170		
Na_2-EDTA	37*	Pyridoxin Hcl	0.5 mg/l
$FeSO_4.7H_2O$	28*	Thiamine HCl	0.1 mg/l
H_3BO_3	6.2	Nicotinic acid	0.1 mg/l
$MnSO_4.4H_2O$	22.3	Myo-Inositol	100 mg/l
$ZnSO_4.4H_2O$	8.6	Agar (Oxoid No 3)	9 g/l
KI	0.83		
$Na_2MoO_4.2H_2O$	0.25		
$CuSO_4.5H_2O$	0.025		
$CoCl_2.6H_2O$	0.025		

*5ml/l of a stock solution containing 5.57 g $FeSO_4.7H_2O$ and 7.45 g Na_2-EDTA per litre of distilled water.

†Medium based on Murashige and Skoog (1962, *see* Section 11.7).

‡The mineral salts may be purchased as a commercially prepared package. The medium should be adjusted to pH 5.7 before the addition of the agar. The agar may be omitted and a filter-paper bridge used to support the culture.

Table 12.5 Examples of varying concentrations of auxin and cytokinin that may be used to induce different types of culture growth

Kinetin† concentration	*Indoleacetic acid concentration†*			
	0	*4 mg/l*	*8 mg/l*	*16 mg/l*
0	1*	2	3	4
1 mg/l	5	6	7	8
2.5 mg/l	9	10	11	12
10 mg l	13	14	15	16

*Number of treatment
†Other auxins and cytokinins may be tried instead of kinetin and indoleacetic acid

12.7 Meristem-tip Culture for Virus Eradication

12.7.1 Culture medium

Media based on Murashige and Skoog's medium, which is detailed in Table 12.4 (*see* also Section 11.3), are suitable for many plant species. If the species to be cultured has already been grown *in vitro* by other workers, follow their procedures in detail. If no information is available on the culture of a particular species, try using the medium detailed in Table 12.4, and carry out preliminary experiments using different concentrations of idoleacetic acid (IAA) and kinetin as shown in Table 12.5. These experiments may indicate the optimum concentrations of auxin and cytokinin for the type of growth required.

Generally high cytokinin concentrations favour multiple shoot formation, and a cytokinin-free medium favours rooting. Other auxins and cytokinins may be substituted for IAA and kinetin.

The mineral salts complex used in the culture medium may be made up in the laboratory, but more frequently it is purchased in a prepared pack which only needs to be dissolved in distilled water. Similarly, the organic constituents of the medium may be purchased in prepared packs, but many workers prefer the flexibility of preparing their own organic constituents, and making fresh solutions of these as required.

One litre of the medium detailed in Table 12.4 may be prepared as described below.

(i) Dissolve one sachet of mineral salts (commercially pre-weighed and mixed to prepare 1 l in quantity), 30 g sucrose and 100 mg myo-inositol in 800 ml of distilled water.

(ii) Add to this solution:
Pyridoxin HCl: dissolve 25 mg in a few ml of N/10 hydrochloric acid and make up to 50 ml with distilled water (500 mg/l). Use 1 ml (0.05 mg/l) of this solution.

Thiamine HCl: dissolve 25 mg in 50 ml of distilled water (500 mg/l). Use 0.2 ml (0.1 mg/l) of this solution.

Nicotinic acid: dissolve 25 mg in a few ml of N/10 hydrochloric acid and make up to 50 ml with distilled water (500 mg/l). Use 0.2 ml (0.1 mg/l) of this solution.

Indoleacetic acid: dissolve 40 mg in a few ml of absolute alcohol and make up to 50 ml with distilled water (800 mg/l). Use 10 ml of this solution (8 mg/l).

Kinetin (6-furfurylaminopurine): dissolve 25.6 mg in a few ml of N/10 hydrochloric acid and make up to 50 ml with distilled water (512 mg/l). Use 5 ml (2.56 mg/l) of this solution.

(iii) Make up the total volume to 1000 ml with distilled water and adjust the pH to 5.7 with N/1 NaOH.

(iv) Add 9 g of agar (Oxoid No. 3 or equivalent) and heat the solution until the agar is thoroughly dissolved.

(v) Add 10 ml of molten solution to 7.0×2.5 cm flat-bottomed, glass tubes. Seal each tube with an aluminium cap or foil.

(vi) Stack the tubes in a wire basket and cover the mouth of the basket with cotton wool and autoclavable paper.

(vii) Sterilize by autoclaving for 15 to 20 min at 15 lb pressure. The sterilized tubes may be stored at 3° to 4°C until used.

As an alternative to agar, the culture explant can be supported on a paper bridge which soaks up the medium from the bottom of the tube. The bridge is folded to the required shape using 9×4 cm strips of chromatography paper (Whatmans No. 1). The dry bridge is placed in the culture tube before the culture solution is added and sterilized as described above.

12.7.2 Heat treatment of the infected parent plant

Incubate the infected parent plant, growing in a pot in compost at 30° to 40°C in an illuminated, constant temperature growth cabinet (16 h photoperiod). The minimum treatment period is likely to be 4 weeks, but up to 12 weeks may be necessary. The plant should be treated for as long as possible at the highest temperature it can withstand. A preliminary experiment will be required to determine the maximum temperature. A pretreatment period at approximately 30°C for 1 week, may help to acclimatize the plant for treatment at a higher temperature.

The meristem-tips must be removed, immediately the high temperature treatment is completed.

12.7.3 Excision of the meristem-tip

Sterilization of the explant

The shoot bearing the bud from which the meristem-tip is to be removed can be surface-sterilized as follows:

(i) Immerse shoot in 70% alcohol for 1 to 2 min and then immerse in 5% sodium or calcium hypochlorite solution containing 0.1% Tween-80 for 5 min.

(ii) Rinse several times in sterile distilled water for 3 to 5 min each rinse.

If the meristem-tip is enclosed within a tightly closed bud, surface sterilization may be unnecessary, particularly if the plant has not been subjected to overhead watering for several weeks prior to the explant being taken. If the shoot is not surface sterilized, great care must be taken to ensure that the dissecting instruments are thoroughly sterilized by dipping in alcohol and flaming, between each excision manoeuvre.

Removal of the tip

Using a binocular stereomicroscope and a fine pair of forceps to hold the shoot, remove the outer leaves of the bud with a sterilized needle or scalpel. Sterilize the needle by dipping in alcohol and flaming between each cut, and remove the remaining primordial leaves until the youngest primordial leaf and meristem dome are exposed. Remove the meristem-tip (*see* Figure 11.2) with a fragment of razor-blade mounted in a handle, and place the tip in a culture tube. The tip taken should be between 0.3 and 1 mm in diameter. It should be emphasized that the smaller the explant removed, the greater are the chances that the plantlet produced will be virus-free (*see* Section 11.3.3).

As the tip is cut from the shoot, the aluminium cap sealing the culture-tube should be removed and the mouth of the tube 'flamed' by passing it through the flame of an alcohol or bunsen burner, for a few seconds. The tip should then be placed on the surface of the agar and the mouth of the tube again 'flamed'. Immediately seal the tube with a 7 cm square of sterilized polypropylene sheet (G.S. 8000 film, available from G. S. Packaging Ltd., Aber Works, Aber Road, Flint, Clwyd, U.K.) held in position with a rubber band. The squares of polypropylene may be sterilized by laying each square separately between sheets of filter paper, enclosed within aluminium foil, and autoclaving them 15 min at 15 lb pressure.

The cultures should be grown at a constant temperature between 23° and 25°C and illuminated for 16 hours per day. Most cultures grow satisfactorily under daylight fluorescent tubes, high pressure mercury lamps or sodium lamps.

12.7.4 Post-culture treatment

Transfer the rooted plantlet to a 4 cm 'Jiffy' peat pot, and gradually 'harden off' the plant in a seed tray covered by a cloche in a glasshouse. The humidity within the chamber should be carefully controlled, and slowly reduced over a two-week period until the plastic cover of the seed tray can be removed. If the plantlet is growing well, after a further few days the peat-pot can be planted *in situ* into a pot containing John Innes No. 3 compost.

The plantlet should then be grown to maturity in sterilized soil in an insect-proof glass or gauze-house. During this period the plant should be indexed for virus at least two or three times (*see* Section 11.5).

Class exercises

(i) Excise meristem-tips from axillary buds of a cabbage 'head' and grow on the medium detailed in Table 12.4. Vary the concentration of auxin and cytokinin in the medium to observe shoot proliferation and rooting (*see* Table 12.5). Also culture pieces of meristematic floral initials removed from a cauliflower curd. These will produce vegetative growth in tissue culture.

(ii) Infect several *Nicotiana rustica* seedlings with cucumber mosaic virus. Four weeks after inoculation (when the apical, systemically infected leaves should be showing mild mosaic symptoms), excise the meristem-tips from the axillary and apical buds and grow into plantlets on the medium detailed in Table 12.4.

Note the size of the meristem-tips taken and the position of the bud on the stem from which each tip was taken. When the cultured plantlets are established in soil, index them for virus by back-testing sap from leaf samples on to *Chenopodium quinoa* plants. Correlate if there is a relationship between the number of virus-free plants obtained, and the initial size of the excised meristem-tips and their position on the stem.

12.8 Other Relevant Information

12.8.1 Useful measurements

Weight	*Metric unit*	*In grams (g)*
	kilogram (kg)	10^3
	gram (g)	1
	milligram (mg)	10^{-3}
	microgram (μg)	10^{-6}

Volume	*Metric unit*	*In litres (l)*
	kilolitre (kl)	10^{3}
	litre (l)	1
	decilitre (dl)	10^{-1}
	centilitre (cl)	10^{-2}
	millilitre (ml)	10^{-3}
	microlitre (μl)	10^{-6}

Length	*Metric Unit*	*In metres (m)*
	kilometre (km)	10^{-3}
	metre (m)	1
	decimetre (dm)	10^{-1}
	centimetre (cm)	10^{-2}
	millimetre (mm)	10^{-3}
	micrometre (μm) (syn. micron (μ))	10^{-6}
	nanometre (nm) (syn. millimicron (mμ))	10^{-9}
	añgström (Å)	10^{-10}

12.8.2 List of CMI/AAB descriptions of plant viruses, virus groups and viroids

Each description contains information on the diseases caused by the virus, its geographical distribution, host range, and detailed information on its biological and physical properties.

Description sets are available from *Central Sales Branch, Commonwealth Agricultural Bureaux, Farnham Royal, Slough, SL2 3BN, England.*

Virus	*Description No.*
Agropyron mosaic	118
Alfalfa mosaic	46:229
Alfalfa latent	211
American hop latent	262
Andean potato mottle	203
Apple chlorotic leaf spot	30
Apple mosaic	83
Apple stem grooving	31
Apple (Tulare) mosaic	42
Arabis mosaic	16
Arracacha A	216
Arracacha B	270
Artichoke Italian latent	176
Artichoke yellow ringspot	271
Barley stripe mosaic	68
Barley yellow dwarf	32
Barley yellow mosaic	143
Bean common mosaic	73
Bean golden mosaic	192
Bean mild mosaic	231
Bean pod mottle	108
Bean rugose mosaic	246
Bean (southern) mosaic	57:274
Bean yellow mozaic	40
Bearded iris mosaic	147
Beet curly top	210
Beet leaf curl	268
Beet mosaic	53
Beet necrotic yellow vein	144
Beet western yellows	89
Beet yellows	13
Beet yellow stunt	207
Belladonna mottle	52
Bidens mottle	161
Blackgram mottle	237
Black raspberry latent	106
Blueberry leaf mottle	267
Blueberry shoestring	204
Broad bean mottle	101
Broad bean necrosis	223

White clover mosaic	41	Wild cucumber mosaic	105
		Wound tumor	34

Virus Group	*Description No.*
Bromovirus group	215
Carlavirus group	259
Closterovirus group	260
Comovirus group	199
Ilarvirus group	275
Nepovirus group	185
Potexvirus group	200
Potyvirus group	245
Rhabdovirus (plant) group	244
Tobamovirus group	184
Tymovirus group	214

Viroid	*Description No.*
Avocado sun-blotch viroid	254
Citrus exocortis viroid	226
Potato spindle tuber viroid	66

12.9 Reference

Pawley, R. R. (1973). Year-round production of test plants for virology. Glasshouse Crops Res Inst Ann R 149–51.

Glossary

Acquired resistance (syn. ***induced resistance*** or ***acquired immunity***): a resistance response developed by a normally susceptible host following a predisposing treatment, such as inoculation with a virus, fungus, bacterium, or treatment with certain chemicals. This resistance is not inherited (*see* ***cross-protection***).
Acquisition feeding time: the feeding time during which a vector feeds on an infected plant to acquire a virus for subsequent transmission (e.g. to become viruliferous).
Acquisition threshold period: the minimum feeding time for a vector to become viruliferous.
Agglutination: a term used in serology to describe an antibody–antigen reaction in which large clumps of reactants are involved, as when antibodies are attached to latex particles before mixing with the virus antigen.
Alate: winged form in the life-cycle of certain insects (e.g. aphids) (*see* ***apterous***).
Allele: one of two or more alternate forms of a gene occupying the same locus on a particular chromosome (*see* ***gene***).
Antibiosis: *see* ***vector resistance***.
Antibody: a specific protein formed in the blood of warm-blooded animals in response to the injection of a protein or polysaccharide.
Antigen: a protein or polysaccharide which induces the formation of antibodies when injected into a warm-blooded animal. The capacity of an antigen to react specifically with an antibody is referred to as *antigenic reactivity*.
Antiserum: the blood serum containing the antibodies when separated from the other blood components.
Antiserum titre: the highest dilution of an antiserum that will react with its homologous virus (*see* ***homologous*** and ***heterologous reaction***).
Apterous: wingless stage in the life-cycle of certain insects (e.g. aphids) (*see* ***alate***).

Capsid: the protein shell of a virus particle.
Capsomere: the morphological sub-units seen on the surface of the

virus particle (virion) in the electron microscope. A capsomere is built up from varying numbers of protein-sub-units (polypeptide chains).
Carna-5 RNA: (syn. ***cucumber mosaic virus RNA 5***) a satellite RNA of CMV which is dependent upon the remainder of the CMV genome for its own replication, but which is not essential for the replication of the CMV particle (*see* ***genome*** and ***satellite***).
Cell manipulation: *see* ***genetic engineering.***
Chelating agent: a chemical such as sodium EDTA which will bind with bivalent and trivalent cations to assist in virus purification.
Circulative virus: a virus which is transmitted by an insect in a persistent manner and which circulates from the insect's digestive tract, through the haemolymph to the salivary glands, before being transmitted in the saliva as the insect feeds.
Clone: a genetically identical group of individuals, originally derived from a single individual by vegetative propagation.
Complementation: occurs when a virus is assisted by another virus (or a strain of the same virus) to replicate.
Constitutive resistance: genetically controlled, inherited resistance.

Decoration test: *see* ***immunosorbent electron microscopy*** (ISEM).
Density gradient centrifugation: a centrifugation procedure in which partially purified virus is further clarified by movement through a gradient. Contaminating components may be separated from the virus particles by velocity centrifugation, usually in a low to high sucrose gradient, which separates components according to their differing sedimentation coefficients, or alternatively, by isopycnic centrifugation usually in caesium chloride or caesium sulphate gradients, which separates the components according to their differing buoyant densities.
Dependent transmission: transmission of a virus (by aphids) that only occurs when the vector feeds on a source plant that is jointly infected by a second virus. The second virus is referred to as a *helper virus*, and the virus that is not transmissible on its own is called the *dependent virus*.
Differential centrifugation: cycles of low and high speed clarification and sedimentation used in the purification of a virus.
Differential host: a plant which gives distinctive symptoms when infected with a specific virus, allowing the virus to be distinguished from others.
Dilution end-point: the lowest dilution in a serial dilution of a virus preparation, that will infect a mechanically inoculated plant.
Dioecious aphid: an aphid whose life cycle alternates between a primary and secondary host plant (*see* ***monoecious aphid***).
Disease gradient: the change in incidence of a disease with increasing distance from the source of infection.

Discontinuous resistance: *see* ***resistance.***
Dominant gene: a gene that is fully expressed in the phenotype of the heterozygote (*see* ***recessive gene***).
Double antibody sandwich: a method in enzyme-linked immunosorbent assay (ELISA) in which the reactants are added to the test plate in the order of antibody, virus, antibody-enzyme complex.
Durable resistance (***durability***): used to describe resistance that is long lasting.

Ecdysis: the moulting of the integument of an insect that occurs between each of its growth stages (*instars*).
Electron microscope serology: *see* ***immunosorbent electron microscopy.***
Electrophoretic mobility: the rate of movement of virus per unit potential gradient. Mobility may be towards the cathode or the anode depending on whether the virus has a net positive or negative charge at the pH used.
Enation: an abnormal outgrowth (often on the leaf) caused by an increase in cell numbers (hyperplasia).
Encapsidation: the enclosure of a virus's nucleic acid genome within a protein shell.
Enveloped virus: plant viruses of the reovirus and rhabdovirus groups which have an outer lipid-protein membrane surrounding the protein shell of the virus.
Enzyme-linked immunosorbent assay (ELISA): a serological test in which the sensitivity of the antibody–antigen reaction is increased by attaching an enzyme to one of the two reactants (*see* ***double antibody sandwich***).
Epidemiology: the study of factors affecting the outbreak and spread of infectious diseases.
Epidermal-strip test: an electron microscope technique by which a virus can be quickly examined in crude-sap extracts. The *quick leaf-dip* and *quick dip* techniques are similar procedures.
Extraction buffer: the buffer used in grinding infected leaves during the initial stages of virus purification.

Field resistance: *see* ***resistance.***
Focus: the site of initial disease infection from which secondary spread may occur (plural *foci*).
Freeze-dried: *see* ***lyophilization.***
Freund's incomplete adjuvant: a substance containing an emulsifier and mineral oil, which is mixed with a virus before it is injected into the muscles of an animal to produce antiserum. The adjuvant allows slow release of the virus following injection.

Gel-chromatography: a molecular sieving procedure, by which viruses are separated from different sized molecules when passed through the pores of gel beads such as agarose. Used for virus purification.
Gel double-diffusion: a seriological test in which the antibody and antigen reactants diffuse towards each other in gel and react to form a visible precipitation line (*see* ***radial diffusion*** and ***immunodiffusion***).
Gene: an inherited factor that determines the characteristics of an organism (*see* ***allele***).

Minor gene: a gene that has small observable effects upon the phenotype.

Major gene: a gene that has large observable effects upon the phenotype.

Gene-for-gene hypothesis: the concept that corresponding genes for resistance and virulence exist in the host and pathogen respectively (*see* ***vertical resistance***).
Genetic engineering: the alteration of the genetic composition of a cell by various tissue culture procedures (*see* ***protoplast-fusion***, ***somaclones*** and ***transformation*** (DNA).
Genome: the nucleic acid component of a virus, which may consist of a single (*monopartite*), two (*bipartite*), three (*tripartite*) or more (*multipartite*) molecular species of RNA (*see* ***multicomponent virus***).
Genotype: the genetic factor which influences the phenotype.
Gradient of infection: *see* ***disease gradient.***

Helper virus: *see* ***dependent transmission.***
Heterologous reaction: a serological reaction in which an antiserum is reacted against an antigen other than the one used in its preparation (*see* ***homologous reaction***).
Heterozygote: possesses two different alleles in a single gene pair e.g. *Ss*.
Homologous reaction: a serological reaction in which an antiserum is reacted against the antigen used for its preparation.
Horizontal resistance: *see* ***resistance.***
Hyperplasia: a malformation caused by an increase in cell numbers. An abnormal increase in the size of an organ is referred to as *hypertrophy*.
Hypoplasia: a malformation caused by a reduction in cell numbers. A reduction in organ size is called *atrophy*.
Hypersensitivity: a reaction in a host plant resulting from virus infection, involving rapid death of the infected tissues. In some instances the area of dead cells is restricted to discrete *local lesions*, and in others the virus may spread rapidly through the plant's vascular system causing systemic necrosis and death. Often considered to be a form of resistance to virus spread (*see* ***lesion***).

Icosahedron: a figure having 20 plane faces, the symmetry of which forms the basis for the arrangement of the protein sub-units of isometric virus particles.

Immune and immunity: *see* ***resistance.***

Immune response: the ability of an animal to produce antibodies as a result of antigens, such as proteins, entering its body either by infection with a pathogenic agent, or by artificial injection. The ability of the antigen to induce this response is referred to as *immunogenicity*, and the substances that are capable of inducing the response are called *immunogens*.

Immunocytological methods: procedures used to study cytopathological disorders in ultrathin sections of virus diseased tissues, using labelled antibodies to diagnose the virus.

Immunodiffusion: a serological procedure in which the antigen – antibody reaction is carried out by allowing the reactants to diffuse in gel (*see* ***gel double-diffusion*** and ***radial diffusion***).

Immunology: the study of acquired immunity in animals and man against infectious disease.

Immunosorbent electron microscopy (ISEM) (syn. ***electron microscope serology***): techniques involving the visualization of the antibody –antigen reaction in the electron microscope. These include:

Trapping: a procedure in which the E.M. grid is first coated with antiserum (referred to as an antibody-coated grid (ACG)), which then attracts virus particles from a virus preparation placed on it.

Decoration: in this technique virus particles are attached to the E.M. grid and then the antiserum is added. Homologous antibodies will react with the particles to coat or '*decorate*' them.

Inbreeding depression: a loss of yield and vigour caused by inbreeding, which may be particularly significant if normally cross-pollinated species are self-pollinated.

Inclusion body: virus induced structures that may occur in the cytoplasm or nucleus of infected plants.

Incomplete dominance: occurs when a dominant gene is only partially expressed in the phenotype of the heterozygote.

Indexing: a procedure for demonstrating the presence of virus infection in a plant.

Induced resistance: *see* ***acquired resistance.***

Infectivity assay: a bioassay using mechanical sap-transmission to quantitatively determine the amount of infectious virus.

Inoculation feeding period (syn. ***test feeding period***): the length of time a vector feeds on a test plant during transmission experiments.

Inoculation threshold period: the minimum feeding period a vector needs on a test plant to transmit a virus.

Instar: a growth phase between moults in an insect's life-cycle.

Integrated control: the combined use of more than one method to effect pathogen or pest control, such as the complementary use of biological and chemical methods.
Isoelectric point: the pH at which a virus particle has a zero net charge.
Isolate: a virus that has been obtained from an infected plant.
Isometric: used to describe virus particles that are approximately spherical in shape.

Koch's postulates: criteria proposed by Koch for proving the pathogenicity of an organism; (1) the suspected causal organism must be constantly associated with the disease; (2) it must be isolated and grown in pure culture; (3) when inoculated into a healthy plant it must reproduce the original disease.

Land-races: stocks of plants selected by farmers on a local basis over many years, which are strongly adapted for local conditions.
Latent infection (***Latency***): infection in a plant without visual symptoms.
Latent period: the period after a vector has acquired a virus before it can transmit it. Often observed in the case of persistent virus transmission.
Lesion: a localized area of diseased tissue often referred to as a *local lesion*. The term *primary lesions* may be used to describe lesions that develop on the inoculated leaves at the initial points of infection.
Longevity end-point: the storage time after which a virus in a crude sap preparation loses its infectivity. Usually determined at 0° or 20°C.
Lyophilization (syn. ***freeze-drying***): a technique by which water is removed under vacuum while the preparation or tissue is frozen. Used to preserve viruses or antisera.

Mechanical transmission: used to describe artificial transmission of a virus in which an infectious preparation is rubbed onto a test plant. May also occur in the field when virus is transmitted from one plant to another by leaves rubbing or root contact.
Meristem-tip: the meristem dome of cells and one or two pairs of primordial leaves (0.5 to 1 mm in diameter), which comprises the explant removed from a bud and grown in tissue culture to produce a virus-free plant.
Metal shadowing: a technique used to prepare viruses for electron microscopy, in which the virus particles are exposed to the vapour of a heavy metal such as gold or platinium. Now replaced by the *negative contrast staining* method.
Modal length: the length that occurs most frequently in a population of virus particles.

Monoecious aphid: an aphid that spends its complete life-cycle on a single plant species (*see* ***dioecious aphid***).
Monophagus: (syn. *oligophagus*) when an insect such as an aphid feeds on a specific type of host plant (*see* ***polyphagus***).
Multicomponent virus: a virus whose genome is divided into two or more parts, each part being separately encapsidated. Hence two or more components are needed to initiate an infection. Note that this is different to a multipartite genome where components may be enclosed in a single particle (*see* ***genome***).
Mutant: an organism that shows one or more discrete heritable differences from a standard type (*see* ***strain***, ***complementation***, ***recombinant*** and ***pseudorecombinant***).
Mycoplasma: a pathogenic agent with a confining unit membrane, but no cell wall, measuring 0.1 to 1.0 μm in diameter and containing both DNA and RNA. They induce virus-like symptoms in infected plants.
Mycovirus: a virus that infects fungi.

Negative contrast staining: a staining procedure used to prepare virus particles for examination in an electron microscope (*see* ***metal shadowing***).
Noisiness: errors that may occur during the replication of a virus's genome.
Non-persistent transmission (syn. ***stylet-borne transmission***): a type of insect transmission in which the virus is acquired by the vector after very short acquisition feeding times, and which is transmitted during very short inoculation feeding periods. The vector remains viruliferous for only a short period unless it feeds again on an infected plant (*see* ***persistent*** and ***semi-persistent transmission***).
Non-preference: *see* ***vector resistance.***

Obligate parasite: a pathogen capable of living only as a parasite and which cannot be cultured on an artificial medium.
Odontostyle: the feeding stylet of a nematode (eelworm).
Oligogenic: a character controlled by a few genes (*see* ***polygenic***).
Ouchterlony test: *see* ***gel double-diffusion.***

Partial fusion line: *see* ***spur precipitation line.***
Persistent transmission: a type of insect transmission in which the virus is acquired by the vector only after a long acquisition feeding period, and in which there may be a latent period following the acquisition feed, before the vector can transmit the virus. The vector remains viruliferous for a long period, often throughout its life span. The virus sometimes multiplies within the vector (*see* ***non-persistent*** and ***semi-persistent transmission***, ***circulative and propagative***).

Plasmodesmata: the cytoplasmic connections between cells.
Polygenic: a character controlled by many genes (*see* ***oligogenic***).
Polyphagus: when an insect such as an aphid feeds on various secondary host species (*see* ***monophagus***, ***primary*** and ***secondary hosts***).
Precipitation (syn. ***precipitin***) ***reaction***: a visible precipitation reaction that occurs when antibodies and antigens react to form an insoluble lattice.
Primary host: the plant on which the sexual forms of an aphid mate and lay eggs to overwinter (*see* ***secondary host***).
Primary symptoms (***infection***) (syn. ***local symptoms***): the symptoms that develop at the site of virus entry (*see* ***secondary*** or ***systemic symptoms***).
Propagative virus: a virus that multiplies within its insect vector.
Protoplast fusion: a tissue culture procedure for somatic hybridization that is used in cell manipulation studies (*see* ***genetic engineering***).
Pseudo-recombinants: new strains of a virus that result from the reassortment of genome nucleic acids during the replication of viruses with divided genomes in mixed infections.
Purification: the separation of virus particles from the host plant and their concentration.

Quick leaf-dip (syn. ***quick-dip***): *see* ***epidermal strip test.***

Radial diffusion: an immunodiffusion serology test in which liquid antigen (or antibody) is placed in a well cut in gel containing the other reactant, and allowed to diffuse out into the gel.
Recessive gene: a gene that is not expressed in the phenotype of the heterozygote (*see* ***dominant gene***).
Recombinant (***recombination***): a new strain of a virus that occurs as a result of the breakage and renewal of co-valent links in a nucleic acid chain, so that the nucleic acids are rearranged in the chain.
Reflective mulch: a polythene or straw layer placed on the soil surface around the crop plant to control air-borne insect vectors.
Resistance: various types of inherited (or constitutive) resistance to virus infection are recognized in plants, but the terms used to describe them have a multiplicity of meanings to different workers. The following terms have been defined according to their most common and accepted usage (*see* ***hypersensitivity*** and ***susceptibility***).

Resistance: a host plant can be considered resistant if it has the ability to suppress or retard virus activity. Resistant is the opposite of susceptible and may be quantitatively identified as high (extreme), moderate or low, depending on the effectiveness of the protective mechanism.

Tolerance: a host response to virus infection that results in negligible or mild symptom expression, but relatively normal levels of virus concentration and movement within the host compared with a susceptible host (*see* ***vector resistance***).

Immunity (***immune***): terms used to describe absolute exemption from infection by a specific pathogen. An immune plant is not attacked at all by the particular virus and is a *non-host* of the virus concerned (*see* ***acquired resistance***).

Field resistance: resistance shown by a host plant under natural field conditions, even though the same host may be susceptible to the virus under experimental conditions.

Horizontal resistance: resistance that protects a host against all genetic variants of a pathogen to a greater or lesser degree.

Vertical resistance: resistance that protects a host against only specific strains of a pathogen (*see* ***gene for gene hypothesis***).

Continuous resistance: a response involving a gradient from severe infection to extreme resistance in a segregating population.

Discontinuous resistance: a response involving distinctive, clear-cut symptoms in a segregating population, which is often controlled by a single dominant gene.

Rickettsia: plant disease agents belonging to the Schizomycetes group of bacteria, which may cause virus-like symptoms.

Roguing: a control procedure involving the removal of diseased plants showing visual symptoms, from a field crop.

Satellite: a term first used to describe a small virus (satellite virus) associated with tobacco necrosis virus (TNV), which is dependent upon the TNV genome for its own replication. Also used to describe certain nucleic acid molecules that are unable to multiply in the host cell, without the aid of other nucleic acid molecules (*see* ***Carna 5-RNA***).

Screening test: a test to observe the response of a range of plant cultivars or types to virus infection.

Sedimentation coefficient: the rate of sedimentation of a virus per unit centrifugal field measured in *Svedberg units (S)*.

Self-incompatible: the inability to produce seed by self-pollination.

Semi-persistent transmission: virus transmission by an insect vector that is intermediate between *non-persistent* and *persistent transmission.*

Somaclones: plants produced by a genetic engineering technique by which single cells or protoplasts are cultured to produce individuals which are genetically variable from their genetically stable parent. The variation induced is called *somaclonal*.

Spur precipitation line (syn. ***partial fusion line***): an antibody–antigen precipitation line formed when two antigenically distinct strains of a

virus are placed in adjacent wells in a *gel double-diffusion* test.
Steckling bed: a sugar-beet seedling bed.
Strain: a variant of a virus that can be recognized by some characteristic of the phenotype. The variant will be serologically related to the type strain, but may have distinct antigens.
Stylet-borne: *see* ***non-persistent transmission.***
Susceptible (***susceptibility***): when a virus readily infects and multiplies within a plant. Susceptible is the opposite of resistant, with low or high levels of susceptibility being recognized.
Svedberg unit: the units used to measure the rate of sedimentation of a virus to determine its *sedimentation coefficient* (S).
Syndrome: the overall development and expression of a disease in a plant.
Synergism (***synergistic***): the association of two or more viruses acting at the same time.

Thermal inactivation point: the lowest temperature at which heating for a limited period (10 min), is sufficient to cause loss of virus infectivity.
Tolerance: *see* ***resistance*** and ***vector resistance.***
Transcapsidation: the encapsidation of the nucleic acid of one virus strain with the protein of another, during simultaneous infection and replication of two strains.
Transformation of DNA: a genetic engineering procedure in which the properties of cells may be modified by the insertion and expression of foreign DNA (*see* ***genetic engineering***).
Transovarial transmission: when virus is transmitted through the eggs of the infected vector to its progeny.
Transstadial: when virus is retained through the moult of its insect vector.
Trapping: *see* ***immunosorbent electron microscopy.***

Vector: an organism able to transmit a virus.
Vector resistance: resistance of a host plant to the vector of a virus. Three basic types of vector resistance are recognized

Antibiosis: resistance in which the growth and multiplication of the vector on the host is inhibited.

non-preference: hosts that a vector does not like feeding on. May cause the vector to probe-feed several times shortly after landing before moving to another plant, which may reduce the transmission of a persistent virus, but could increase the spread of a non-persistent virus.

tolerance: the ability of a host plant to withstand a vector's attack without suffering severe damage. This type of vector resistance does not control virus spread (*see* ***resistance***, ***tolerance***).

Vertical resistance: *see* ***resistance.***

Virion: the complete virus particle consisting of the nucleic acid and protein shell.

Viroid: a pathogenic agent consisting of ribonucleic acid of low molecular weight without a protein coat.

Virulence: the relative capacity to cause disease.

Virulent: strongly pathogenic.

Viruliferous: a vector that carries or contains virus.

Virus cryptogram: a descriptive code summarizing the main properties of a virus.

Virus inactivator: a chemical that inactivates a virus causing loss of infectivity.

Virus inhibitor: a chemical that prevents virus transmission without inactivating the virus. Inhibition of transmission may be temporary and can sometimes be avoided by dilution. Inhibitors frequently act on the recipient host rather than on the virus itself.

Viviparously: a method of a sexual reproduction occuring in aphids by which the young are born alive and active.

Volunteer: a plant from a previous season's crop that regenerates in a subsequent crop e.g. a potato tuber.

Index